The Food System
A GUIDE

Geoff Tansey & Tony Worsley

EARTHSCAN
Earthscan Publications Ltd, London and Sterling, VA

The opinions expressed here are those of the authors and do not represent in any way the policies and opinions of CSIRO-Australia

First published in the UK and USA in 1995 by Earthscan Publications Limited

Reprinted 1996, 1997, 1999 and 2000

A catalogue record for this book is available from the British Library

ISBN: 1 85383 277 4

Typesetting and figures by PCS Mapping & DTP, Newcastle upon Tyne
Cover design by Andrew Corbett
Printed and bound in Great Britain by Biddles Ltd, *www.biddles.co.uk*

For a full list of publications please contact:

Earthscan Publications Limited
120 Pentonville Road, London N1 9JN
Tel: +44(0)20 7278 0433
Fax: +44(0)20 7278 1142
email: earthinfo@earthscan.co.uk
website: http://www.earthscan.co.uk

22883 Quicksilver Drive, Sterling, VA 20166–2012, USA

Earthscan Publications Limited is an editorially independent subsidiary of Kogan Page Limited and publishes in association with the International Institute of Environment and Development and WWF-UK.

This book is printed on chemical chlorine free paper from sustainably managed forests.

Praise for 'The Food System'

'Frank, devastatingly honest and expertly researched, this is the essential guide for anyone seeking to understand the global food jigsaw. It reveals the lunacies of marketing, the greed, the corruption, the hidden deals that control the way the world is fed. If you want to understand why the world feeds itself in the way it does, you must read this book.'

Derek Cooper, The Food Programme, BBC Radio 4

'This is a book we have been waiting for! It opens the door and guides the reader into the complicated web of resources, resource management and the often conflicting interests of consumers, producers and markets constituting the food system in the rich world – which also directly or indirectly affects the poor. As a university teacher I am especially eager to try it out with students in nutrition… The book is organised in a new and original way, and is highly readable, thanks to people who know how to serve their words in appetising dressings!'

Wenche Barth Eide, Nordic School of Nutrition, University of Oslo; formerly Technical Adviser in Nutrition to the International Fund for Agricultural Development

'This extremely valuable book is the first in the field. It gives an excellent overview of the food system today and policy issues surrounding it. It will make an excellent textbook for many specialist courses – home economics, consumer studies, business studies, nutrition and dietetics, health and social policy to name but a few – which need the perspective provided by this book. Some students will need it to refer to, others to use as a general tome, and still others requiring a general insight.'

Tim Lang, Professor of Food Policy, Thames Valley University

'This is the best text I have found to support my course because it shows the reality of food policy issues from an international perspective. It contains excellent coverage of all the major issues and I particularly like the case studies incorporated throughout. I have no hesitation in recommending it as a prescribed text to students.'

Mary Mahoney, Course Director, Faculty of home Economics, Science and Humanities, Bath College of Higher Education

'This book provides a well balanced and informed source of reference as well as a good read. The structure of the book is very helpful. It draws together scientific, economic and political material, stresses their interrelated nature and discusses each of them in a language which is not too technical for the ordinary reader to follow.'

John S Marsh, Professor of Agricultural Economics, Director of the Centre for Agricultural Strategy, University of Reading

'This will be an extremely useful addition to the food studies area, and will definitely be on the essential reading list for students. The book fills a gap in the market, and will form the basis of reading for students at a wide range of levels. It can be used as a general introduction to the complexity of food today. It provides useful and stimulating material for those at higher levels, and the examples given are elaborated sufficiently well to allow further investigation.'

Stella M Walsh, Consumer Services Management, Leeds Metropolitan University

Geoff Tansey is a member of the Guild of Food Writers, Visiting research Fellow at the University of Bradford, and was the founding editor of the journal *Food Policy*. He has also worked as a consultant for the UN and other international organisations.

Tony Worsley is director of the Food Policy Research Unit of CSIRO, Australia, and was Professor of Social Nutrition at the University of Otago, Dunedin, New Zealand before moving to the National Centre for Epidemiology and Population Health at the Australian National University.

Contents

List of Illustrations

Boxes

Tables

Figures

Glossary

ACC/SCN	Administrative Committee on Coordination/ Subcommittee on Nutrition	IGD	Institute of Grocery Distribution
ADAS	Agricultural Development and Advisory Service	IMF	International Monetary Fund
AFRC	Agricultural and Food Research Council	IOCU	International Organization of Consumers Unions
BECU	Bureau of European Consumer Unions	IUF	International Union of Food and Allied Workers Associations
CA	Consumers' Association	MAFF	Ministry of Agriculture,
CAP	Common Agricultural Policy		Fisheries and Food
CGIAR	Consultative Group on International Agricultural Research	MMPA	Marine Mammal Protection Act
COMA	Committee on Medical Aspects of Food Policy	NACNE	National Advisory Committee on Nutrition Education
COT	Committee on Toxicity	NAFTA	North American Free Trade Agreement
CSE	Consumer Subsidy Equivalent	NFA	National Food Authority
CSIRO	Commonwealth Scientific and Industrial Research Organization	NFF	National Farmers Federation
		NGO	Non-governmental organization
ECOSOC	Economic and Social Council	NHS	National Health Service
EU	European Union	OECD	Organization for Economic Cooperation and Development
FAC	Food Advisory Committee		
FAO	Food and Agriculture Organization	OPEC	Organization of Petroleum Exporting Countries
FDA	Food and Drugs Administration	SEA	Single European Act
		UN	United Nations
FPRU	Food Policy Research Unit	UNICEF	United Nations Children's Fund
GATT	General Agreement on Tariffs and Trade	USDA	United States Department of Agriculture
GDP	Gross Domestic Product		
GRAIN	Genetic Resources Action International	VALS	Values and Lifestyles
		WDPAN	World Declaration and Plan of Action for Nutrition
IDA	International Development Association	WFP	World Food Programme
IFAD	International Fund for Agricultural Development	WHO	World Health Organization
		WTO	World Trade Organization

Preface

Chance meetings can change your life. Ours did. We first met one lunchtime at the Food Policy Research Unit in Bradford University in the autumn of 1988; little did we realize the effect. Tony suggested that Geoff should let him know if he was going to visit New Zealand and – to cut a long story short – Geoff gave a lecture or two at the University of Otago, Dunedin, New Zealand, in 1989. Tony saw the basis of a book in Geoff's lecture on food policy and a publisher expressed an interest.

The book has been an ambitious undertaking for several reasons. First, there is the scope of it. We have tried to synthesize many of the isolated facts and figures which emerge continuously from the many industries, organizations and individuals associated with food. We have had to be broad ranging rather than deep. We have also tried to be entertaining and to make the book easy to read.

Secondly, we have had to bridge the gaps in our very different backgrounds – Geoff in soil science, history of science and technology, agricultural extension, and freelance writing and journalism, and Tony in psychology, consumer behaviour, research institutions, and university research and teaching – to cover the whole range of the food system. This has not been helped by the that fact we live 18,000km apart – which has kept the electronic mail, fax and phone circuits busy!

We have each written some chapters (Tony Chapters 4 and 7, Geoff Chapters 2, 3, 5 and 6) and contributed varying amounts of the rest. In pulling the manuscript together, Geoff has tried to give it one voice and to integrate the material. We have both gone over the whole script twice, incorporating each other's comments as well as those of various people who have been kind enough to review drafts.

Perhaps our biggest challenge is you, the readers. You could come from any part of the food system and each have vastly different experiences, interests and concerns. Indeed, if you are a farmer, food technologist, nutritionist or any one of a number of professionals in the system, or training to be so, you will have far greater knowledge of your specific area than is covered in this book. What we aim to provide, however, is an overview which will help in understanding the whole system, what the various actors in it are up to and to see the relationships between what might seem, at first, to be disconnected areas.

Acknowledgements

We are indebted to an enormous number of people who have spent their lives working on specific aspects of the food system. They have provided the raw material for us to draw on in producing this book. Our major challenge has been what to leave out. We are grateful for the support of the University of Otago, Dunedin, New Zealand, in the very early stages when Geoff gave the lecture which Tony said would make the basis of a book. We also thank the National Centre for Epidemiology and Population Health, Australian National University (ANU), Canberra, Australia, which provided a one month visiting fellowship for Geoff to work with Tony on developing the book's outline in October – November 1990. We thank, too, the CSIRO in Adelaide for its support to Tony. In Britain, we thank the University of Bradford and the Food Policy Research Unit in the Department of Biomedical Sciences for their award of an honorary visiting research fellowship to Geoff. This has enabled him to research the book more easily and to keep in electronic touch with Tony through the Janet worldwide computer network that links academic institutions.

We also thank those who have read the whole or part of various drafts of the book, in particular Charlie Clutterbuck, Felix Fitzroy, Anne Hobbiss, Tim Lang, Maureen Ludlam, John Marsh, Mary Mahoney, David Roberts, Kathleen Tansey, Roy Taylor, Stella Walsh, Neil Ward and Verner Wheelock. We thank, too, the staff and students of the Food Policy Research Unit of the Universities of Otago, ANU and CSIRO, for their input through our discussions with them. Thanks are due, too, to the representatives of many different organizations which have provided materials for the boxes and checked them over. We are grateful to the Institute of Grocery Distribution Research Services and to Seymour Cook Food Research International for supplying us with some market research reports. We are responsible for any inaccuracies that remain, although we have done our best to avoid them.

We gratefully acknowledge the permissions given by the publishers of the various extracts used throughout the book. Each of these is clearly referenced and we thank the publishers and authors. We also thank the staff at Earthscan. Finally, we thank our long-suffering families who have supported us in this venture.

This book is dedicated to the memory of

David Gunn McLeod

26 July 1955–25 July 1993

1

Introduction

The man and woman in the street know *that their world is changing and worry about it. ...Nothing is certain except that we face innumerable uncertainties; But simply recognizing that fact provides a vital starting point, and is, of course, far better than being blindly unaware of how our world is changing.*

Kennedy, Paul (1993) *Preparing for the Twenty-first Century*
Harper Collins, London, pp344 and 348.

This book is a guide to the *entire* food system. However, we focus on the rich, industrialized world where the global food system is being developed and promoted. Food issues are linked to the exertion of power, influence and control by the different actors within the system.

Of course, no guide can go into detail about all the topics it covers. In trying to provide a broad sweep across the system in a few hundred pages, we can only give an introduction – each chapter could be a book in itself. We hope to encourage readers to look further and with better informed eyes into the issues and areas that concern them.

Who is it For?

The book will be relevant to you if you are:

➤ a student on a food-related course – from nutrition to dietetics, home economics to catering, agriculture- horticulture to retail management – or a student of the behavioural and social sciences;
➤ concerned about the balance between the public good, private interests and long-term sustainability in the production and consumption of food;
➤ an ordinary citizen interested in aspects of the food you eat, such as its safety, healthiness, and its social and environmental effects.

The Food System

We use the idea of a food system in discussing the how and why of what we eat – ie, how food is produced and reaches our mouths and why we eat what we do. It subsumes the terms 'food chain', which is too linear a model for today, and 'food economy', which is too narrowly economic. The idea of a system implies that there is an interconnection beneath the surface of things which, as we shall see, is very much the case when we look at any aspect of food today. The food system links three different aspects of life:

1

➤ Biological: the living processes used to produce food and their ecological sustainability.

➤ Economic and political: the power and control which different groups exert over the different parts of the system.

➤ Social and cultural: the personal relations, community values and cultural traditions which affect people's use of food.

The modern food system has really come together since the Second World War. As The *Oxford English Dictionary* defines a system, it is a 'set or assemblage of things connected, associated or interdependent so as to form a complex unity, a whole'. The food system reflects the prevailing social and economic influences around the world and is a system largely developed, run and promoted worldwide by economic institutions in the rich and powerful industrial nations.

The triumph...

Food and water are our most basic needs. Food is such an obvious need that it often escapes our close attention until it is not there or until we consume an unhealthy mix of it. One of the greatest achievements of this century is the feeding of very many more human beings than ever before. In the industrialized countries, food is available in great variety when and where people want it. For an increasing number of developing countries in Latin America and Asia, famine and food scarcities have become a thing of the past.

...and challenge

Yet, as we write this book:

➤ millions of people still face starvation in Africa owing to conflicts, drought and inadequate policies, while Europe's farmers are having to set aside land because they produce too much;

➤ the USA and the European Community, after coming to the brink of a trade war over agricultural subsidies, agreed a compromise in the General Agreement on Tariffs and Trade (GATT) world trade talks which made many farmers in rich and poor countries fear for their future;

➤ a large British firm, part of one of the world's largest multinational food companies, is fined for what trading standards officers reportedly called 'creeping adulteration' – ie, putting soya in a product labelled 'pure ground beef';

➤ governments struggle to balance consumer and health demands for better food labelling with the more conservative approaches of some industry groups;

➤ food safety remains an issue for consumers in rich countries as reported food poisonings increase;

➤ the decline in biodiversity is speeding up, causing concern about current farming practices and the long-term sustainability of food supplies;

➤ the World Declaration on Nutrition, agreed at the United Nations' first International Conference on Nutrition in December 1992, states that 'about 780 million people in developing countries still do not have access to enough food to meet their basic daily needs' and 'at the same time, chronic non-communicable diseases related to excessive or unbalanced dietary intakes often lead to premature deaths in both developed and developing countries';

➤ the world's human population is forecast to double in size to 10 billion in the next 50 years, yet every year more fertile land becomes desert or is built on.

Clearly, there is a long way to go before the food system serves the needs of *all* the world's people, either now or in the future. From farm to mouth, the health, safety, honesty, availability, equity and environmental impact of today's food system is ringing some alarm bells.

But we must look beyond today. The present food system serves best people in the rich industrialized countries – that 20 per cent of humanity that uses about 80 per cent of the world's resources. The questions are: can such a division of the world be sustainable and will the food system meet the needs of ten billion people in 2050? To answer such questions, we need to see the connections that exist, and how change in one aspect of the food system can affect others. As we discuss what the major actors in the food system are currently doing, these are the questions to bear in mind. A much greater debate is needed about our food future, and it is one to which everyone, either as a worker in the food system, or as a politician or consumer, can contribute.

A changing world

The 1990s are bringing major and rapid changes. Six societal changes world-wide will have great influence on the food system:

1. Increasing longevity – people are living longer – and the numbers of people are increasing (see Figure 1.1). Current trends will reduce the diversity of species and put greater strains on the earth's ecosystems – from increasing desertification and deforestation of land to pollution of the atmosphere and the seas and destruction of the ozone layer.

2. Increasing urbanization – Europe and North America are no longer the only major urban regions of the globe. It is becoming an urban world. The number of large conurbations has increased rapidly throughout Latin America (Mexico City is the world's largest), Asia and Africa. Urbanization has extended the food chain, linking farmer to eater, and the spread of supermarkets has been spectacular. In Japan, for instance, in the early 1980s only 10 per cent of consumers shopped at supermarkets; in 1992 the proportion rose to 60 per cent. Urbanization fuels the rise of a consumer monoculture – urban people buy food produced for and carried to cities from all parts of the globe. Centralized buying by supermarket chains can lead to smaller producers being squeezed out of the market as they are unable to produce the quantities required.

3. Globalization of the food market – large companies have expanded to control larger and larger shares of trade in agriculture,

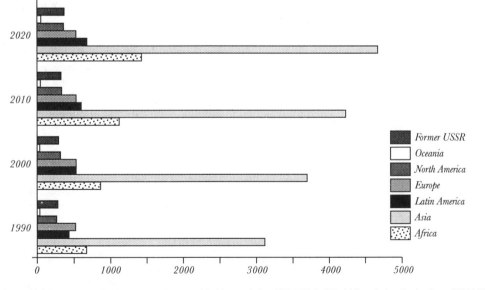

Source: UN, World Population Prospects 1993, quoted in Nygaard, David F (1994), 'World Population Projections, 2020' 2020 Brief 5, IFPRI, Washington DC.

Figure 1.1 UN population projections, 1990–2020 by region, medium variant

manufacturing or retailing, with smaller businesses from farms to grocers' shops being squeezed out of business. The effects of the Third World debt crisis, the dominance of free-market ideologies, plus the collapse of communism in Eastern Europe and the former Soviet Union are helping the globalization and marketization of economies all over the world, drawing even the most isolated self-provisioning peasants into a global market. Business life has never been as corporatized.

Global companies owe scant allegiance to particular nation states. Frequently, profits made in one country end up in the parent company's country; national taxes can be avoided through creative accounting schemes, such as transfer pricing or repayment of loans, so that real profits are turned into paper losses. Co-operation between nation states or their citizens, however, does not match the size and power of these global companies, so their activities are largely of their own choosing. The concentration of economic power into a smaller number of boards of management and the limited power of shareholders to control executives are leading to a concentration of food power of concern to governments and citizens.

4. Increasing technological change – major changes, particularly in biotechnology and information technology, offer different actors varying kinds of control over their parts of the food system. Genetic engineering, for example, may allow plant and animal species to be redesigned, and patent laws permit the patenting of seeds, plants and animals, thus transforming previously public goods into private property. These and other possibilities are generating much controversy over what should be done and how, and who should receive the benefits and bear the costs of change.

5. Changes in attitudes and values – market researchers in countries like the USA, Germany and Australia report that increasing numbers of citizens and companies are becoming more interested in community values such as honesty, 'substance' and environmental sustainability. The 'Me' generation may be rethinking its self-centredness. There is also an attempt to remove overt discrimination in many societies, shown by, for example, the increased sensitivity over the use of non-sexist and non-racist written language, and Equal Opportunity and Affirmative Action legislation in various countries. Such changes are reflected in food advertising – for example, one Australian advertisement for lean meat now focuses on the advantages of meat for women ('plenty of iron') rather than for men, as found in older advertisements. There is some debate about the desirability and effect of these changes. Are they cosmetic or do they represent real changes in the power balance between social categories?

6. The decline of the traditional 'housekeeping' role – an increasing number of women are entering the job market or returning to full- or part-time paid employment after their child-bearing years. The traditional role of the housewife, which was largely a time-consuming occupation concerned with the preservation, preparation and serving of food and requiring many skills to process basic food ingredients, is declining. The wider range of prepared foodstuffs, better distribution and availability of fridges, freezers and so on, has reduced drastically the time and skills necessary for such food preparation, but also it requires more money to exercise this choice. Working in the labour force has given women relatively more economic influence, especially over domestic purchases such as food, and has also contributed to the demand for convenience foods and labour-saving devices such as the microwave. Indeed, it is predicted that by the year 2000, 60 per cent of retail food products in some countries will be prepared meals. People in paid employment have less time to perform household tasks, so that increased incomes often mean a lack of time. People may no longer have the time, energy,

gy, motivation, or even the skills to prepare food from scratch.

Criteria for a good food system

All these factors present major challenges for the future of our food. We believe that the many different concerns about this can be summed up as a challenge for the development of policies and practices for a food system that will ensure a safe, secure, sufficient, sustainable and nutritious diet for all, equitably. Such a diet will also allow people to enjoy and take pleasure from food in culturally appropriate ways. Thus, the criteria to use when looking at changes in the system and evaluating policies are: is it safe? is it sufficient? does it enhance food security? is it sustainable? does it promote a nutritious diet? does it enable all to meet their food needs?

An individual's role

Individuals play a part in the food system in two ways. First, as an individual, you must eat food to live. But food is often used to satisfy non-nutritional needs, such as those for social acceptance and influence. Secondly, you may be an actor in some part of the food system – a farmer, shopkeeper, checkout till operator, dishwasher, consumer or scientist. Food is the world's biggest business. About a quarter of all workers are involved in the food industry, not including those in the home who do much of the work of transforming basic ingredients or bought meals into the foods we put into our mouths.

The food system is dynamic and changing. What matters is the nature and direction of the changes. Obviously, food is part of the broader economy and society and what happens in food mirrors the values, activities and trends in the larger whole. By the same token, finding ways to make the food system work more effectively and to deal with the challenges it faces, could hold lessons for the rest of the economy and society.

Global changes, individual limits

The food system is changing as part of the global economic restructuring of the 1990s. Food businesses in the industrialized world face the dilemma that comes in all markets where the basic needs of those with effective demand (money) have been met. But that demand is limited because people can only eat so much food. No matter how tempting the products, people can only increase the amount they eat by some degree and this does have physical consequences – obesity. The poor may actually need more food and be unable to afford it, but those who can afford it do not need it.

It is in the nature of the competitive market economics of the 1990s that businesses cannot stand still but must grow to survive and maintain and increase profits for their shareholders. With food, if this cannot be done by increasing sales of basic foodstuffs, it can be done by turning basic foodstuffs into other, more expensive, products. Hence, the variety of snack foods and the invention of new products, and even the introduction of 'fat' spreads with little or no nutritional value.

Anyone involved in farming, food technology, dietetics, nutrition, marketing, catering, food retailing, importing, exporting, or any other aspect of today's food system will face some of these issues and have a hand in meeting the challenges they present. We hope this guide will help you to understand those challenges. Although there are differences in different places, there is a similarity in the processes that affect them today

A Guide to the Book

The book is in three parts. The first part, after this introductory chapter, deals with the background basics of the food system. In Chapter 2, we focus on the biological and ecological aspects of the present food system, which underpin the rest. In Chapter 3, we briefly look at the historical roots of the food system and in Chapter 4 we examine how food relates to culture and individual human needs.

In Part 2, we move into the economic area and look at groups of people as key actors in the food system and their various roles. These key actors have competing interests. Usually, each group of actors acts in ways that fit their particular needs and argues for policies that suit them. However, within each group there are also differing interests. Often, those with the most need have the weakest voices – the poor, small farmers and consumers being the weakest and least well-organized of those involved.

In Chapter 5, we look at farmers, food workers and traders. These groups provide the inputs, labour and raw materials that those in Chapter 6 – the manufacturers and processors, retailers and wholesalers, and caterers – transform and sell. In Chapter 7, we look at individual citizens as consumers of food.

Part 3 examines various control mechanisms in the food system. Chapter 8 describes how various actors try to influence and control nature and production processes through science and technology, at the uses of information and the media, and at management techniques. In Chapter 9, we examine first the controls society puts in place through laws, rules and regulations to influence the distribution of benefits and costs throughout society from the actions in the food system. Then we look at the elements that make up food policy and what is needed to make effective policies before moving into the concluding chapter.

For us, this is a logical way to proceed, but it is not the only way. Initially, you may want to skip the background in Part 1 and go straight to the key actors in Part 2. You can repackage the order as you wish, but we suggest that you cover it all to get the whole picture.

Throughout the book we use profiles, stories and quotes from a range of sources as illustrations. By referring to these and the short guide to further reading at the end of the book, readers can pursue their own interests further. Owing to the very rapid changes in the food system, some of the examples may date quickly – people are buying and selling food companies with great rapidity – but this does not matter. The point is to put specific people, actors and institutions into context. Furthermore, the forces affecting change are similar, even if the circumstances in which they operate differ.

Global reach

Although we focus on the industrialized countries, the food system we describe is engulfing the developing countries, too. An understanding of what is happening in the market-leading countries and their boardrooms is vital for developing countries if they are to develop appropriate policies. As the global economy becomes more interdependent, finance, communications, science and technology, and private trading and production institutions, all play a growing role in the global market.

The food system has changed rapidly this century. The pace of change is likely to increase with new tools, such as biotechnology and information technology, coming into use. Changes will affect different groups in different ways. Some will benefit, some will not. Governments and legislators will have to decide how far to react to these changes and which controls are necessary to safeguard individual, social and environmental interests.

Global forces have local effects, from terms and conditions for workers to agricultural practices to consumer desires, and vice versa. How far can we control the activities of sectional, private interests for the public good? What sorts of institutions are needed? Food policies cannot be carved in stone but must change as the actors change. Policies are required which enable sectoral interests to prosper in harmony with the common good. These are the kinds of issues and questions we trust you will bear in mind in reading the book and which we deal with in Chapter 10. Here, we examine the challenges and opportunities for new alliances, directions and policies in the food system as we approach a new millennium so that, at last, it fulfils its prime objective – to provide a safe, secure, sufficient, sustainable and nutritious diet for all, equitably.

Part 1

Food Matters

Food is so basic to our well-being that it is almost too obvious; it is taken for granted. Yet today's relative abundance in the rich world is a new phenomenon and civilizations that have achieved such abundance have not always managed to maintain it. The modern food system is an historical creation and a wide range of forces have shaped it. These forces could have balanced out differently and, if they had, the abundance of food enjoyed by the wealthy nations today may have been a different story.

In Part 1, we see briefly how the food system is rooted in the biological processes supported on the planet. These, we are coming to realize, are being stretched increasingly by human activities. Then we provide a potted history of food, and look at the changes that have shaped the food system today. Finally, we look at the complex role of food in fulfilling human needs – nutritional needs as well as many others. In recognizing the various needs that food fulfils, we equip ourselves to deal with the many appeals the various actors in the food system make to us as consumers and citizens.

2

Food and the Biosphere

Humanity has the ability to make development sustainable – to ensure that it meets the needs of the present without compromising the ability of future generations to meet their own needs. The concept of sustainable development does imply limits – not absolute limits but limitations imposed by the present state of technology and social organisation on environmental resources and by the ability of the biosphere to absorb the effects of human activities. (p8)

The world produces more food per head of population today than ever before in human history. In 1985, it produced nearly 500 kilograms per head of cereals and root crops, the primary sources of food. Yet amid this abundance, more than 730 million people did not eat enough to lead fully productive working lives. There are places where too little is grown; there are places where large numbers cannot afford to buy food. And there are broad areas of the Earth, in both industrial and developing nations, where increases in food production are undermining the base for future production. (p118)

World Commission on Environment and Development (1987) *Our Common Future* Oxford University Press, Oxford, by permission of Oxford University Press

Sustainability has become something of a 'buzz' word in the 1990s. It reflects, however, a growing concern for the environment of which we are a part, the impact we have on it and its ability to continue to sustain us. Nowhere is this more crucial than in the food system, which is first and foremost a global biological system. It relies on the thin film of life contained in the biosphere – an area stretching about 10km above and below sea level, where most living processes take place. Much of the biosphere is a product of those life processes, but other events – from seismic to cosmic – also influence this fragile film.

Policy makers often seem to assume that human technical ingenuity will find ways to sustain us whatever we do to the biosphere or however high the human population and its material demands become. This may be realistic, but we doubt it. Overconfidence could have catastrophic consequences. Unless we get our relationship with the biosphere right, the sustainability and physical security of food supplies cannot be guaranteed. In this chapter we look at the biological basics of the food system.

The Biosphere

In any set of conditions, the biosphere has a limited carrying capacity for life. Unfortunately, we do not know what that capacity is. When conditions change, the forms of life and the balance between them adjust, but in that adjustment what was the dominant species can be swept away. The rapid disappearance of the dinosaurs some 65 million years ago seems to have been due to a major change in planetary conditions. Human beings are relative newcomers to the earth, with a history going back a few million years at most. Although our species did not create the conditions for humankind to evolve, it is now having a major impact on the conditions which enable it to survive.

The key question about the survival of the human race is: does it matter whether people are in balance with the rest of life on the planet or not? If it does not, then we can destroy all other life as it suits us and rely on our ingenuity to support ourselves. If it is so, then how we affect the biosphere and its ecosystems is significant.

The major eco-systems on earth – the forests, grasslands, oceans, marshes, estuaries, lakes, rivers, tundra and desert – are run by the energy used by the biosphere. This is only about 1 per cent of the total solar energy that reaches the earth's surface. Solar energy is absorbed by the earth, oceans and atmosphere, and is reradiated by them; the variations in this cycle drive the weather patterns.

At its simplest, the biosphere depends upon plants to use solar energy to transform carbon dioxide into organic compounds and to produce oxygen during photosynthesis. Respiration by plants and animals in turn uses up oxygen and releases carbon dioxide. The sun also powers life. Green plants are the primary producers. In photosynthesis, green plants use light energy with carbon dioxide and water to create organic compounds, such as carbohydrates like sugars and starches, and release oxygen. Over half of the energy captured (or fixed) in this process is used later for respiration but some is stored. This stored energy, in leaves, seeds etc, forms the basis of food chains that support most other life, from the micro-organisms that break down the plants to the animals that graze or browse on them.

Green plants both photosynthesize and respire (breathe), but other organisms only respire. Without green plants, the carbon dioxide and water produced in respiration would not be reused to make more plants which then provide food for animals, fungi and bacteria.

'Diversity is strength' is one of the key rules that govern the stability of living systems. Diversity reduces the biosphere's vulnerability to change. Yet one result of humankind's activities is to reduce the biological diversity of life

on the planet by replacing a vast range of life forms with ourselves and a narrow range of plants and animals that are moulded to serve us. As human numbers have increased, the number of other species has decreased. How far this diversity will decline before it causes such instability that conditions will change in ways hostile to human life, is an open question.

Until recently, the biosphere was expected simply to absorb the effects of human action rather than be changed by it, but now the scale of that action is such that it is no longer the case. The major life-sustaining cycles – nitrogen, carbon, oxygen, water, energy and minerals – are interconnected, complex and feed upon each other. They are repeated time and time again, recycling life's building-blocks. Change that affects one cycle will affect the rest in turn. Briefly, we will summarize what those cycles do.

Cycles for life

Water

Water is essential for life. The weather system moves water around the planet. Water transports nutrients in plants and animals. Oceans act as a heat sink, storing up vast amounts of solar energy. Frozen water at the poles reflects back solar radiation and helps to keep the planet cool.

The water cycle involves the continual evaporation and precipitation of water. It evaporates directly from water surfaces and through transpiration in plants. Thus, water is drawn up from the roots through the plant, transporting nutrients with it, and is transpired through the leaves. As the water vapour enters the atmosphere it is blown about in weather systems and is redeposited as rain or snow on land where plants and animals use it, and where it plays an important role in soil formation and transport. The destruction of tropical forests could have a major effect on weather patterns.

In a crop producing 20 tons fresh weight, 2000 tons of water will pass into the roots. At harvest, maybe 15 tons is in transit in the plant. The dry weight of 5 tons includes 3 tons of water which was fixed by photosynthesis, when it provided the hydrogen atoms to make up the carbohydrates and released oxygen. But the energy fixed in the dry matter of the crop will be less than 1 per cent of the total solar energy received by the crop – nearly 40 per cent will have been used to evaporate the transpired water (see Box 2.1).

Carbon

In the carbon cycle, carbon is transformed from the carbon dioxide in the air into carbohydrate in plants by photosynthesis, and is returned from carbohydrate to carbon dioxide by plant and animal respiration and the decomposition of dead organisms. Plants convert a part of this carbohydrate into other nutrients, such as protein, and use a part for energy, releasing the carbon dioxide produced through leaves and roots. Plants are consumed by animals which digest them as food, respire and release carbon dioxide. After plants and animals die, micro-organisms in the soil break them down and release carbon dioxide (see Figure 2.1). A similar path is followed in the sea among marine plants and animals.

The rate at which plants fix carbon dioxide – by producing lots of vegetation – varies greatly from a high value in the tropical forests, where 1–2kg of carbon are fixed per square metre of land surface, to as little as 1 per cent of that in the Arctic tundra or most desert areas. The forests and fields in the middle latitudes assimilate between 0.2 and 0.4kg per square metre.

By far the largest amount of carbon on earth is stored in sedimentary rocks like limestones. Some is stored in fossil fuels like coal and oil which have been used as fuel for the past 150 years, thus sending the carbon back into the atmosphere. This is leading to an increased concentration of carbon dioxide and threatening to cause a greenhouse effect, which will raise the global temperature and alter patterns of weather and plant growth.

Box 2.1
Water Use and Water Problems

Each person needs about 5 litres of water each day for drinking and cooking and another 25–45 litres a day for hygiene and health. According to the World Resources Institute 3.4 billion people get by on about 50 litres per day. In rich countries, however, people use much more water. The average American uses around 350 litres a day and the average Australian 570 litres a day. This personal use represents only one tenth of what nations like Australia use for industry and agriculture.

Fresh water is probably the most precious liquid on the planet and it is becoming scarcer. Human activities are polluting fresh water at an alarming rate. Even in the Australian state of Victoria which is not particularly infamous for its water pollution problems:

➤ 30 per cent of natural wetlands have been lost;
➤ more than half of the shallow and deep freshwater marshes have disappeared;
➤ nutrient and salinity levels are killing plant and animal life in many rivers;
➤ 31 of 46 species of native fish are endangered, two are extinct;
➤ drinking water throughout the state does not meet the WHO's goal of zero E. coli contamination suggesting sewerage pollution.

Sources: CSIRO, *Water* no 18, CSIRO Research for Australia Series, East Melbourne, CSIRO Publication, undated and 'Down the Drain' *The Helix*, 1991, no 21, pp 6–9

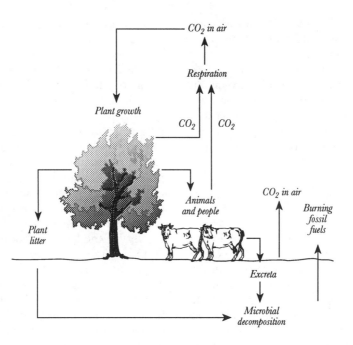

Figure 2.1 Schematic diagram of the carbon cycle

Oxygen

Our oxygen supply comes from plants. Oxygen absorbed from the atmosphere powers the high rate of energy release needed in higher organisms. It does this through complex chemical pathways which rely on oxidation-reduction (redox) reactions in cells between different carbon compounds. These reactions form the basis of biological energy supply and demand.

Oxygen is one of the three fundamental building blocks of carbohydrates (sugar and starch molecules consisting of carbon, hydrogen and oxygen atoms in various ratios) which are produced by photosynthesis. Water (H_2O) and carbon dioxide (CO_2) with light energy combine in photosynthesis to produce carbohydrates and free oxygen, the water molecule being the source of the free oxygen. These primary products of photosynthesis undergo a vast number of transformations in living cells, many driven by redox reactions.

The biological oxidation of molecules like carbohydrates produces carbon dioxide, water and energy. The oxygen atom from the organic molecule appears in the carbon dioxide and the molecular oxygen accepts the hydrogen atom from the carbohydrate to form water. In biological oxidation, the main way that organic molecules are oxidized is by dehydrogenation, – ie, by the removal of hydrogen.

The biosphere is constantly exchanging water vapour, carbon and oxygen with the atmosphere and hydrosphere (seas and oceans). It takes about 2 million years for all the earth's water to be split by plant cells and reconstituted by plant and animals cells, about 2000 years for the oxygen to be recycled and only 300 years for the carbon dioxide.

Nitrogen

Although nitrogen makes up 79 per cent of the atmosphere we breathe, most living things cannot use it directly. Yet it is an essential part of living organisms, forming part of amino acids, the basic compounds in proteins, which are essential building blocks for life. Nitrogen is 'fixed' or captured by specialized organisms or, since the Industrial Revolution, by industrial processes, to produce nitrogen fertilizers.

To be usable by plants, atmospheric nitrogen has to be transformed into one or more compounds of nitrogen, such as nitrates or nitrites, which plants can absorb. Various kinds of bacteria, some living free in the soil and others living in nodules on the roots of legumes, can do this. The fixing of nitrogen requires energy as it has to be oxidized. These bacteria contain enzymes which catalyse the reaction, allowing it to take place at much lower energies than is required in industrial production. In this, nitrogen has to be passed over a catalyst at high temperatures (400–500°C) and pressures (around 200 atmospheres) in the Haber process to produce ammonia, a first step in producing nitrogen fertilizers.

The nitrogen compounds produced in plant root nodules or in soil or added as fertilizers can be absorbed by plant roots and used to make amino acids and proteins. If leguminous crops are planted, they leave a residue of nitrogen compounds in the soil which can help to fertilize other crops that are unable to fix nitrogen.

Micro-organisms decompose dead plants and animals into their component amino acids. Assuming that oxygen is present, these can then be converted into carbon dioxide, water and ammonia, which can be reabsorbed by plants. A few micro-organisms can convert the ammonia into nitrite and nitrate ions while other bacteria, denitrifying bacteria which work in oxygen-free (anaerobic) conditions, convert these back into nitrogen gas.

Unfortunately, as more and more nitrogen fertilizer is used, the natural denitrification process may become unable to cope, with the excess nitrogen compounds running off into the waterways. The levels may be too high for human consumption, or may lead to algal blooms. In rivers, lakes and coastal waters, high nutrient levels can cause excessive growth of algae and plants which use up the oxygen in the water, leading to eutrophication with fish dying for lack of oxygen (see Box 2.2).

Minerals

Carbon, hydrogen, nitrogen and oxygen are the four main constituents of living matter, but much smaller amounts of many other miner-

---

Box 2.2
Over-Fertilization and Eutrophication in The Netherlands

For over 20 years, about a quarter of the 500,000–600,000 tonnes of phosphates used in or transported (in the waters of the River Rhine) to The Netherlands each year has accumulated in the environment. About ten per cent of the four to five million tonnes of nitrogen also accumulates. As a result, even if no more surpluses were added The Netherlands would feel the effects for years to come.

To deal with this problem of nutrient over-supply, the government aims to reduce nutrient discharges to 70–90 per cent of 1985 levels by the year 2000. An integrated approach, involving agricultural measures, more effective sewage treatment plants and treatment of industrial wastewater, and control of phosphates in detergents is being used. Since agricultural practices account for about 75 per cent of both nitrogen and phosphorus accumulation, combating fertilizer discharges is a major part of the programme. The government aims to have nutrient inputs and outputs in balance by 2000. Since nitrates threaten the quality of groundwater, a phased reduction in the use of manures and chemical fertilizers is a vital step.

The cost of dealing with over-fertilization was ECU 130 million in 1989 which was expected to rise to ECU 390 million in 1994 and ECU 560 million in the year 2000.

Source: OECD (1991) *The State of the Environment*, OECD, Paris, p181

als are needed. Sulphur is an essential part of some amino acids and is fixed by a few bacteria operating in anaerobic conditions, while phosphorus is essential in the key energy reaction in living cells and derives from the minerals making up soil. Magnesium lies at the heart of the chlorophyll molecule which allows photosynthesis to take place. Potassium, calcium, phosphorus, silicon and many other minerals are also essential. These, too, have their cycles with the water-soluble elements being absorbed by plant roots from the soil water. In nature, plants have used and adapted to whatever elements are present in the soil; various animals eat the plants and humans eat both. Modern agriculture, however, adds them in the form of fertilizers and feed supplements, and so do some humans as vitamin and mineral supplements.

From cycles to lines

These cyclical, continually repeating and recycling processes of the biosphere are quite different from the ethos of today's dominant industrial society – itself barely 200 years old. This is based on mining, extractive and industrial practices where processes are linear: they have a beginning, middle and end, use up the resources involved and generate waste at each stage.

The economics that have developed alongside industrial civilization mirrors it, discounting and treating as external those costs that do not relate directly to the process involved and may not easily be measured. Such costs are left to society, future generations or the biosphere to deal with. This is one reason for the growing debate between environmentalists and economists over the adequacy of economics to deal with the real world. It is a debate that is vital for future food security, as is the debate over technological supremacy versus the balance of life approach.

Overall, we have hardly begun to come to terms with these ideas, although we recognize that our use of fossil fuels may change the climate via a greenhouse effect. The introduction of new substances into the environment may also have adverse effects. For example, DDT which accumulated in living tissues affected birds and CFCs are damaging the ozone layer. However, we still know too little about the operation of the biosphere to say how what we do will affect its working, although our knowledge about particular parts of it is growing. It is a concern of especial relevance when it comes to food. Table 2.1 lists some of the possible adverse effects on human health that may result from global environmental change. Many arise because of their effects on the food system, at the foundation of which lies agriculture.

Table 2.1 Possible adverse effects on health due to possible global environmental change

Environmental damage	Manifestation	Type (direct, indirect) and timing* (early, late) of adverse effects			
		Direct, early	Direct, late	Indirect, early	Indirect, late
Enhanced greenhouse effect	Global warming and climatic change	Heat-wave related illness and death		Altered distribution of vector-borne infectious diseases	Reduced viability of edible fish in warmed oceans
		Natural disasters: cyclones, floods, landslides, fires		Food shortages due to altered agricultural productivity	

Table 2.1 continued

Environmental damage	Manifestation	Type (direct, indirect) and timing* (early, late) of adverse effects			
		Direct, early	Direct, late	Indirect, early	Indirect, late
	Sea-level rise	Increased risk of flash floods, surges	Inundation, social dislocation, sanitation breakdown, farm loss	Consquences of damage to foreshore facilities, roads, etc	Destruction of wetlands, decline in fish stocks
Stratospheric ozone depletion	Increased UV-B flux at earth's surface	Sunburn, photo-kerato-conjunctivitis; suppression of immune system; increased risk of infection; cancer	Skin cancer; ocular effects; cataracts, pterygium		Impaired growth of food crops and marine micro-organisms (base of aquatic food web)
Acid aerosols (from combustion of sulphurous fossil fuels)	Acid rain (and other precipitation)	Possible effects on respiratory system		Killing of aquatic life, reduced food, impaired crop growth	Impairment of forest growth, reduced ecosystem productivity
Land degradation, over-intensive agriculture and excessive grazing	Erosion, sterility, nutrient loss, salinity, chemicaliz-ation, deserti-fication	Decline in agricultural productivity	Rural depression, migration to fringes of cities (shanty towns); (see also final entry)	Exposure to higher levels of pesticides and fertilizers may also lead to toxic algal blooms in waterways	Consequences of silting up of dams and rivers
	Depletion of underground aquifers	Lack of well-water for drinking and hygiene	Decline in agricultural productivity		
Depletion of plants and animals, loss of biodiversity	Destruction of habitat	Deforestation, disruption of local culture and health	Shortage of edible species		Deforestation, greenhouse enhancement
	Loss of genetic diversity (species and strains),			Loss of medicinal chemicals and other health-supporting	Greater vulnerability of plants and livestock; decline in

Table 2.1 continued

Environmental damage	Manifestation	Type (direct, indirect) and timing* (early, late) of adverse effects			
		Direct, early	*Direct, late*	*Indirect, early*	*Indirect, late*
	weakening of eco-systems			materials	vitality of eco-systems
Other effects of overpopulation, particularly in poor countries	Proliferation of crowded urban slums and shanty towns (due to migration and high fertility)	Infectious diseases, malnutrition, antisocial behaviour	Effects of breakdown of social organization		Various consequences of overload of local eco-system

Note: * The designations 'early' and 'late' are notional only, indicating the relative time of occurrence.

Source: McMichael, A J (1993) 'Global environmental change and human population health: a conceptual and scientific challenge for epidemiology,' *International Journal of Epidemiology*, vol 22, pp 1–8

Agriculture

Agriculture is the bedrock of the food system. In developing agriculture, humankind has modified and continues to modify the workings of the biosphere. Agricultural techniques were developed in place of hunting and gathering, to provide more managed, secure food supplies, as outlined in Chapter 3. They gave rise to many complex farming systems worldwide which, while reducing the diversity of species, still sought to operate on a cyclical basis. However, where soil or water resources were overexploited, disaster ensued.

Nowadays, fishing is the only form of hunting that still provides a significant part of our diets, but here, too, fish-farming is becoming more important. Overexploitation of wild fish stocks (a resource), however, is threatening fish stocks world-wide, as more sophisticated technology, larger boats and industrial fishing enable large quantities of fish to be caught.

In agriculture – which we use broadly to include horticulture and forestry – farmers prepare, tend and harvest animals and plants, using what knowledge they have under particular ecological, economic and cultural constraints. The basic cycle starts, for plants, with land preparation. This requires labour and tools. Changes in these have had a dramatic effect on what can be done – for example, the area that could be ploughed in a day increased from a fraction of a hectare to many hectares as farmers moved from human to animal power and then to tractors using fossil fuels.

Land preparation makes the soil suitable to receive seeds, although today the texture and nutrient content that were produced in the past by fallows (leaving the land bare for a year or more) and leys (planting with grass or other cover crops that were subsequently ploughed in) are now produced with machines and fertilizers. After planting, farmers wait for the crop to grow and ripen, apart from applying occasional treatments: weeding, where manual hoeing is being replaced by herbicide spraying; fertilizer application, where animal and green manures are being replaced by artificial fertilizers; watering, if irrigation is required and feasible. When the crop is ripe, it is cut or harvested. Farmers sell or store the valuable part. The rest is left to refertilize the fields, to feed animals, or to be burnt off.

It is a similar process with animals: breed or buy, fatten or crop (eg, for milk or wool) and sell. Animals, however, require daily attention and there may always be animals present on a farm at different cycles of their lives. Most farming used to be mixed, with farmers keeping some animals and raising crops, but now these activities are often separated.

Today, an industrial approach to agriculture, often with similar techniques and crops, dominates in the rich countries and is spreading world-wide. In this process the diversity of plant and animal life is greatly reduced. This may increase the vulnerability of the food-production systems we have created and managed to sustain through continued use of human-added inputs – fossil fuels for energy, artificial fertilizers to provide soil fertility, pesticides to deal with weeds and pests, seeds specially bred for high yields, and ever more complex and expensive tools to reduce the need for labour.

Agriculture involves using three classic factors of production – capital, labour and land – with the first two affecting what can be done with the third. The tendency has been to reduce degrees of uncertainty – for example, those caused by climate or natural variations, by intensifying production using more inputs of capital and labour to modify the land, plants and animals. Increased capital expenditures, in particular, have allowed increased energy inputs into farming. In this chapter, we focus on the land itself, as the basic biosphere ingredient in agriculture. Capital and labour are discussed in Part 2, where we look at how the actors in the food system use these to control agricultural production through provision of inputs like fertilizers or pesticides, or in the technology of the suppliers who provide seeds and breeding animals.

Land and soil fertility

How agriculture uses land is crucial, for the land must be able to sustain production for ever unless alternative land or methods are available for us to produce the food and other products that we need. The key to this production is the soil – that thin layer above the bedrock in which plants grow. Some primitive plants, like lichens, live directly on rock, secreting substances to break down the rock and release nutrients, but most plants need soil. Soil is not sexy. It is dirt. But it is the most complex dirt you are likely to pick up.

Soil forms a space-time continuum. Even if you stand still on the same spot, the soil underneath you will change with time owing to complex changes within it, and if you walk away, the soil underneath your feet will change, too. It is a complex biological, chemical and physical system.

Soil forms from the rocks below it and the plants above it. The rocks weather away, breaking down into small particles; the sand, silts and clays that make up soil. Plants provide the organic matter which binds these particles together, producing soil that provides homes and sustenance for a huge variety of microscopic and larger animal life. Just a small spoonful of soil contains millions of organisms, such as bacteria and nematodes, organic and inorganic matter, water and water vapour, that undergo complex, far from fully understood, chemical reactions. The solids, liquids and gases in the soil are the physical environment which supports plant roots, trees, and even buildings. Getting society's relationship with soil wrong can help to destroy a civilization, as the ancient Romans found when they destroyed their granaries in North Africa through over intensive cultivation.

Under most conditions, it takes between 3000 and 12,000 years to build enough soil to form productive land, depending on the climate, base rock, and so on. Some soil is washed down sloping ground to valleys, making them more fertile, and into rivers. This soil may be deposited on the flood plains around rivers, like the Nile in Egypt where, for thousands of years, floods brought new soil.

Like most natural systems, soil can recover from damage, over-grazing, over-intensive cultivation, or too widespread use. This happens in traditional slash-and-burn agriculture where the soil fertility is exhausted after a few years. If it is then left for a few decades for vegetation to regrow, it will regenerate.

But if soil is used over-intensively, it is lost. Land degradation, soil erosion and salination are of major concern over much of the world's agricultural land – from Australia to Africa, from the mid-west of the USA to East Anglia in Britain. Only about 11 per cent of the earth's land surface – some 1500 million hectares – is cultivated. A similar amount could be cultivated with sufficient investment in drainage and irrigation, but 100–140 million hectares were lost over the last 20 years to land degradation. In the mid-1990s, land in 99 countries is slowly turning to desert, affecting 900 million people. This desertification is one of the most serious environmental problems according to the United Nations Environment Programme. The UN estimates the global income lost annually to desertification to be US$42 billion, half in Asia (see Figure 2.1).

Land degradation is an insidious process and hard to see. Salts build up slowly in badly irrigated and drained land, poisoning it. Land is damaged by over-cultivation or over-grazing, and is eroded by wind and water. In the 1930s, a great dust-bowl developed in the USA, where the soil structure had been so broken down and farming practices had left the soil exposed to winds, that enormous amounts of soil blew away in huge dust storms. Today, in the Kenyan highlands, steep hillsides are cultivated and subject to deep gully erosion, while in Australia, salination and rising water tables threaten large areas (see Box 2.3)

These effects are not inevitable. They derive from farming practices and economic pressures that cause practices leading to degradation. As practices have changed, as the areas cultivated have increased, more people have been fed. But farming methods that intensify production by mining the soil's fertility without replenishing it, or by exposing it to erosion

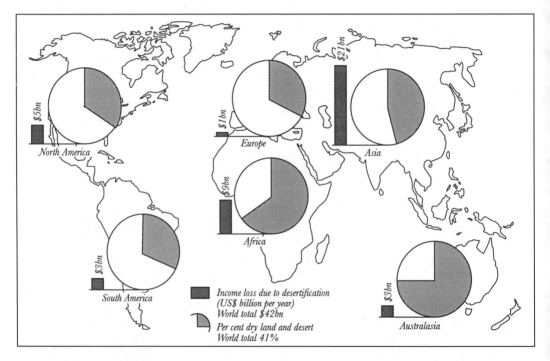

Source: Panos (1994) Media Briefing no 8, London; percentage figures from 'A New Assessment of the World Status of Desertification', *Desertification Control Bulletin*, no 20, 1991

Figure 2.2 Per cent of land which is desert or dry land and annual forgone income due to desertification.

Box 2.3
Landcare

For most Australians the country's number one environmental problem – land degradation – is hard to see unless it blows into the cities in a dust storm or turns the snow red. But wind and water erosion, salination, and soil acidity threaten the future of Australian agriculture, which accounts for around 40 per cent of Australian export earnings.

Over 55 per cent of land is degraded in the arid zone and in the non-arid zones 68 per cent of the extensive cropping areas need treating to combat degradation, according to Rod Roberts, author of *Land Conservation in Australia – A 200 year Stocktake* (Soil Conservation Service of NSW, 1989).

Annual losses of agricultural and pastoral production due to land degradation are about A$600 million according to a study by the Department of Primary Industries and Energy.

Treatment will cost at least A$2 billion, probably more, believes Roberts who bases his estimates on the only national assessment of land degradation carried out in the mid-1970s.

'Our environmental problems are very different from the European Community's', says Philip Eliason, the National Farmers Federation's (NFF) deputy director. 'The EC's seem to stem from overuse of inputs. In Australia it's more to do with the system – groundwater movements, salination, deforestation and destabilization of soils.'

These problems led the NFF, which represents about 170,000 farmers through 23 affiliated organizations, into an unusual alliance with the Australian Conservation Foundation (ACF), which has over 22,000 members. Early in 1989, they lobbied the Federal government to increase funding to voluntary groups to tackle land degradation.

The result was a 10-year commitment of A$340 million to a Landcare programme by the national government – a name borrowed from the state of Victoria, which started Landcare in 1986. In Victoria in the early 1990s, about 80 groups involved some 3700 landholders with 1.8 million ha – more than the total area of National Parks in the state. Nation-wide, about 15,000 landholders are in about 600 groups according to a report on the first year, with group numbers expected to peak at around 1500 by 1995.

Miriam Wallis is secretary of one of the oldest Landcare groups at Whiteheads Creek Catchment Area, about 50 miles north of Melbourne, Victoria, and a member of the Seymour environment group. She believes no government could fix all the land and wants landholders to 'own' their land problems.

After using inappropriate English farming methods in Australia for over a century, 'now the chickens are coming home to roost,' she says. 'We tried to turn this country into an English parkland. All this was thick forest, and we burnt it down.'

Group chairman, Tom Newton, agrees. He shepherds on a motorbike and his sheep dog rides with him until they reach the paddock. 'It is easier on his feet' explains Tom, who has over 6000 sheep on some 3000 acres. The ironstone soil is rock hard, due to heavy stocking with hard-hoofed animals rather than the soft-footed native marsupials. Tree clearing has led to salination in some places and earlier burning and grazing also led to changed pastures with annuals rather than perennials, causing erosion problems.

The group, which was established in 1987 and has over 40 members, is busy planting trees. The work is slowing run-off, reforesting hill tops where water infiltrates into the groundwater table causing salinity by raising the water table, and preventing gully erosion, but there is a long way to go.

Traditionally, farmers did not interfere with what a neighbour did on his farm or even discuss it, according to a senior national official, 'but with Landcare they go for walks on other farms and actually ask why are you doing that?' He believes Landcare groups are 'the most significant sociological change in rural Australia', as they are breaking down barriers between farmers.

The programme aims to 'catalyse not subsidize' groups. 'The US government provides over $1 billion for soil conservation and farmers see it as a government responsibility' says an official, 'We don't want that. Soil conservation must be part of farming, they must own it'.

Landcare groups alone will not reverse land degradation, emphasizes Andrew Campbell, author of the report. It is also necessary 'to have viable technical solutions available and a favourable socio-economic framework' like new rating systems which promote land improvements and do not penalize them.

Voluntary action may not be enough to deal with land degradation as only the most aware farmers are likely to join the groups. Drawing an analogy with pollution, where the polluter pays penalties, Rod Roberts calls for any unreasonable or negligent use of the land to be outlawed.

Others say Landcare does not go far enough. Since the problem lies with the farming system, the basic need is to change the production system, according to David Dumaresq, Lecturer in Agroecology at the Australian National University.

Source: Revised version of article in *Financial Times*, 27 February 1991; see also Alexander, Helen (1993) *Lessons in Landcare — Australia's Model for a Better Farming Future*, SAFE Alliance, 38 Ebury St, London

without safeguarding it, are a long-term threat to sustainability.

There are limits to the amount of cultivable land, with those areas in the developing world with the largest population growth projections having the least amount of land in reserve for future use (see Figure 2.3). The population of the world is expected to grow to around 10 billion by 2050, so the existing soil resources – or expansion into tropical forest areas, which have poor soils and are essential for climate balance, or problem soils requiring irrigation, terracing and other inputs – have to produce more food, fibres and beverages.

Land and water management go hand in hand. Failure to manage the land properly will allow water to erode the soil. Irrigating land that lacks water, or draining land with too much, increases its productivity until another limiting factor such as soil fertility comes in. Much of the post-war, world-wide major investment in agriculture has gone into irrigation (see also World Bank in Annex 1). But if irrigation is not done properly, with sufficient drainage, it can lead to soils becoming salty and unusable. Throughout the world's arid zones, half a million hectares of irrigated lands become desertified each year – roughly equal to the area newly irrigated each year. India has the second largest area, after China, of irrigated crop lands in the world – 40 million hectares – but one-fifth of this area is now affected by salinity and waterlogging which can cause desertification.

Building dams to capture surface water and drawing on underground water for irrigation have both environmental and political implications with many areas of the world facing water shortages if present practices continue. More efficient and effective techniques are needed. Much could be learnt from the long-standing water management techniques that are practised in various parts of the world, from Sri Lanka to Bali to Yemen. These involve social organization as well as basic technical skills (see Box 2.4).

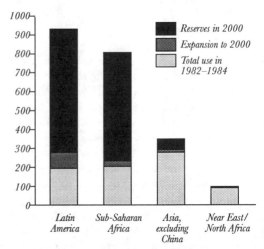

Source: Alexandratos, Nikos (1988) *World Agriculture: Towards 2000*, FAO & Belhaven Press, London

Figure 2.3 Land use and reserves (million ha, arable land)

Box 2.4
Yemen's Hanging Gardens

While nearby countries like Sudan and Ethiopia suffered famine in the 1970s and 1980s, the similarly drought stricken Yemen Arab Republic survived thanks to a centuries-old farming system which uses a highly developed terracing and water capture system. This uses the tiny amounts of rain once or twice a year to extraordinary effect, but now it is threatened with collapse.

On the 3000 metre high plateau that runs the length of Yemen are fertile terraces. Each field is levelled, with a bank around it, and culverts are built down each gully so that when it rains water is spread out along the contours and held in the soil before passing on to the next field.

The controlled run-off flows down to the foothills. Here, farmers in the wadis (valleys) rely largely on water and top-soil from the highlands. They bank their fields to trap and spread water for an agreed period. A 'water master' ensures Islamic principles of water allocation are observed and arbitrates in disputes.

Coupled with this are various husbandry techniques based on tillage tools like chisel ploughs, and methods to maximize moisture conservation and soil fertility – limited household livestock production, use of household and animal wastes as fertilizers and the controlled coppicing of trees.

Now the mountain terraces are crumbling as they are stripped of trees and shrubs for firewood and walls are not maintained. Huge chunks of fertile farmland are being swept downhill in increasingly violent floods. In some areas, 80 per cent of cultivable land has been washed away or, downstream, covered by up to 4 metres of rubble. The irrigation systems at the very bottom are becoming clogged with topsoil.

The problem lies with the kind of development in Yemen over the past 20 years, according to Anthony Milroy, who first started working in Yemen in the 1970s and is director of the Arid Lands Initiative (ALI). Attention focused not on the highlands but on market-based irrigated monocropping of high value crops to promote exports.

The 10 per cent of Yemeni farmers in the lower, flatter areas with access to groundwater got most aid. Subsidized imported grain, greater educational opportunities unrelated to rural life, and the focus on irrigated farming, disrupted the farm economy for 90 per cent of farmers.

To avoid disaster, Milroy believes development efforts must be reversed. But above all, listening and learning from farmers running a system that has fed people for at least 2000 years is required.

Proven technologies of use in semi-arid areas exist, he says. For example, using Australian rapid tree sowing techniques farmers can re-establish vegetation 50 times more cheaply than using traditional forestry industry methods. The means to reach communities that need them also exist, for example private sector methods like franchising, sales agencies, and profit incentives.

But farmers need control over development finance. If a package of finance and proven appropriate technologies can be brought together with organized end users, then the private sector can help solve the problems, believes Milroy.

In Yemen, new crops and less labour-intensive tools are needed. But farm researchers must go out with farmers and learn from them. The ALI has developed a simple and locally made gab iron – reinforced mesh and wire boxes to put stones in – which can be used to re-establish more permanent diversion structures and help reclaim the wadis. On the highland terraces another simple gab-iron enables terraces to be built more quickly and last 20 years instead of five.

Source: Revised version of feature in *Financial Times*, 5 April 1990

Technically, the UN's Food and Agriculture Organization (FAO) is sure we can produce enough food, even with existing technology, for the foreseeable future. Yet for the most part it is not a technical question, but as the irrigation example shows, social, economic and political, in which using good techniques for sustainable production becomes an overriding concern.

Sustainable farming systems mean sustainable soil use: the soil cannot be mined and treated as an industrial feedstock or raw material, to use and discard. If an industrial analogy is needed, then soil is better seen as something of a catalyst – we use it to produce useful products, but it is not used up in the process and at the end it must remain in the same condition.

The science begun in the nineteenth century into how the soil works has led to a much greater understanding of how we can influence it to produce more. But this has often focused on one particular aspect of soil. The chemical, as provider of nutrients which could be supplemented by fertilizer; the physical, as a structure, which should not blow away in the wind or be washed away in the rain; the biological, as a living system – all of these aspects combined together are necessary to produce sustainable soil fertility.

The Challenge of Sustainability

There is a growing critique of the type of intensive agriculture that has developed in the industrial world, which has reduced the biological complexity of farming systems and seeks to control production by increasing capital and energy inputs. From a short-term economic viewpoint, it is productive, but the measure of productivity itself may be flawed, as critics like Vandana Shiva suggest:

> *Monocultures of the mind generate models of production which destroy diversity and legitimise that destruction as progress, growth and improvement. From the perspective of the monoculture mind, productivity and yields appear to increase when diversity is erased and replaced by uniformity. However, from the perspective of diversity, monocultures are based on a decline in yields and productivity. They are impoverished systems, both qualitatively and quantitatively. They are also highly unstable and non-sustainable systems. Monocultures spread not because they produce more, but because they control more. The expansion of monocultures has more to do with politics and power than with enriching and enhancing systems of biological production. This is as true of the Green Revolution as it is of the gene revolution or the new biotechnologies.*

Shiva, Vandana (1993) *Monocultures of the Mind –*

Perspectives on Biodiversity and Biotechnology, Zed Books and Third World Network, London, p7

Today's highly input-responsive agricultural systems rely on high energy inputs which use fossil fuels, which are old solar energy stored. Without these inputs, to power machinery and to spread artificial fertilizers and pesticides, yields would drop, unless more efficient ways of converting solar energy to food, via plants, were found. Inputs like fertilizer and pesticides also leak out of the agricultural environment and affect other ecosystems, as happened with the pesticide DDT and with fertilizer run-off into lakes which kills fish.

Moreover, these simplified systems rely on monocultural practices – growing the same kind of crop, with the same genetic make-up in the same area or producing meat or milk from animals kept in factory farming conditions. Such monocultures predominate in modern agriculture, but require high levels of inputs to ensure that unwanted weeds, pests and diseases do not decimate the crops or animals. They also tend to focus on the production of particular parts of the crops, such as the grain and treat the rest as waste. This is unlike the older, mixed farming methods, many of which are still practised in developing countries which use the whole range of outputs from the land for different purposes – grain for eating, straw for feeding animals, mixed crops to reduce insect pests, and so on.

Although agriculture depends on the land's physical state, which can be classified according to the soil type, slope, climate and other factors into types suited to different uses, this can be changed and agricultural productivity increased by investments in drainage and other inputs. This depends on the terms by which they are available, which in turn are determined by the socio-economic forces in play in the food system. We look at this in more detail in Part 2 and the ways in which societies seek to meet their food needs. First, though, we look at how the current food system has developed in the potted history of food in the next chapter

3

Modern Food – where did it come from?

For perhaps 100,000 years Homo sapiens were successful hunters and gatherers, living in small bands, part of larger social and political alliances. Their material worlds were surely limited, but their mythic worlds undoubtedly were rich, and these treasures passed from generation to generation. Then, between twenty thousand and ten thousand years ago, people began to organise their practical lives differently, sometimes exploiting plentiful food resources in a way that allowed less mobility, more stability, perhaps more possessions. Finally, from ten thousand years onward, food production – as against food gathering – became more common, villages sprang up, small towns, cities, city-states, and eventually nation states. What we call civilisation had arrived, founded on generations of slow cultural changes.

Leakey, Richard and Lewin, Roger (1992) *Origins Reconsidered – In Search of What Makes Us Human,* Little, Brown and Co, London, pp339–40

As supermarket customers from Aberdeen to Adelaide, Washington to Wellington, facing 15,000 or more items in a store, we may find it difficult to relate the food on the shelves in the packets, tins, jars and chill cabinets to the struggle to produce the basic foodstuffs on the farm. But to understand the food system, we must. In this chapter, picking from a myriad of changes, we present a potted history of food.

The modern food system is not inevitable but has deep historical roots which are bound up with humankind's various attempts to control the biological, socio-economic and cultural aspects of food. The interplay of the forces involved has shaped the food system, producing food shortages and surpluses, hunger and overnutrition, technological brilliance and junk foods in the same world.

Today's food world is a mixed up world. We may associate coffee with Brazil, cocoa with Africa and potatoes with Ireland, but that is because these crops have been shifted around the world from their native habitat. The story of food is the history of how different people have roamed the world, conquered it, mixed it up and controlled it for their own benefit.

Humankind first gained control over the production of food through an agricultural revolution that lasted for millennia. This in turn both required and permitted great changes in the way people lived and paved the way for the development of a variety of civilizations. Exotic foodstuffs and flavourings were highly prized by the rich and powerful and much effort was spent on securing them. New technologies were developed that helped to increase productivity and supported growing populations.

The expansion of the European powers led to huge changes in food production and consumption and helped to produce a trading system that linked many different parts of the world. With industrialization, the need to control food supplies and produce foodstuffs for the growing number of workers in towns and cities led to farm produce being seen increasingly as raw material for an industry that processed it into foods. Industrial processors needed the right ingredients at the right time and place to enable their factories to run smoothly and a distribution system to deliver their products to their buyers. New technologies for processing and transporting foodstuffs allowed mass production and marketing.

Thus, an increasingly complex, interlinked system developed, for food production, processing, marketing, distribution and consumption, in which change in one area causes change in many others.

Changing the Environment and People's Behaviour

For most of human history, people lived off nature – hunting animals and gathering fruits, nuts, berries and insects. They lived in small groups and moved around, following animals and seasonal changes in vegetation. In these hunter-gatherer societies, a sexual division of labour probably existed, with the younger males hunting and the elders, women and children gathering plants and small animals. Their diet depended on their location and the season. Life was short: half of the population died by the age of 20 and 90 per cent by the age of 40.

Probably over 100,000 years ago, early humans discovered the use of fire to cook some of what they ate – at first they probably roasted meat, then perhaps they cooked roots in the embers. People maybe also discovered the keeping properties of dried meat, and that meat tasted better if it was eaten after rigor mortis had worn off – ie, after it was 'well hung'.

From hunter-gatherer to herder-grower

Around 12,000 years ago, as the ice sheets retreated from much of the northern hemisphere and the climate warmed, people started to tame and herd animals and to grow plants for food. With this major climatic change, the first agricultural revolution began. It changed how food was obtained and spearheaded humankind's push to control its environment. Starting probably in West Asia, people settled around the rich wild grain areas and then progressed from simply collecting grain to planting it. Over perhaps 2000 years, they developed methods and tools to help

them cultivate what they planted. Thus, agriculture has shaped today's environment, for by and large we see a farmed landscape – whether in the American Great Plains, Soviet steppes, European or Indian valleys, or African landscapes.

Farming animals and plants brought major advantages. Greater quantities and more regular and secure food supplies were possible. Herding animals was a more attractive and assured source of meat than hunting them and over several thousand years a number of species were domesticated (see Table 3.1). The first domesticated animals were probably helpmates such as the dingo for Australian aborigines. When cattle were domesticated they provided both milk and a power tool for pulling ploughs, which made crop cultivation more productive.

Some regions around the world harboured particularly wide varieties of different plants – wheat in the Middle East, maize in Mexico, potatoes in the high Andes in Peru. These 'Vavilov' centres of genetic diversity, named after the Russian who did most work on them in the early part of the twentieth century (see Figure 3.1), are the sources of most of today's cultivated species.

Initially, people harvested fields of wild grain; then they realized that if they planted seeds from these plants they could be more certain of getting their food. In time, in West Asia, wheat developed from crosses between primitive emmer and einkorn wheats and goat grasses. It became one of humanity's megacrops, along with rice, maize and potatoes. Travellers and conquerors spread wheat cultivation to poorer climates in Europe and Asia. In some places weeds in wheat, notably rye

Table 3.1 Approximate date of domestication of animals

Animal	Approximate date of domestication
Dingo	40,000BC
Wolf (became the dog)	11,000BC
Sheep and goats	9000BC
Pigs	7000BC
Cattle	6000BC

Source: Tannahill, Reay (1988) *Food in History*, Penguin, London

and oats, flourished and in turn became cultivated in their own right.

In other parts of the world, the native plants and animals were different. Between 7000 and 5000BC in the Tamaulipos mountains of Mexico, for example, people grew summer squash, chilli pepper and the bottle gourd. Maize was cultivated in the Tehuacan

valley. Mexico was also home to the turkey and dogs were bred for the table. Beans were grown in South America by 5680BC and potatoes by 3000BC. Rice was cultivated in northern Thailand by 3500BC and in India the jungle fowl was domesticated to become the chicken.

Tools and techniques

Throughout this prehistoric period, people experimented with techniques and tools to help them farm more effectively and passed on their experience. In about 2500BC, the Sumerians left the first written records of farming techniques, codifying thousands of years of trial and error experimentation into good farming practice. Earlier generations had developed technologies that increased productivity – harrows, rakes and mallets to cultivate the soil, and the hoe which evolved into a scratch plough, initially pulled by a man, to make a groove in light soils. When people learned how to replace

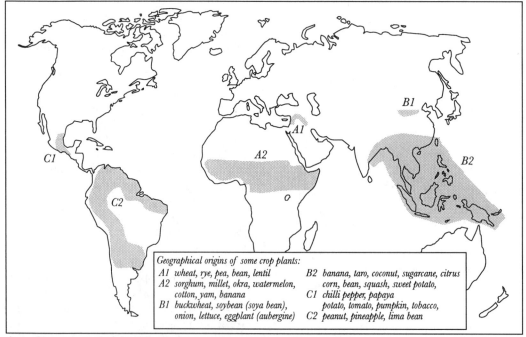

Geographical origins of some crop plants:
A1 wheat, rye, pea, bean, lentil
A2 sorghum, millet, okra, watermelon, cotton, yam, banana
B1 buckwheat, soybean (soya bean), onion, lettuce, eggplant (aubergine)
B2 banana, taro, coconut, sugarcane, citrus corn, bean, squash, sweet potato,
C1 chilli pepper, papaya potato, tomato, pumpkin, tobacco,
C2 peanut, pineapple, lima bean

Source: Chrispeels, Maarten J and Sadava, David (1977) *Plants, Food and People*, W H Freeman, San Francisco, p 122. Redrawn from Harlan, J R (1971) 'Agricultural Origins' Centers and Non-Centers' *Science*, 174, p 472

Figure 3.1 Centres of genetic diversity

human energy by harnessing oxen to pull the plough in the third millennium BC, they could use heavier ploughs and make deeper cuts, exposing more soil and thus providing more nutrients for the crops.

Techniques and technologies for harvesting, winnowing and grinding crops followed. The sickle – at first a curved blade of horn or wood with flint teeth – made the harvest possible. It was labour-intensive work.

Cultivation changed the environment. Trees and shrubs were felled. The early farmers practised slash-and-burn shifting cultivation, knowing that after a few years the land they had cleared would become desert and would have to be left unused for decades to recover. In Classical Greece, the hillsides were denuded of trees to provide houses, ships' timbers and charcoal for metal working. Rains washed away the thin soils, with floods in the valleys. There has been environmental degradation ancient and modern.

From societies to civilizations

Cultivation also changed human culture and societies, especially after about 5000BC when people in West Asia discovered irrigation – that watering their fields produced more crops more reliably. Diverting water from streams into the bordering fields greatly increased their yield, and the number of people living on the flat lands suitable for irrigation also increased. An administrative system was required to manage the canals and allocate water. However, irrigation also led to desertification, as, without drainage, irrigated lands became saline. As deserts spread, the ancient civilizations began imperial expansion and the search for new lands to control.

People became tied to the land they farmed, and forms of administration and organization developed to deal with this. Agriculture produced a more settled and secure food supply which allowed population growth and the development of civilizations. These had spare capacity which enabled them to support substantial ruling classes, priests and philosophers with the excess production. Even so, most people lived on the land.

Civilizations have since risen and fallen; cities have relied on supplies from the countryside, and the privileged and well off have relied on an ever widening circle of supply, secured by either wealth or military power, or both.

Agriculture also helped to change the division of labour, with women seemingly having increased burdens in cultivation, harvesting and food preparation. Precisely what happened is unknown, however. With managed food production, human numbers increased over thirtyfold in the first 7000 years of agriculture, from about 3 million to 100 million by 3000BC. Since then, our numbers have grown over fiftyfold to more than 5 billion by 1990.

Diverging diets for rich and poor

What people have eaten has depended, for most of human history, on what was available in their fairly immediate surroundings. For hunter gatherers, it was what they could catch or glean (see Table 3.2). For early farmers, it was what they could grow or raise. Nevertheless, agriculture led to a dietary revolution. The staple foods – those which make up the bulk of the diet – became cereals or starchy root crops. In West Asia and North Africa, these were supplemented, according to wealth, with beans, lentils, chickpeas, meat and fish.

The staple grains were eaten as pastes and various kinds of breads. Fermentation was discovered and mead, ale and wine were known when the first civilizations appeared. In Ancient Egypt, people found that bread, too, could ferment – raising or leavening it – to produce a lighter product, although a special wheat was needed to make well-leavened bread.

For the Ancient Egyptians, the River Nile provided a fresh supply of fertile soil each year, after it flooded, giving them a measure of food security. But drought could pose problems. Normally, the country was self-sufficient and it had a well-developed system of government. And from the earliest times, one of the concerns of government has been with food security for the population it administers.

Table 3.2 A rough guide to eating patterns throughout history

Civilization	Staples and everyday food	Special or rich people's food	Drink
Neolithic hunter-gatherer	Mammoth, musk-ox, bison, shellfish and fish near the sea or rivers, edible roots, greenstuffs, berries, nuts, small animals, wild beans, lentils, chickpeas, seeds of grass-like plants	Dependent on season and location	Water
Sumerian	Barley pastes or bread, beans, onions	Fish, beef, veal	Ale
Egyptian	Bread, onions		Beer
Greek	Grains, olives, figs, grain and pulse pastes, goats' milk cheese	Fish, herbs, spices	Wine
Roman	Grain pastes, course bread, millet porridge, olives, raw beans, figs, cheese	Spices, herbs, liquamen, silphium, kid, eggs, asparagus, grapes, pears, apples	Water, wine
Medieval peasant	Breads, seasonal roots, herbs, eggs, occasional meats	Pot-au-feu	Ale, mead, water
Asian nomads	Meat, blood, mares' milk, curds, yoghurt, cheeses		Fermented milk, milk
Chinese	Rice, bean curd soup, seasonal vegetables, fish, soy sauce	Imported delicacies	
Aztec (Mexico)	Maize as porridge or tortillas, beans, sauce of tomatoes or peppers	Turkey, dogs, wild game, human sacrifices	
Incas (Peru)	Maize, potatoes, squash, beans, manioc, sweet potatoes, peanuts, tomatoes, avocados, chilli peppers	Game, dogs, ducks	
Nineteenth-century UK industrial worker	Bread, potatoes, jam	Bacon or cheap meat on Sundays	Tea, beer

Source: Tannahill, Reay (1988) *Food in History*, Penguin, London

From Egypt, one of the earliest attempts by government to ensure food security is reported in the biblical account of Joseph; the Pharaoh's dream of seven bountiful and seven lean years resulted in the subsequent development of stores to provide grain for the lean years, some of which was sold abroad.

Much later, in the eleventh century AD, the

Chinese emperor Ying-tsung tried systematically to ensure food security by establishing regulatory granaries. Stock was bought up when harvests were good and was released when they were poor so that prices were not too high for ordinary people to afford.

Not everywhere could produce food as easily as the Nile delta of Egypt, and in different regions of the world some regional trade developed, in which foodstuffs were exchanged for other goods in some form of barter. Food was harder to produce for the Ancient Greeks, for example. In the sixth century BC, the only agricultural export allowed from Greece was olive oil. This, like all the legal frameworks governments establish for agriculture, affected what was produced. In Ancient Greece, it led to the gradual replacement of other remaining trees by olive trees.

After the Greeks adopted money in 625BC, trade moved from a barter system to a money-based one – peasants could not borrow grain but had to borrow money to pay for it. The Greeks came to depend upon trade for the necessities of life. As the country prospered from its exports of precious metal, olive oil and, later, wine, the diets of the rich and the poor, of the peasant and the city dweller, diverged. The availability of and accessibility to different foods changed. The poor continued to eat mostly barley pastes, gruel and bread with olives, figs and goats' milk cheese, with the occasional salt fish as a relish, while the rich ate more fish and meat, drank more wine, and ate prepared dishes with exotic, imported ingredients.

In Ancient Rome, a complex food chain supplied the city, factories produced some essential products and divisions between what the rich and poor ate grew (see Box 3.1). The Roman Empire fell to peoples who had been partly driven from their lands by invaders from the East, partly by the need for year round grazing for their herds, and partly by growing populations. They ate frugally, with milk, cheese and meat featuring in their diet. Relatively little is known about the diets of the rich or poor in Europe from the fall of the Roman Empire until about the twelfth century. Their diets probably consisted of bread, green and root vegetables, supplemented with pigmeat, rabbit, and perhaps fish.

Cooking technology limited what could be done with the food that was available. Around the Mediterranean, for example, the charcoal stove permitted frying. Further north, the cauldron over the fire was the norm, forming the basic stockpot that provided something hot to go with bread.

Varied cuisines, regular shortages

Different regions of the world developed different staples, sauces and treatment methods for foodstuffs – from freeze-drying for potatoes in the Andes to air-drying meats in China. The flocks of the central Asian nomads supplied most of their needs – meat, usually mutton, milk, blood, curds, cheese, yoghurt and a fermented mare's milk. The Mongols also produced dried milk, according to the thirteenth-century traveller Marco Polo:

> *First they bring the milk to the boil. At the appropriate moment they skim off the cream that floats on the surface and put it in another vessel to be made into butter, because so long as it remained the milk could not be dried. Then they stand the milk in the sun and leave it to dry. When they are going on an expedition, they take out about ten pounds of this milk; and every morning they take out about half a pound of it and put it in a small leather flask, shaped like a gourd, with as much water as they please. Then, while they ride, the milk in the flask dissolves into a fluid, which they drink. And this is their breakfast.*

p100 from *The Travels of Marco Polo* translated by Lathem, Ronald (1958), Penguin Classics, London copyright © Ronald Latham, reproduced by permission of Penguin Books Ltd

Most staples are rather dull on their own, and tastier sauces or flavours were used, as in ancient Rome, to increase their palatability. One mark of rank, power, privilege and wealth

Box 3.1
Ancient Rome's Bread and Circuses

In Ancient Rome, the key to survival for the rulers became 'bread and circuses' for the masses, ie basic, reliable food supplies and entertainment. The city lay at the heart of an enormous empire. It required produce from around that empire to survive as well as the smooth functioning of Roman technology. The milling and baking system was an essential part of this.

Until about the fifth century BC, grains had been ground on saddle querns with a backward and forward motion, a movable stone being pushed and pulled over the grain on the fixed stone. Then it was found that the motion was easier if a lever was used to move the rubbing stone and, somewhat later, that a rotary motion with the lever would grind the grain.

This allowed the development of a mass production technology – a couple of donkeys turning a large mill stone over a base stone – and professional millers. Rome's millers also became bakers in the second century BC. Bread baked in large ovens, preferably made from flour as refined as possible, and eaten by the rich, was regarded as the best, especially since most flour was very course with much chaff in it.

The city's rulers found they needed to subsidize food for its poorer citizens and free grain was provided for 40,000 adult male citizens in 71BC. By the time of Augustus, 320,000 citizens were receiving grain. Later bread was given away, then wine too.

The sauce 'liquamen' was essential in much of Roman cooking to flavour most dishes. Such was the demand that it was also mass produced in factories. The Romans also used a strong herb, silphium, to flavour foods until it disappeared around AD70, probably through over-exploitation from its source in Cyrene in North Africa. It was replaced by asafoetida, another pungent herb, from Persia.

Spices also featured prominently in cooking. Well-established trade routes existed before Roman times to transmit these high value products (see Figure 3.2) to the wealthy around the Mediterranean. Such was the demand for spices in Rome that the traditional suppliers, the Indians, who supplied via Arab merchants, had to look farther afield and find sources from what is now Indonesia.

Paying for all these imports proved a drain on the Roman economy; spices accounted for 44 of the 86 classifications of goods imported to the Mediterranean from Asia and the east coast of Africa in the first century AD. After Rome collapsed in the fifth century AD, people returned to more of a barter economy and to the land. Even in classical times, however, about 90 per cent of the population was still engaged in agriculture.

was access to more varied and exotic foodstuffs such as spices. People in different regions and societies also developed detailed rituals and methods for dealing with foodstuffs and deciding which were not to be eaten, – for example, the Jewish law and prohibitions on pork and the Hindu rules on beef. Different religious traditions played a prominent part in developing laws for their societies to follow, and food laws and rituals often demonstrated 'belonging' and 'sanctity' to the religious group.

Despite managed food production through agriculture, food supplies were still subject to the vagaries of climate. Everywhere at the end of winter, food was relatively scarce.

In bad years, after poor harvests due to climate or disease, people starved.

Ergotism, a disease of rye, caused both famines and illness in Europe in the Middle Ages. Arab invaders brought with them the barberry bush which harboured rusts which attacked wheat and led to harvest failures. But the Arabs, who had been the middlemen traders in spices since the seventh century BC, also brought irrigation as well as citrus fruits, almonds, rice, sugar and saffron to Europe.

In Asia, where nomads had herded animals for millennia, the people at the western end became stockbreeders and developed intricate, complex organizational systems to

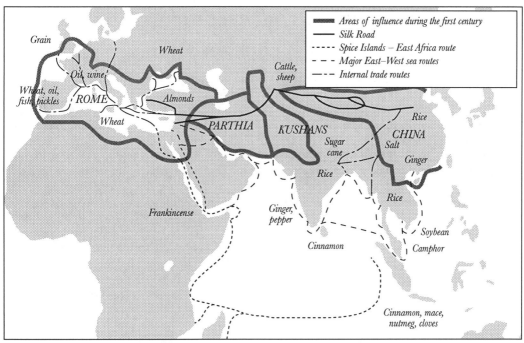

Source: Redrawn from original map by E H Holloway

Figure 3.2 Ancient trade routes

regulate stock movements and the use of grazing lands. But when their herds were threatened by drought, they moved both west and east – at various times sweeping into Europe or clashing with China.

In China, most people ate mainly cereals and vegetables, while the feudal lords had a different cereal-based diet that was high in meat and fish. The cooking style that developed, of stir frying for short periods, may stem from the time when peasants moved to temporary accommodation from spring to harvest time when fuel would have been scarce. The Chinese also used soya beans to provide a strongly flavoured sauce to liven up their basic staple, rice. Soya beans also provided high protein curds, milk and bean sprouts.

Rich Chinese were keen to try imported delicacies which could travel along trade routes like the Silk Road, which ran from China to the Mediterranean. One such imported luxury was 'stone honey', which was made from the sun-dried boiled juice of sugar cane.

Sugar cane had been introduced into India by the fifth century BC, most probably from Indonesia or New Guinea. The Aryans, who invaded India in the second millennium BC, seem to have introduced the liking for dairy products, like ghee (clarified butter) and curds, and these made up an important part of the ordinary diet, along with vegetables and grains. The basic grain varied according to the part of the country, with wheat in the north, millet in the drier areas, and rice on the plains. Sauces made from a variety of spices, lumped together under the umbrella term 'curry', played an essential part in making these interesting and palatable.

A Second Agricultural Revolution

Already by the Middle Ages, much agricultural technology had been developed and very great movements of food sources, flavourings and tastes had taken place. High-value, low-weight, flavour-enhancing spices were widely traded and some civilizations, like the Roman, had developed major import-export capacities, notably for grain. These were restricted to particular regions, however, and most people still lived in the countryside. They ate what they grew or herded around them, traded with neighbouring communities and supplemented their staple, bulk foods with flavour enhancers ranging from soy sauce to pepper. The food movements, however, were relatively slow and limited compared with those still to come, in which Europeans played the central role.

Europe changes

Technological changes underlay socio-economic changes in Europe where religious, trading, administrative, craft and other groups developed. These had differing interests and expectations which were increasingly divorced from those of the peasants who produced their food and over whom they ruled. Between about the sixth and ninth centuries, three new technologies – the plough, horseshoe and crop rotation – brought about a second agricultural revolution in northern Europe and set the scene for European agriculture for nearly a thousand years.

For millennia, peasant farmers had used versions of the scratch plough invented by the Sumerians, which simply scratched a V-shaped furrow in the soil in which seeds were planted. But this could not plough heavier land. It was the mouldboard plough – developed first in China between 200BC and AD200 and introduced into Europe by the Slavs in the sixth century – that allowed more and heavier land to be cultivated. This meant that more people could be fed and allowed population growth. But the mouldboard plough needed eight oxen to pull it and it was

expensive to make. Only richer farmers could afford to own one; others had to co-operate to run one plough and also had to merge their land into larger fields.

Soon after adopting this plough, European farmers found that they could make land productive for two out of three years by switching from a two-field to a three-field rotation. Instead of growing a crop in one field one year and leaving the other fallow, one field could be planted with wheat or rye at the end of one year; peas, chickpeas, lentils, broad beans, oats or barley could be planted in another field in the spring of the next year and the third field was left fallow.

Then, with the development of horseshoes and the use of a new collar-harness, which also originated in China, farmers found that they could reduce the number of draught animals needed, as horses could pull much more than oxen. When they were not tied to the plough, horses also offered farmers freedom of movement.

Urban growth and urban needs

The more productive technology underpinned the growth in the number and size of European towns in the Middle Ages. Controlled, peaceful markets were needed to supply townspeople who increasingly used money rather than barter to obtain what they wanted. Even so, the towns produced much of their own foods, as the former peasants who now made up the townspeople grew what they could and kept animals. The markets needed regulation to protect the customers from fraud, and various administrative mechanisms were developed for this (see Box 3.2).

The cities' demand for cereals led to greater production throughout Europe, but when the Germans opened up the Slav lands in Eastern Europe in the twelfth century, the plains there produced abundantly and water transport allowed the produce to be carried west easily and cheaply. Subsequently, many areas in Western Europe went out of cereal

Box 3.2
Food Fraud and Adulteration

Food fraud of all kinds has a long history in Europe's towns and cities and was widespread in medieval times. It persists today. Some medieval merchants and suppliers sold underweight or adulterated products like wine, ale, flour and oil. Guilds grew up to protect and self-regulate different trades and skills, while the authorities also tried to regulate both their activities and market trading. Inspectors were appointed by authorities throughout Europe – from the French langueyeurs who inspected pigs' tongues for ulcers to Italians who checked the fish stalls to destroy stale fish.

In 1327, a London baker, John Bird, was caught cheating his customers by using a small trap-door cut on the table where they laid their dough to be baked in his oven. One of his staff cut pieces off the dough while he talked to the customer before he put the trimmed loaf in the oven. He used the stolen dough to make bread for public sale.

Adulteration became more acute as people piled up in cities, especially after the Industrial Revolution. The food chain got longer with handling, transport, storage and marketing needed, and food suppliers dealt with more people. Cheating was particularly lucrative for scarce, expensive products. The answer, for the unscrupulous, was to bulk them out with whatever seemed suitable. Some, such as mustard husks, pea flour, or juniper berries which were added to pepper, or ash leaves added to green China tea, were relatively harmless. Others, like the black lead tinted used leaves added to Indian tea, or poisonous copper and lead salts used to colour sweets, were less so. Still others could lead to widespread acute food poisoning.

The growth of scientific analysis in the nineteenth century brought laboratory tests that could show just how much food was adulterated. The results came as a shock – bread with alum in it, beer with iron sulphate, coffee with chicory or acorns, and cocoa with brick dust.

The demand grew for food purity laws and in 1860 the first British Food and Drugs Act was passed, to be revised and strengthened in 1872. The Cooperative Society made much of their impact at this time by selling wholesome, unadulterated products and providing the education to make this worthwhile.

Adulteration, however, is not a thing of the past. Our favourite tale is of a man convicted in Italy in 1969 of selling grated umbrella handles as Parmesan cheese. In 1991 a survey of pure, unsweetened orange juice showed that three-quarters of the 21 samples from Britain's leading suppliers were adulterated with pulp wash, corn syrup and/or sugar; this too required new analytical techniques to detect the adulteration.

Other kinds of fraud also continue. In the EU today, for example, the European Commission is trying to stem multi-million dollar frauds by those making false claims from the various subsidies and export refunds available through the EU's Common Agricultural Policy.

---

production. Some moved to vines, as in parts of France, and stockbreeding increased.

The towns were dirty, smelly and rat infested – the black rat lived off refuse and its fleas spread typhus to human lice which infected people. Townsfolk tended to be malnourished and were vulnerable to food shortages. After bad harvests, landlords and peasants kept more of what they had, townspeople went short and prices rose. But, in famines, the rural people had to look after themselves, as any reserves of grain tended to be distributed in towns. In many places, grain reserves were kept in specially built granaries by guilds and city administrators to provide some security of supplies – one of the achievements of Tudor and Elizabethan England.

Sheep, which provided meat, milk and wool, were the preferred animal, but they had to be moved from summer to winter grazing, which caused conflicts with crop farmers (see Box 3.3). Sheep's milk was consumed in England in the thirteenth century, but 300 years later cows' milk and its products were preferred.

Keeping animals over winter in the Middle Ages was difficult owing to the lack of feed. Many were killed in the autumn and some means of preserving them was needed. Salting was the chief means. Either a dry method was used in which the meat was packed in a granular bed of salt (but it was expensive to produce the necessary fineness); or the meat was cured in brine, – ie, immersed in a strong salt solution. However, salting by either method could be too expensive for many, and only a fat carcass would make it worthwhile – which may be why relatively little mutton was salted.

The technology available to process and cook foods in the home was also very limited. In the town, there was communal use of bakers' ovens, and shops of various kinds developed which sold ready-made foods. The idea for such fast foods dates back millennia to the Mesopotamians, but in Europe it took a monetary economy and sufficiently sizeable towns for it to spread from the 1100s onward. A leg of roast mutton in England in 1363 cost a day's wage for an agricultural labourer – one penny. Drying was the other main preservation process used during the Middle Ages. In much of Europe it was used more for fish than for meat but in other parts of the world, it was used to preserve fruit, vegetables and meat. Although the rich may have been able to afford fresh produce for most of the year, for most people

---

Box 3.3
Can Growers and Herders be Friends?

Herding domesticated animals brought about a conflict that persists in places today – that between livestock herders and crop farmers. Sheep, cattle and goats moved where the pastures were and, in the more settled transhumance system, to higher land in summer and lower land in winter. This could mean crossing farmland given over to crops. Elaborate conventions developed in various societies to minimize the conflicts.

Even today in Turkey, for example, animals are moved up to mountain pastures in summer and down again in winter. Flocks and herds are always accompanied by a shepherd whose job it is to ensure they do not graze the growing cereal fields and to find pastures.

States, however, may want to develop their agricultural resources and settle nomadic herders. The dislocation caused to herders when the great Cukurova plains in Turkey were closed off from them for irrigated crop production formed the basis of a Turkish novel:

Beyond Aladag Mountain is a long valley, densely wooded, with hundreds of springs bubbling forth everywhere, bright cool pebbly springs bedded in mint and heather. It is light that flows from these springs, not water, but a tinkling brightness. Since time out of mind this valley has been the summer pasture of the wandering Turcomans, the Yöruks and the nomads of the Aydinli tribe. Just as the Chukurova plain has always been their wintering place, so the long valley beyond Aladag has been the summer pasture....

In 1876 a battle took place between the Turcoman nomads and the Ottoman rulers. The Ottomans wanted to settle the nomads, to tie them to the earth, to make them pay taxes and enrol them in the army. The Turcomans refused to be yoked. They resisted fiercely, but were beaten in the end and compelled to settle.

Kemal, Yashar (1976) *The Legend of the Thousand Bulls*, William Collins, London

Modern intensive animal rearing systems avoid moving animals to the food by keeping them in one place and growing or manufacturing food which is taken to them. These rearing methods have already spread to pigs, poultry and cattle and may spread to sheep next.

the salted meats were probably cooked with dried peas or beans, or whole grains which together with the spices, masked the saltiness. Boiled salted meats went well with bland, thick mixtures, which were usually based on grain, crumbs or almond milk and rice.

By and large, poor people's food consisted of dark bread with something from the stock pot, cheese and curds, all washed down with ale or wine. For the richer European, who lived in the country or was a town merchant, a meal was structured rather like a Chinese restaurant meal is today, with a range of dishes being served and people picking from a few of them.

Cooking and eating implements were limited. A dagger-like knife and a spoon were the norm in northern Europe until the eighteenth century, as forks spread from Byzantium to Greece to Italy and then further north only slowly. Some dishes needed no implements – for example, pasties which had the meat, sauce and plate all in one package.

Expanding Empires and Shifting Foods

One of the most crucial issues for the wealthy traders in Europe's medieval cities was control of trade and access to items valued by the wealthy in these different societies, for example spices. The spice trade was dominated by the Italian city states. The Arab conquests from the seventh century had disrupted the older trade routes, with spices, for example, now being routed via Baghdad to Trabzon, Constantinople and thence to Venice. The Crusaders took the taste for new foods and spices back to Europe, and the Venetians, who had diverted the fourth Crusade to sack Constantinople in 1204, took a monopoly grip on the spice trade.

The Venetians' grip on the extremely profitable spice trade was a spur to the great exploration from Europe which led to the integration of the world's foods. It is unclear how profitable this was in the fifteenth century, but Reay Tannahil in *Food in History* quotes one estimate in 1621 that suggested that 3000 tons of spices could be bought in the Indies for £91,041, but by the time they reached Aleppo the price had reached £789,168.

The Portuguese and Spaniards wanted to get a greater share of the profits by finding their own sources of spices. To do this they had to find a way around Africa to the East or sail west around the world. The Portuguese founded an empire stretching from Brazil to Indonesia in the process. The Spanish found Mexico and South America. Later, the Dutch, French and British all explored the world to ensure that they, too, could reach the spices and find other riches, including slaves.

Europeans began to spread food (and drink, see Box 3.4) from the Americas around the world. The Spanish found manioc (cassava) in Cuba but the Portuguese transferred it to Africa in the sixteenth century, since when it has become a staple African food crop. Manioc can resist locusts and stay edible when left in the ground for over two years, although it does need careful processing to remove the toxins before consumption.

Maize was carried from America across to the Mediterranean and, possibly with Ferdinand Magellan, on to the Philippines before the middle of the sixteenth century. The Portuguese introduced it to Africa where it grew more rapidly than other grains, but over reliance on it as a foodstuff in Africa and parts of Europe led to disease – pellagra. Maize lacks the vitamin niacin, which the native Americans obtained by eating maize with tomatoes, capsicum peppers and fish. But most people ate very little fruit and vegetables in Europe and Africa at that time, so when maize became the staple foodstuff, diseases caused by dietary deficiency resulted.

In time, the native Mexican tomato and capsicum pepper became ubiquitous in Spanish cooking. In Peru, the Incas ate mainly vegetarian food. Maize was the staple in the lowlands, while potatoes and another tuber, the oca, were eaten in the highlands. The native peoples

Box 3.4
Mixing Drinks

Beverages are an important part of our diet, although they make a varied nutritional contribution. These too have changed greatly since the Middle Ages. In northern Europe, grain-based beers and mead, even at breakfast, were the main drinks until tea, coffee and chocolate became popular. Spirits were widely produced by the sixteenth century following the discovery by alchemists in the twelfth century of how to separate liquids from one another. Water was often contaminated so fermented drinks could be safer and also made a considerable nutritional contribution to the diet.

Tea found its way to Europe from China, possibly by way of Japan, but it had arrived in Holland and Portugal by 1610 and the first public tea sale occurred in England in 1657. Coffee travelled from Ethiopia to the Arabian peninsula to Mecca, Cairo, Damascus, Aleppo and Constantinople and on to Europe. Oxford boasted its first coffee house in 1650. In 1720 the Dutch introduced the plant into Java and then Ceylon, and the English took it to the West Indies. One plant arrived in the north of South America and later four arrived in Brazil. For many years the whole of the South American crop came from these plants. In Europe, both tea and coffee remained drinks of the well-off until the Industrial Revolution.

learnt how to preserve the tubers by freeze-drying. After harvest, the crop was exposed and left overnight in the cold. Next day men, women and children trod out the moisture. This process was repeated for about four or five days and the tubers were then dried and stored.

The Spanish used potatoes as basic ships' stores and so took them back to Europe. By the 1570s, they seem to have been common in Spain and by the end of the sixteenth century, they were grown in Italy, Belgium and Germany and, to some extent, as food for livestock in England. They proved to be a most useful crop, with great masses of people able to subsist cheaply on them.

The Spanish also took another Peruvian crop, the groundnut or peanut, half-way around the world, although they had first met it in Haiti. It went to the Malay archipelago and then to China in the early seventeenth century.

The cocoa tree could not be grown in Spain or in the rest of Europe. It only grew in the Spanish and Portuguese colonies and for over a hundred years they maintained a monopoly over its production and over the method for preparing it for consumption (see Box 3.5).

It was another crop, however, that had the most dire consequences for people half-way around the world – sugar cane. The Spanish

began to cultivate it in the Greater Antilles, which include Cuba, at the start of the sixteenth century. Labour became short as the local populations were decimated, largely owing to illnesses against which they had no natural immunity. At home, slavery was common and the Spanish turned to slavery to fill the plantations, as did other European powers.

Gold and silver became more attractive for the Spanish, and the Portuguese saw the profitable opportunity offered by the use of slave labour to grow sugar in their huge colony in Brazil. They exchanged slaves for cloth, hardware, spirits and firearms in West Africa and took the slaves to Brazil. The five sugar plantations in 1550 grew to over 350 by 1623. The Dutch, English, French and Danes followed suit into the sugar and slave trades, so that by 1600 up to one million Africans had been moved to the Americas; this rose to two and three-quarter million in the seventeenth century and seven million in the eighteenth century. Deaths in passage were in the millions, too.

Sugar became very important in the trade and diets of the Europeans, so much so that the Dutch gave up New Amsterdam (New York) to the English for the sugar lands of Surinam. The French gave up Canada for Guadeloupe where they could grow sugar. It arrived on the scene as supplies of the tradi-

Box 3.5
Chocolate — from Montezuma to Modern Consumer

For the Mayan Indians who first used cocoa around AD600, cocoa beans were money as well as the basis of a bitter, spicy cold drink they called 'chocolatl'. The Aztecs of Mexico learned about cocoa and chocolatl from the Maya who paid part of their tribute (a form of taxation paid in goods) to the Aztecs in cocoa beans.

When the Spanish conquistador Cortes confronted the Aztec emperor Montezuma at the Aztec capital Tenochtitlan (now Mexico City) in 1519–20, cocoa beans were sold in the market and chocolatl was drunk in the palace. When Cortes returned to Spain in 1528, he took some cocoa beans and chocolate-making equipment with him and for the next hundred years the Spanish court kept the secret of chocolate to themselves. But they dropped the chilli pepper from their version of chocolate and just added cinnamon, nutmeg and sugar.

Chocolate spread to France and Britain in the seventeenth century as a high class drink. In London, White's Chocolate House opened in 1693. The Europeans took cocoa cultivation to other tropical areas. Slowly supplies increased and prices came down. In the early 1990s, it was grown in 33 countries with the main suppliers being in West Africa and Asia.

By the nineteenth century, cocoa was a popular drink made with cocoa powder or pieces of cocoa blocks made by many firms. In Britain, several Quaker businessmen — the Cadburys, Frys and Rowntrees — promoted it as an alternative to alcoholic drinks.

The processing method left quite a lot of fat in the cocoa — the bean is over half fat — so many firms added various ingredients to make it more palatable. Some were innocuous like potato starch and sago flour, perhaps treated with treacle, but others were unpleasant, like brick dust and iron oxide, added by more unscrupulous firms to stretch the amount of cocoa powder in the packet and so make it cheaper.

In the late twentieth century, however, chocolate means a solid sweet melt-in-the-mouth snack, not a drink, to most people. The change began in the nineteenth century with improved technology for preparing cocoa. In 1828 a Dutchman (van Houten) invented a press which squeezed much of the fat (cocoa butter) from the beans. This produced a purer cocoa and made additives to make the drink more palatable unnecessary.

The waste product, cocoa butter, is essential in making eating chocolate. When cocoa butter is mixed with sugar and added back to the cocoa liquor, it enables plain chocolate to be set into moulds easily and accounts for the melt in the mouth quality. The cocoa butter is solid up to 31 °C but melts at body heat of 36 °C.

In 1875 a Swiss manufacturer (Daniel Peter) produced the first bar of milk chocolate using powdered milk, although since 1849 Cadbury brothers had sold a milk chocolate drink. George Cadbury took a van Houten press to Britain in 1866 and began experimenting with it. Following the success of milk chocolate, Cadbury's developed their own milk chocolate bar, made from fresh milk, and launched it in 1905.

The firm made sure that its products were marketed and distributed. It developed its own transport fleets in the nineteenth century, using canal boats, the railway and horse-drawn, then motorized vehicles. Machinery was used in the factory for the key processes, but many thousands of people were employed to check, wrap and pack the products. Craftsmen from many trades — such as canister- and box-making, machine construction, sheet metal work, printing, carpentry and joinery — worked in-house, but these skills were gradually phased out in the 1960s and the work was sub-contracted out.

Recently, new machinery was introduced which reduced the need for labour still more. A worker, who worked in the Bournville factory for over 40 years, recalled 11,500 employees in the factory in the 1950s, but by the early 1990s just 2000 ran the production.

Box 3.5 *continued*

Confectionery had become big business. In the UK in 1992, spending on sweets and chocolate was about £100 per second, and total sales reached almost £4.2 billion – 71 per cent on
chocolate and 29 per cent on sugar sweets. The big three manufacturers in the UK – Cadbury,
Mars and Nestlé–Rowntree – accounted for over three-quarters of all chocolate sold there.

World-wide, chocolate accounted for over half of the US$73.7 billion confectionery market in
1991, according to Euromonitor. Almost 60 per cent of these sales are in Europe, 18 per cent in the
USA, with Asia and the Pacific accounting for 14.5 per cent. The top three chocolate markets by
value in 1991, reports Euromonitor, were Germany (US$5.8 billion), the UK (US$4.6 billion) and
the USA (US$3.9 billion), out of total chocolate sales of US$37.7 billion. The top three consumers
of confectionery as a whole (including chocolate, sweets and chewing gum) by country in annual
per capita terms are the Germans (16kg), British (14kg) and Belgians (13kg), whereas the Brazilians,
a large consumer in terms of developing countries, eat only 2kg per capita per year.

In Britain, Cadbury has managed to turn part of its factory into a tourist centre where up to
3000 visitors per day pay to hear the story of chocolate and see Cadbury products promoted.
Advertising has played a major role in this sector and displays include the company's advertisements since the 1950s. A recent advertisement for a box of chocolates took over six months to
make. In 1992, all confectionery advertising in Britain was estimated to have cost £95 million.

New technologies enable new products to be produced and these are carefully, and expensively, market tested before they are launched nationally. The brand image is being stretched further
too, with the name Cadbury being franchised out to makers of biscuits, cakes, ice cream and deserts
who use Cadbury chocolate in their products.

Cadbury is now part of Cadbury–Schweppes, which operated through over 50 companies
world-wide in confectionery and beverages (soft drinks) at the end of 1993. It had sales of £3.7 billion in 1993, made £416 million before tax and spent £425 million on marketing its wares. The
industry is becoming more international with companies seeking to globalize their brands. Some 34
of the world's 40 leading confectionery companies are from North America and Europe and there
are just six leading players in the international market, according to Euromonitor.

tional sweetener, honey, were dwindling, partly owing to the decline of monasteries as a
result of the Reformation in Britain.

Other plants taken from the Americas
included tobacco and rubber, both of which
had a major impact on agriculture elsewhere
in the world. However, the only significant animal to travel from the Americas was the
turkey, although European domesticated animals were imported to the Americas. Europe's
plants, too, were transported overseas, including the staples wheat and barley. The
discovery of the vast fishing banks of
Newfoundland in the sixteenth century gave
impetus to deep-sea fishing, with large fleets
from Europe trawling there.

As the Europeans explored and colonized
as much of the world as they could (which also
posed problems for the sailors, see Box 3.6),

they took their animals, plants and agricultural
methods with them. North America was settled at first by the British, French and Dutch.
The settlers developed farming and gradually
expanded across the continent, taking grain
production and cattle rearing with them. Later
in Australia the introduction of European animals caused major changes which are now
viewed as an ecological disaster – for example,
with rabbits, goats, cats and foxes, swept
through the continent and colonized it. In New
Zealand, European animals had a devastating
effect on the unique flightless birds that had
developed on the islands over tens of millions
of years, and on other animals that could not
compete with the introduced species. The
country's native vegetation, too, was largely
wiped out and replaced by plants introduced
by the immigrant farmers.

Box 3.6
Limeys

The long sea voyages of the explorers and traders meant there was a growing demand for food for sailors and travellers. The first priority of crews on reaching land after a long trip was fresh water and food. Often they had to survive on worm-infested ships' biscuits, and many died from scurvy which was later understood to be the result of vitamin C deficiency – half of Vasco da Gama's crew on the voyage to find India died from scurvy.

Although green herbs and citrus fruit were known to cure scurvy by about 1600, they were not provided for the average sailor until over two hundred years later, probably because of cost. By the end of the eighteenth century, the British Navy was giving out citrus juice, usually mixed with the rum ration, after crews had been at sea for five or six weeks. This cut the mortality rate dramatically. When lime juice from the West Indies was used from the mid-nineteenth century, British sailors became known as 'limeys'.

From Agriculture to Industry

In Europe, the introduction of new crops and animals from the colonized lands changed the landscape there, too. But other changes in European society also affected food production. The Reformation shifted power between the church and state, altering landowning patterns. Changing demands for capital from landowners in response to social and economic pressures brought demands for greater income, and consequently for greater productivity from the land. These helped to fuel a third agricultural revolution.

Many of the individual attempts at what became known as 'improvement' were pulled together in the Low Countries – Belgium and Holland. Little land was available there, so quite intensive production developed with the use of organic fertilizer to replenish the land's fertility. A complex sevenfold rotation system was developed there in the eighteenth century along with specialized dairy farming.

Frederick the Great turned to Holland for advice in developing Prussian agriculture and produced a modified fourfold rotation system. These rotation systems were able to produce animal feeds so that cattle could be kept over the winter. Jethro Tull developed a seed-planting drill, which reduced waste and improved yield as a result of even sowing in lines. The growing industrial revolution in the late eighteenth and early nineteenth centuries allowed the mass production of machinery.

Enclosing Britain, altering the world

The effects of these changes, coupled with other developments, had a decisive impact in Britain, which in turn helped to change the world. British farming changed dramatically from the mid-seventeenth to mid-nineteenth centuries with the new methods, the introduction of new crops, seed drills and rotations. Flax, potatoes and sheep all played a part. The potato, introduced from South America, grew well. It enabled a much greater yield of food to be obtained from the same area and allowed more people to subsist off less land. Some land could thus be released for more profitable use. Flax was more profitable than food crops in eastern England. In the Scottish Highlands, grazing sheep were more profitable than tenant farmers producing largely for subsistence, who were cleared off the land.

In England, the enclosure of common land – an early form of privatization – and the consolidation of fields changed the social structure and prevented many people from farming. The former feudal lords became landowners who wanted higher incomes to fuel new, more affluent lifestyles. They rented land to a small number of tenant farmers rather than to the many peasants who formerly farmed the land. These peasants became agricultural labourers or landless and, in Britain, the peasantry ceased to exist. Decision making about farming the land shifted from the peasantry to the landowners who invested through their tenant farmers for income. Landless peasants also had their uses – as settlers in overseas colonies, as producers in domestic manufacturing, for example of cloth, and then as workers in the factory systems of the nineteenth century.

Major political events also promoted agricultural change. The Napoleonic Wars in the early nineteenth century, for example, brought pressure for a more intensive use of land in Britain, which was subject to an economic blockade.

With the expansion of European control of much of the world, the production of the colonies was moulded to suit the needs of the colonial power. The British turned parts of West Africa and the West Indies into plantations – large monocultures – to supply raw materials for the soap or sugar industries in Britain. Later, other exotic products for the growing middle class, who could afford to spend more of their income on non-essential items like tea, chocolate and bananas, could be imported in quantity. British concern and agricultural research in its colonies was almost exclusively geared to export crops, and local food production was neglected.

Industrial revolution – the British experience

Alongside these changes, and itself influencing them, was a development that has done most to shape food today. It was the first Industrial Revolution. With industrialization, a new, more linear approach to production developed, with many more actors and interests involved in what became an increasingly industrialized food system. It was a revolution not simply of technology, although technological change and innovation were essential, but also of the social and economic relations between people and their work. Two important, if often neglected, economic elements

were the development of banking – with the creation of credit money by lending what one person deposited to several people, thus providing finance for invention – and the limited liability company. The latter removed much of the risk to private individuals and entities in developing new products and methods. Instead, society as a whole is left to deal with the consequences of effects that are not allowed for in the economics governing company operation, such as pollution or poverty due to low wages.

During the first Industrial Revolution, which began towards the end of the eighteenth century, production became focused in centralized units – ie factories. Workers came to factories to produce rather than work in their homes, as was previously the case with cloth, for example, or in small craft workshops. Towns and cities expanded greatly, with the population both drawn to the towns and pushed off the land.

In the nineteenth century, the growing urban populations needed food and the employers wanted them fed cheaply, which meant bringing in food from outside Europe and, for some, reliance on a single crop – potatoes – for the bulk of their food intake. A steam and transport revolution allowed the transport of grain, and new preservation and freezing techniques developed during the nineteenth century – canning, freezing and chilling – which permitted a greater range of foodstuffs to be moved long distances.

Industrialization brought a great increase in productivity, but the increased production had to have markets, which in turn fired up the desire for the Empire, especially in Britain. The farming and production practices in whole countries were reoriented by the colonialists. They produced raw materials in the colonies, shipped them to the industrial heartlands, first to Britain and then to the rest of Europe, and exported processed and finished products back to the colonies. The Indian cotton industry, for example, was decimated, with raw cotton exported to the mills of Lancashire and finished cloth sent back to India. Plantations of rubber, cocoa, coffee, tea and jute spread around the Europeans' overseas conquests. More and more of the national wealth, measured in terms of the gross national product, came not from agriculture and processing its products but from manufactured goods of an ever increasing variety.

In Britain, the new rich of the towns, the industrialists, had different interests from the traditional power holders, the landed gentry and landowners. They wanted cheaper food to be made available from the colonies that had been opened up in the Americas and Australia and New Zealand, so that wages did not have to rise; they did not want a local market that was protected from overseas competition. 'Free trade' therefore became their economic doctrine. The growing middle class, with greater disposable incomes, could afford a greater variety of foods.

The power of the new industrialists prevailed over the landowners in Britain when the repeal of the Corn Laws in the 1840s removed the protective barriers from wheat produced in Britain. This opened the door for cheaper American and Canadian wheats from the vast prairies. British land prices fell, but industrial workers could be fed more cheaply than they could from home-produced food.

The flood of imported grain affected all of Europe's agriculture (see Figure 3.3). Farmers from Norway emigrated in large numbers, the Danes turned more to bacon and butter production, and in England much land went out of grain production and into fruit, vegetables and livestock. In Germany and France, where the peasantry still existed, agricultural interests were politically powerful enough to be able to defend themselves through protective measures and tariffs. In North America, land was plentiful for the settlers, labour was short, and this gave great impetus to the development of labour-saving farm machinery.

These changes took time to work through and for a few decades after 1840, there were attempts to increase agricultural productivity in Britain. Government help, in the form of low interest rates, was given, notably for improved drainage. Fertilizer use was growing, which relied partly on the import of phos-

phates from guano deposits off the coast of Peru. Mass-production techniques allowed manufacturers to produce large quantities of new agricultural equipment, like chilled iron ploughshares, reapers and threshers.

Towards the end of the nineteenth century, however, investment was heading abroad. The central policy in Britain was to obtain cheap food from the colonies and other suppliers world-wide. There was a great agricultural depression in Britain and arable land was converted to permanent pasture. This required fewer workers and more people left the land. Land prices fell and by 1914 only a quarter of Britain's food was produced at home. Investments abroad in transport – railways and refrigerated ships – enabled grain from North America, bananas from the West Indies and refrigerated meat from Australia and New Zealand to arrive in Britain. New technologies for the mass production of white flour, and instant foods like biscuits, chocolate, and so on made them increasingly available at prices most British people could afford.

The new European meat markets led to a rapid expansion of cattle farming in the USA from 1870 to 1890, with cattle taking over the Great Plains from the buffalo and the native American Indians. Australia and Argentina also supplied beef, while New Zealand became a mutton supplier. But the preservation, packing and transport costs were such that the price could not be brought down low enough for the poor of Europe and production exceeded effective demand. This, coupled with climatic disaster in the USA in the 1880s, led to many large meat packing concerns going out of business there and farmers becoming a major political lobby. In Australia, sheep farmers came into conflict with gold-diggers, while in Argentina, conflict between the locally-owned ranching interests and largely foreign-owned packing interests continued until the 1930s.

For poor people, crowded in often appalling conditions in the industrial cities of the nineteenth century, life was hard and their diet insufficient. They subsisted on white bread, potatoes, tea, jam, some cheese and cheap meat or a bacon joint on Sundays. Scurvy, rickets and

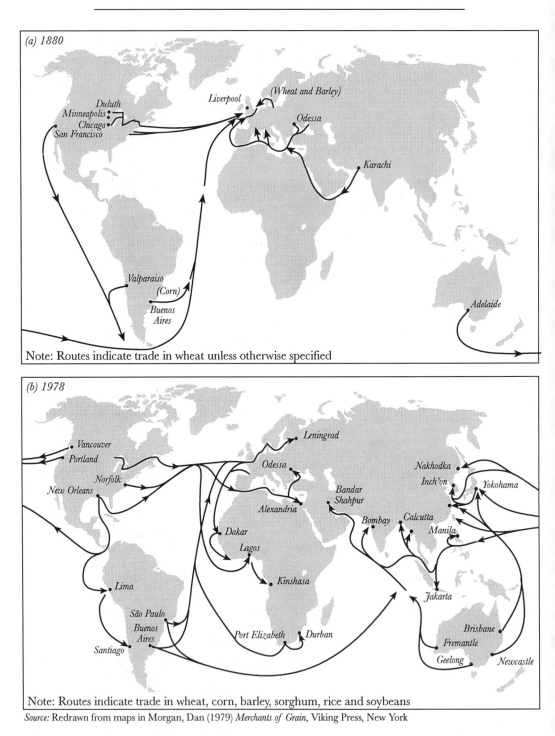

Figure 3.3 Major world grain routes, 1880 and 1978

tuberculosis plagued the poor. The rich ate plenty, although their diet was changing. Dishes were becoming grouped together, first in France, then in Britain. This produced the three-course menu which is now considered traditional – with a starter, main course and dessert. For the middle classes a new range of foods, such as bananas and foods preserved in new ways, such as tinned meats, peaches and apricots, became available for their tables.

Science, technology and change

Science and technology were essential elements in the socio-economic changes of the Industrial Revolution. In post-Reformation Europe there was a growing sense that people could both understand the world and shape it. This, coupled with powerful political and economic interests, gave rise to increasing scientific and technological exploration. Eventually, scientific inquiry into understanding nature came together with the work of inventors tinkering with it to find new machines, techniques and tools to use, which continues until today.

Improvements to agricultural production technologies were the first to emerge, as discussed above. However, during the nineteenth century, scientific experimenters began to focus on how plants and animals grew and how farming might be improved further. In 1848, Lawes and Gilbert started the first experimental husbandry farm at Rothamsted in southern England and looked at the response of plants to fertilizers.

New distribution and processing technologies were also needed to release the productive potential of all the lands open to European colonists, so that these lands could be used profitably to feed industrial workers in Europe. Bulk food distribution of grains required the development of railways overseas which allowed grain to be transported cheaply to ports and then shipped to Europe. British capital invested widely in overseas railways and its former and existing colonies became its offshore farms and markets. The price of home produced wheat fell from over 50 shillings in the 1850s to less than 30 shillings in the 1890s.

Processing and distribution technologies

Canning, freezing and chilling technologies transformed preservation and distribution in the nineteenth century. A Frenchman, Nicholas Appert, took the first step in canning as he preserved meat, fruit and vegetables in glass bottles, which were heated and sealed. Then, in England in 1812, Brian Donkin used tin cans in 2–6lb sizes. The treatment was effective in these, but when contractors used larger 9–14lb sizes for canned meat for the British Admiralty, the heat treatment was ineffective, leaving bacteria still active in the middle which caused the meat to putrefy, so that it was unfit for human consumption. This did nothing for the reputation of canning.

At that time, people believed that the preservation effect was due to the exclusion of air. It was not until after the 1870s, when Louis Pasteur's work on bacteria was gaining acceptance and the Massachusetts Institute of Technology worked out the appropriate heat and time treatments for different sizes and products, that canning became convenient and safe.

Meat was the main item canned until the late nineteenth century. Australia supplied the UK market with 16,000lb of canned meat in 1866 and 22,000,000lb in 1871 at half the price of fresh meat in England, but the USA took over as lead supplier after its Civil War ended in the mid-1860s.

Then better methods – freezing and chilling – were introduced for meat preservation. For industrial workers, canned fruit and vegetables had the greatest effect, because new foods were introduced into their diet year round. After meat, fruit and vegetables were the second most popular preserved foods.

In the 1830s, ice-making machines were invented, based on a compression-evaporation principle, and were improved in the 1850s with an ether compressor that allowed better refrigeration. Sea refrigeration had to wait for development of the ammonia compressor which allowed refrigerated ships to take frozen meat from Australia to Britain. The first shipment arrived in 1877. Chilled meat could be imported from the USA as the journey was shorter. Fishing, too, benefited from the development of ice-making machines and faster steam-powered trawlers which allowed fresh fish to be taken to the shops.

Substitute foods

Dependence on overseas suppliers for food, however, could mean reduced food security, not least as a result of the wars that beset the continent of Europe. The blockade of France and Europe by Britain in the Napoleonic Wars, for example, almost cut off supplies of sugar. This led to a search for local substitutes for foreign supplies of sugar cane and the development and production of sugar beet in Europe. The first sugar beet processing factory was opened in Silesia at the start of the nineteenth century and by the 1840s there were nearly 60 factories in France alone.

Many other raw materials have also been replaced, or at least joined, by locally developed substitutes, some natural others artificial, from synthetic fibres instead of wool and cotton to margarine – at first called butterine – which appeared in the nineteenth century but took longer to develop. In some places, the development of margarine was delayed by powerful agricultural lobbies, who were keen to secure the place of dairy farmers.

Cooking technologies

The quality of cooking was just as important as the ingredients in determining the palatability and desirability of food. Both the available ingredients and cooking technologies were changing in Europe owing to three inter-related changes in the nineteenth century. First, a wider range of raw materials was marketed to a wider range of people; secondly, better cooking equipment such as built-in ranges was available; and thirdly, books that

instructed primarily the middle classes how to cook were appearing.

The mass-production techniques for meat and grain, with supplies coming largely from abroad, led smaller farmers in Britain to turn to dairying and fruit and vegetables, which meant that a wider range of foodstuffs became available. Industrialization produced a substantial middle class of people who could afford to emulate the rich and eat a wider range of foods. Apart from wealthy households, which employed servants to do the cooking, poor people had few implements to cook with, limited fires and ovens and working wives had no time for elaborate preparation.

In the late eighteenth century, Count Rumford, better known for his work on cannons, invented a more fuel-efficient stove. This was developed into the iron range which was found in most middle class homes by the 1860s. It was slowly superseded by the gas stove, an innovation which was linked to the development of the coal gas supply. This gave adjustable heat and allowed a precision in cooking not seen before.

Changing diets and health

Technical developments in food processing permitted the mass production of foods previously eaten only by the rich, such as white flour and milled rice. Steel roller mills, invented in 1834, could grind grain finer and produce white flour on a large scale and these replaced the small-scale mills using stone wheels during the latter half of the nineteenth century.

A new way of preserving milk, patented in the UK but developed by Gail Borden in the USA in the 1850s, was sweetened, condensed milk. This was popular among the troops in the American Civil War and demand continued after the war. But the cheaper varieties were often made from skimmed milk which lacked the fat and vitamins D and A found in whole milk.

These and other developments led to dietary changes. Unfortunately, these often harmed the poor, because the new processed products deprived foods of some ingredients that had made up an essential part of their diet. These essential ingredients were obtained from other foods by richer people who ate a more varied and more expensive diet.

Just what the different ingredients in foods necessary for a nutritious diet were, was not known, although carbohydrates, fats, proteins and minerals had been identified as essential parts of foods by the late nineteenth century.

The problems that could arise from over-reliance on a refined staple food were most clearly visible in Asia where polished white rice replaced relatively unmilled brown rice towards the end of the nineteenth century. A disease called beriberi, which attacks the nerves, heart and digestive system, caused so many deaths in the Dutch East Indies, that in 1886 the Dutch sent out a team to investigate the problem. It took them over 15 years to show that beriberi was not an infectious disease but was a condition caused by a reliance on polished rice for the bulk of the diet. Christian Eijkman found that the rice germ contained essential substances that became known as vitamins.

In Britain, the widespread use of these refined products contributed to deteriorating health in the general population. Although it had been noticed in the nineteenth century, the scale of the problem became clear during the First World War when the British carried out a mass medical examination of 2.5 million men in 1917–18. They found that 41 per cent of men supposedly in their prime were totally unfit for military service, mostly as a result of under-nourishment. This finding prompted a great interest in nutrition research (see Chapter 4).

Modern times

After the First World War, the scene was set for the development of today's food system. Industrialization had brought new actors on to the food scene and a more linear approach to food production. Scientifically, there was interest both in the effects of food on people

and in making agriculture more productive. Food processors wanted better methods for preservation and new products to make from the existing ingredients. Others were developing substitutes for tropical products in temperate zones and substitutes for natural products in factory processes. Traders were adept at moving large volumes of foodstuffs across the globe using the growing rail and shipping transport infrastructure.

In Britain, the emphasis on cheap food from abroad continued, apart from a brief period of support for agriculture and home production during and shortly after the First World War. There was a slow, undramatic shift of power from Europe to the USA. Flawed peace settlements and old rivalries helped to precipitate a slump and depression that provided fertile ground for the genesis of the Second World War. New powers wanted part of the colonial spoils and to extend their control over resources and peoples.

By the Second World War, the importance of a good quality food supply was recognized in Britain. Food self-sufficiency before the war was only about 30 per cent – ie, Britain produced less than one-third of the food consumed by its people. As much food as possible had to be grown in the country during the war and it had to be well distributed if the population was to be adequately nourished. Unprecedented measures were taken to ensure that a basic, nutritionally sound diet was available to all in the population – itself a splendid achievement. Such state direction of the food system was unusual in the West.

After the war, there was a major reassessment of the state of agriculture in Europe. In Britain, for example, the Agriculture Act of 1947 set about to remedy the neglect of farming that had persisted for over a hundred years. It recognized that government support was essential for the survival of the agricultural sector as a major domestic food producer. Policies were designed to increase farm incomes, using direct price support through deficiency payments when prices fell short of those needed to secure adequate farm incomes, and to increase production. Farmers' incomes were supported out of taxes through the Treasury.

In mainland Europe, massive US aid, under the Marshall Plan, was used to rebuild farming and the economies, and a whole set of new international institutions was established, mostly under the UN, to finance reconstruction and development and to monitor and regulate trade – the General Agreement on Tariffs and Trade (GATT), although agriculture was not covered by GATT. Many of these international institutions have had a major impact on agricultural development in poorer countries (see Annexe 1). Large international companies also developed, with skills in different areas of the food system.

In the 1950s, former enemies France and Germany joined together with Belgium, Italy, Luxemburg and The Netherlands to form a Common Market, at the core of which lay a food security policy – the Common Agricultural Policy. By the early 1990s, this European Community, called the European Union after ratification of the Maastricht Treaty in 1993, had expanded to 12 members. Three more states joined in 1995, with still more queuing up to join, including former East European states, one of which, East Germany, had already become a member following German reunification.

Most significantly, in the 30 years after the Second World War, colonialism broke down and most colonies fought for or were given their political independence. This happened in the context of a deeply divided Europe and America, with a Cold War between capitalism and communism colouring the actions of both East and West to change anywhere. However, despite political independence and an ability to play off the East and West superpowers, the newly independent colonies remained locked into a production and economic system designed by their former colonial rulers to produce raw materials for them.

The debt crisis

The interdependency of economic relations in the post-war world was illustrated when the industrial world's command of cheap energy in the form of oil, which underpinned post-war economic developments, was dealt a severe shock in 1973. The producing nations got together in the Organization of Petroleum Exporting Countries (OPEC) to ration production and greatly put up the price. In an attempt to recycle the petro-dollars deposited by OPEC countries in Western banks, bankers proffered huge loans in the 1970s to the leaders of many 'Third World' governments, partly in the hope of industrializing their countries. Then, interest rates were low.

By 1983, interest rates had risen dramatically and the price of primary agricultural and mineral commodities on which these countries depended for their foreign incomes had dropped. Mexico was the first to try to suspend repayments. The debt crisis began. Various devices were tried to maintain the repayments, such as the temporary suspension of interest payments, and the International Monetary Fund (IMF) and World Bank supported Structural Adjustment Programmes. Gradually, Third World Debt became manageable for the banks until, by the early 1990s, many bankers believed that there was no longer a crisis.

However, within many debtor countries there has been massive suffering and social disruption. The countries of Africa, for example, pay in total about $1 billion per month in interest payments. This has crippled health, education and social services. Despite interest and capital repayments, the level of debt grew by over 60 per cent.

According to the OECD, between 1982 and 1990, total resource flows to developing countries amounted to $927 billion... Much of this inflow was not in the form of grants but was rather new debt, on which future dividends or interest will naturally become due in the future.

During the same 1982–90 period, developing countries remitted in debt service alone $1345 billion (interest and principle) to the creditor countries. For a true picture of resource flows, one would have to add many other South-to-North outflows such as royalties, dividends, repatriated profits, underpaid raw materials and the like. The income–outflow difference between $1345 and $927 billion is thus a much understated $418 billion in the rich countries' favour. For purposes of comparison, the US Marshall Plan transferred $14 billion 1948 dollars to war-ravaged Europe, about $70 billion in 1991 dollars. Thus in the eight years from 1982–90 the poor have financed six Marshall Plans for the rich through debt service alone.

Have these extraordinary outflows at least served to reduce the absolute size of the debt burden? Unfortunately not: in spite of total debt service, including amortisation, of more than 1.3 trillion dollars from 1982–90; the debtor countries as a group began the 1990s fully 61 per cent more in debt than they were in 1982.

George, Susan (1992) *The Debt Boomerang: How Third World Debt Harms Us All*, Pluto Press, ppxv-xvi

The level of debt has pitted one nation against another in a desperate attempt to earn foreign currency from commodity exports, which has also been encouraged by the major international financial institutions which advised all developing countries to go for exports, often commodity crops. Production rose, prices fell. The encouragement of single commodity agriculture – for example, coffee plantations and cattle ranches weakened the ability of many poor countries to feed themselves by making them over-reliant on world commodity markets for the income required to import basic foodstuffs.

Three Latin American countries now rely on the production of an illicit export commodity, cocaine, to maintain their economies. It is far more profitable to grow cocoa for cocaine than it is to grow conventional crops such as cotton. These and other responses to debt, affect adversely the societies of Europe and North America. They boomerang, according to Susan George, in environmental destruction, drugs, cost to taxpayers, lost jobs and markets, and immigration pressures, and heighten conflicts and war.

Many of these internationally supported schemes have also brought about considerable environmental damage – for example, cattle farming in southern Africa and the destruction of the rainforests of Brazil. This has helped to increase the desertification of previously fertile land and has reduced the economic potential of the land. Consequently, there has been a growing concern about the environment since the 1970s, when some scientists began to warn of unforeseen and potentially damaging results for the environment from the practices of industrial production.

Summary

What has emerged today is a food system for those who can afford to buy, which is largely the people living in the Organization for Economic Co-operation and Development (OECD) countries and the richer parts of the world elsewhere – that is, around a quarter of the world's population. It uses an industrial approach to agriculture and food production, is highly productive in response to high inputs and overcomes seasonality for all foods. It draws on produce from around the world and, by using a mixture of trading and preservation techniques, enables a wide range of foodstuffs always to be available.

In the development of this food system, foods became more and more like commodities, rather than matters of life and death, or of religious and cultural meaning. Commodities are produced, traded and transformed, bought and sold, in a market whose reach has extended from a largely local level to an increasingly global stage. It is a market in which the actors seek to control their costs, their production or marketing practices, as closely as they can. They want to minimize their uncertainties and costs and maximize their returns. It is a market in which each actor is thrown into competition with others, both within their areas of operation and outside them. Individuals trying to function within this, to meet their food needs, or to make a living, have increasingly less control of how they do so, with their choices constrained by market forces beyond their control.

The new food system has also produced diets very different from the earlier consumption of relatively little processed food, with much more cooked and highly processed foodstuffs available in an ever widening range of dishes drawn from cuisines around the world. It has also contributed to increasing human longevity.

What it has not done is brought an end to hunger in the poorer countries, or provided diets that are sound and healthy for the affluent, or yet demonstrated its long-term sustainability for a population twice that of the present world. Concerns about these basic issues are helping to fuel an increased debate about our food future. In the next chapter, we look at the human needs which food meets and which the system should fulfil, and the many roles food plays in people's lives and culture.

4

Food, Culture and Human Needs

Give me neither poverty nor riches,
feed me with food convenient for me.

Proverbs, ch 30, v 8

As individuals, we need sufficient, safe, nutritious food for a healthy life. But what do we eat and why? Obviously, that depends partly upon external factors such as our economic status, but there are also deeper internal forces at play. In this chapter, we consider the basic psychological, social and cultural needs of individuals that food meets. Understanding our individual relationships to food, like our species' relationship with the biosphere, is important in developing an overall picture of the food system and of how the different actors in the system use the needs of individuals to promote their specific interests and economic benefits.

Basically, a food is a part of the environment that we eat. Unlike nutrients, human foods are defined culturally. In most cultures, people only ingest a small number of the available, potentially nourishing substances. Much human socialization relates to the development of forms of disgust towards objects which could provide some nourishment – for example, cockroaches and other insects.

Nutrients, however, are the biochemical compounds found in foods which sustain our bodies' biological processes. Macronutrients, which we need in large amounts, are the proteins, carbohydrates and fats; micronutrients, which we need in much smaller amounts, include vitamins and minerals. In addition, foods contain many hundreds of non-nutrient compounds which may be important for the body's biochemical processes.

Human beings value food for far more than its nutrient content. For example, expensive, difficult to prepare foods may be served to show guests how much they are valued by their host; busy executives may invite employees to breakfast to save time; lovers may share special meals together in 'special' restaurants. We use these symbolic properties of foods to meet our psychological and social needs – for example, to express love and self interests.

Our Bodies, Food and Health

Sufficient, safe, nutritious food is an essential ingredient for good health. Despite much progress, an enormous number of people are hungry and malnourished (see Box 4.1). To overcome hunger and malnutrition, the fundamental challenge is to enable people to produce or buy sufficient food for a healthy diet. For those able to do so, the challenge is to avoid overnutrition or unhealthy diets. The threats come from getting the nutritional balance wrong or from microbiological or chemical contamination of food.

Box 4.1
Hunger and Malnutrition

Hunger and malnutrition remain the most devastating problems facing the majority of the world's poor. In spite of general improvements in food supplies and health conditions and in the availability of educational and social services throughout the world, malnutrition in its various forms persists in virtually all countries. Although it is impossible to measure accurately the total number of malnourished individuals, the latest data indicate that about 20 per cent of the developing world's population – over 780 million individuals – suffers from insufficient food intake and over 192 million children are suffering from protein-energy malnutrition (PEM). Many more people, approximately 2000 million, are also affected by various micronutrient deficiencies. In addition, chronic diet-related non-communicable diseases, such as obesity, cardiovascular diseases and some cancers, are emerging public health problems in most countries.

Reprinted from *International Conference on Nutrition: Nutrition and development – a global assessment – 1992*
(Revised edition), FAO and WHO, Rome, 1992.

Fifty years ago the meeting that led to the establishment of the Food and Agriculture Organisation (FAO) in the mid-1940s declared that 'The first cause of hunger and malnutrition is poverty'. This is still true today. It is the poor who go hungry, the better off generally do not. Undernutrition, as the International Conference on Nutrition in 1992 made clear, is a continuing scandalous tragedy of enormous proportions. Undernutrition, as defined by the FAO, is not eating enough food to provide the minimum energy requirement for a healthy body and a healthy active life.

It is one form of malnutrition, a broader term that means a person's diet does not contain the right mix of foods to provide the nutrients needed for healthy living. Some diets can lack micronutrients, like vitamins A and C, or minerals like iodine. These deficiencies can debilitate, blind or kill.

Another form of malnutrition is overnutrition, in which people eat too much of the wrong mix of foods. This can lead to non-communicable diseases like heart disease and some cancers, which also lead to early death and poor quality of life.

All these forms of malnutrition can be found in the same country. In the Philippines, for example, for the 48 per cent of the population below the poverty line who spend at least 60 per cent of their income on food, chronic undernutrition is a threat. Yet 5 per cent of the population suffer from the diseases of overnutrition and still others suffer from iodine-deficiency diseases.

Chronically undernourished people, especially children, are much more likely to fall ill and die from infectious diseases. Since they are generally poor, they also live in poor conditions, lacking safe water and proper sanitation, which increases the likelihood of illness.

Undernourished women are themselves disadvantaged and tend to produce underweight children, which greatly disadvantages them in life. In fact, one of the main guides to the nutritional status of the population is the prevalence of low birth-weight babies.

Globally, there would be enough food to go round if the distribution were more equitable. Average food availability rose from 2290 calories per person per day in 1961–63 to 2700 calories in 1988–90, despite the world's population increasing by some 1800 million.

The starving child has become the most common symbol of malnutrition. Yet it misrepresents the problem. Visible malnutrition, usually the result of exceptional circumstances, affects only about one or two per cent of the world's children. Avoidable micronutrient deficiencies are a major, less publicized, cause of malnutrition that stunt the mental and physical growth of millions of children.

Vitamin A deficiency blinds an estimated 250,000 children a year in the developing world, and leaves tens of millions more susceptible to the three leading causes of child death – diarrhoeal diseases, measles and pneumonia. Small amounts of fruit or green vegetables in the child's daily diet can prevent it.

Over 200 million children are affected by the most visible symptom of iodine deficiency disorders (IDD) – the swelling of the thyroid gland that produces goitres in the neck. Because their mothers lack iodine, at least 30,000 babies are stillborn and over 120,000 are born as cretins. Millions more grow up stunted, listless, mentally retarded and incapable of normal speech or hearing owing to IDD. The problem can be solved inexpensively, as it has been in most industrialized countries, by adding iodine to common salt. Iodized bread and water have also been used for the general population and iodized oil injections for specific groups.

Source: UNICEF (1993) *The State of the World's Children 1994*, Oxford University Press; UNICEF (1993) *The Progress of Nations 1993* UNICEF, New York; Hetzel, Basil S (1989) *The Story of Iodine Deficiency – An International Challenge in Nutrition*, Oxford University Press

The development of nutrition science

Nutrition science seeks to understand what effects the nutrients in foods have on our bodies. It grew in response to health problems in the industrialized cities of Western Europe and North America in the late nineteenth century and early twentieth centuries. Poverty was endemic among the working classes. People did not get enough to eat. They did not get enough energy or enough of what later became known as micronutrients – vitamins and trace elements – from their diets.

Parasitic and infectious diseases were prevalent. Of course, some people died of heart disease and cancer, but what was striking were the numbers of infants, children and adults who were swept away by periodic epidemics of measles, dysentery, diphtheria, scarlet fever and influenza. Being well-nourished was known to provide people with some degree of resistance to infectious and parasitic diseases, so the search was on for the nutrients which conveyed this protection.

This research yielded great success – for example, in discovering vitamins and combating deficiency diseases such as beriberi, as discussed in Chapter 3. When combined with civil engineering works which provided better sewage treatment, waste disposal and cleaner water, infectious disease rates dropped substantially. This decline in infectious disease rates happened before the discovery and distribution of antibiotics, and public health authorities and nutritionists can take a great deal of credit for it. One longer-term effect was increased longevity. The vast majority of parents could expect their babies to live to adulthood. From the mid twentieth century in the industrialized countries, most grandchildren were alive to meet their grandparents. Adult men began to live to around 65–70 years and women even longer.

Scientists had studied what was missing from the diets of the poor. Knowledge about energy metabolism and the roles of vitamins and minerals increased. A few investigators also surveyed various social classes, looking for the reasons for the lower nutritional status of the poor. One favoured solution was nutrition education, since it was believed that poor people – ie mothers and wives – simply did not know how to buy, prepare and cook the right sorts of foods. American nutritionists in particular encouraged mothers to feed their children more fat, carbohydrates (starch and sugars) and other sources of energy. This is a far cry from today, but people were far less sedentary at the turn of the century, or even 30 years ago! The problems occurred in industrialized cities, partly because when people moved into them they had to give up their old eating habits and growing food for themselves.

From the 1930s to the 1960s, nutritionists expanded their knowledge of energy and protein malnutrition through studies of colonial populations in Latin America, Africa and Asia. But by mid-century, the impetus of the earlier phase of nutritional enthusiasm had largely exhausted itself. In industrialized countries, longevity had increased and infant mortality had decreased. Nutritional status had been markedly improved through government actions, such as education, and mother and baby clinics at which foods and vitamin supplements were administered, and better living conditions in general. All that had to happen, it seemed, for global Utopia to be achieved was for 'undeveloped' countries to industrialize.

By the 1960s, infectious diseases killed and maimed far fewer people in the West than previously, but other diseases were affecting adults, shortening people's lives by years. These were the cardiovascular diseases such as coronary artery disease, strokes, adult onset diabetes, as well as various forms of cancer such as lung, bowel and breast cancer. Men as young as 30 years of age were dying of heart disease. Narrowing of the coronary arteries, in particular, seemed to have increased since the Second World War. Autopsies of young adult American soldiers killed in Vietnam revealed more signs of atherosclerosis, caused by deposits of plaque narrowing the arteries, than similar autopsies in the Korean War in the early 1950s. Scientists wanted to know why (see Box 4.2).

What do we know now?

There are now well over 40 recognized nutrients. Annex 2 lists the main ones and their bodily functions. In addition, many hundreds of substances exist in food which may play useful physiological roles, but little or nothing is

---

Box 4.2
Risk Factors and a Town called Framingham

A key study began in the late 1960s, in a small town called Framingham in Massachusetts. This study has had a massive influence on public health policy. Several thousand men were medically examined and then followed to see which diseases they eventually died from. From this developed the notion of 'risk factor' as the investigators found that a number of biomedical factors put the men at risk of premature death. Cigarette smoking, elevated blood pressure and elevated serum cholesterol levels were the three most important factors, along with a number of lesser factors such as obesity and so-called Type A Behaviour Pattern (a key feature of which was the expression of hostility). These findings implicated diet and lifestyle in the causation of heart disease, which, in the context of 1970s and 1980s America, put a great deal of emphasis on individual change rather than on changes in the food supply and distribution.

At about the same time, a study conducted in seven countries showed that countries which had high average serum cholesterol levels tended to have a greater prevalence of heart disease. In clinical studies, the serum cholesterol level was shown to be related to the total fat intake of the individual, being increased twice as much by saturated fat intakes as it was reduced by polyunsaturated fat intakes.

In general, subsequent work has supported the Framingham findings, extending the links between habitual food intake and risk factors for other diseases such as bowel cancer and adult onset diabetes. However, with all the work has come confusion and controversy, perhaps increasingly so as different types of cholesterol have been identified with different properties.

Sources: Kannel, W B, Cordon, T and Castelli, W P (1979) 'Obesity lipids and glucose intolerance. The Framingham Study' *American Journal of Clinical Nutrition*, vol 32, pp1238–5; Keys, Ancel (1970) 'Coronary heart disease in seven countries' *Circulation*, vol 41 pp1–211; Truett, J, Cornfield, J and Kannel W (1967) 'Multivariate analysis of the risk of coronary heart disease in Framingham' *Journal of Chronic Disease*, vol 20, p511

known about their physiological effects. A cup of coffee, for example, contains hundreds of chemical compounds but probably only around 20 have been examined in scientific studies, most notably caffeine.

With rising affluence, diets tend to contain more sugars, more animal fats and fewer sources of carbohydrates (such as fruits, vegetables and cereal grains). Unfortunately, such affluent diets place people at greater risk of contracting a variety of diseases such as diabetes, heart disease and some forms of cancer. Men and women, however, may be affected quite differently. For example, if we look at the prevalence of one of the major 'diseases of affluence', heart disease, and the male and female death rates in many countries over 30 years, we find that in affluent countries in which a lot of saturated fats are consumed, men are at greater risk from heart disease (see Figures 4.1 and 4.2), although women's risk rises substantially after menopause.

Being male is thus a major risk factor in itself for these diet-related diseases of affluence. Although women's mortality rates from heart disease are fairly similar across the countries of the world, male mortality rates vary greatly. The highest rates are found in Finland, Scotland and other European countries; the lowest rates are found in non-urbanized countries and South-East Asia, and are lowest of all in Japan. Japanese men have rates as low as those for Japanese women. Clear explanations for this sex difference are not yet available but it seems related to the ways that men and women metabolize cholesterol and the ways that their endocrine systems work. Men seem to be more sensitive to the fat (and other nutrient) content of their diets than women.

Obesity is certainly a factor. The sort of body fat deposition that is typically male on the upper trunk ('apple shaped'), is far more of a risk factor than fat which is deposited on the thighs and lower trunk, which is typical of women ('pear shaped'). Obesity seems to be dependent on the balance between energy

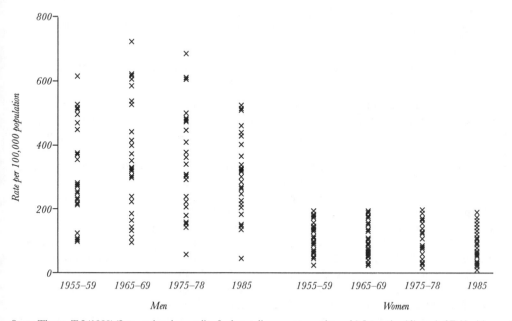

Source: Thorn, T J (1989) 'International mortality for heart disease: rates and trends' *International Journal of Epidemiology*, vol 18, pp520–28.

Figure 4.1 Death rates over 30 years from coronary heart disease in men and women aged 45–64 in 27 countries, selected data years, 1955–59 to 1985

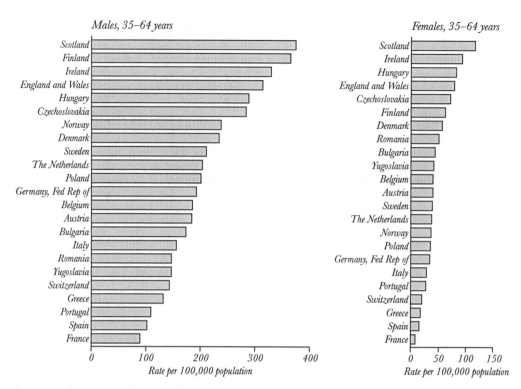

Males, 35–64 years   Females, 35–64 years

Rate per 100,000 population   Rate per 100,000 population

Source: James, W P T in collaboration with Ferro-Luzzi, A, Isaksson, B and Szostak, W B (1988) *Healthy nutrition: preventing nutrition-related diseases in Europe*, WHO Regional Office for Europe, Copenhagen (WHO regional publication, European Series No 24)

Figure 4.2 Death rates for ischaemic heart disease in Europe (standardized mortality rates per 100,000 population, 1980)

intake and energy expenditure. People today expend much less energy during their working or leisure lives, but they still consume roughly the same amount of energy as 30 years ago. Hence, the increasing prevalence of obesity. Ironically, it is women who do most to reduce their weight, although obesity is more dangerous to the health of men.

Elevated blood pressure (hypertension) is still a major disease risk factor, although its causation is not clear. For a population, the level of sodium (salt) intake is a major predictor of blood pressure and hypertension, although other nutrients such as calcium also appear to be involved. However, it now seems that only some people are salt sensitive and this sensitivity appears to be a genetic predisposition. Across the board, reductions in the salt content of processed foods such as bread

and processed meats could save many thousands of lives each year in Western countries such as Britain. Sodium intake is generally beyond the control of individual consumers unless they go to great lengths to avoid such common foods as bread and processed meats. Without the co-operation of government and the food industry, people will continue to die needlessly or become chronically ill because of hypertension-related stroke and heart disease.

Premature deaths, however, are only the tip of the health care iceberg. Morbidity (suffering and disability) due to illness costs vast amounts. Already in countries like Australia, which are experiencing increasing longevity, for every year of extended life *two* years of morbidity-burdened life accompany the longevity increases.

Finally, we look at cholesterol. During the

late 1970s, it was realized that cholesterol could be divided into at least three types: high density lipoprotein (HDL); low density lipoprotein (LDL), and very low density lipoprotein (VLDL). HDL seems to be the healthful fraction that mops up substances which are integral to the arteriosclerotic process. LDL and VLDL, however, are the villains in the heart disease story. High levels of these substances help to develop plaque in the artery wall.

Work on dietary fibre has shown that some forms of water-soluble fibre form volatile fatty acids in the gut and these can remove cholesterol fractions from the blood. Other research has shown that so called unsaturated fats, such as those found in certain fish (the 'omega three' fats), can form part of the cell membrane and affect its workings. Low intakes of these unsaturated fats are believed to affect the composition of cardiac cell membranes, which affects the electrical rhythm of the heart. This can lead to cardiac arhythmias and sudden deaths which may occur in the absence of any narrowing of the coronary arteries. In addition, the ways we respond to high fat diets vary according to our genetic make up.

It is small wonder that the public is confused, especially when the term 'cholesterol' is used ambiguously to refer either to serum cholesterol or to dietary cholesterol. When cholesterol was first linked to heart disease, people assumed that it was largely absorbed together with animal fats in their diets. So the age of cholesterol-free food was born. However, our bodies make most of the cholesterol we have and cholesterol in the diet now seems to produce only a small part of heart disease risk. The liver regulates the level of cholesterol fractions in the blood, cutting down production if we take in a lot through our food and vice versa, so scientific interest is now centred on this aspect.

Cholesterol is only one of several risk factors. Additionally, instead of the simple certainty of 30 years ago – cholesterol as a risk factor for heart disease – we now have many lipoproteins. Attempts to reduce serum cholesterol levels in populations have met with little success, apart from the well designed Oslo Heart Study which reduced total mortality rates and heart disease deaths in subjects who cut their fat intakes and ceased smoking.

Nutrition guidelines

Many findings emerge from the burgeoning nutrition science industry every year. They show that food intake is closely, if complexly, related to the ways our bodies function. Nutritionists have suggested guidelines for public food and health policies, and dietary guidelines and goals have been developed by several countries and adopted recently by the World Health Organization (WHO).

The first set of guidelines in the heart disease era were the US Surgeon General's dietary guidelines and goals for Americans in 1977. They have recently been revised, but basically their advice remains the same for the 1990s, and they have been adopted and adapted by many other countries.

Basically, the guidelines (see Box 4.3) are about the variety of foods we eat. They suggest that healthy diets are achieved by eating mainly from four groupings of foods (some would add dairy foods or their equivalents): cereals, fruit, vegetables and meat and meat substitutes. They also caution about the overconsumption of fats, sugars and salt and the need to keep our body weight within healthy limits. Essentially, they are a criticism of diets based solely on highly processed foods and emphasize the need to choose a wide variety of foods from relatively unprocessed sources.

The guidelines have been the focus of opposition from producers whose products are inconsistent with them. The US Department of Agriculture (USDA), for example, devised a well-researched dietary pyramid to help promote the guidelines (see Figure 4.3).

Until the late 1980s, there was a general belief that dietary guidelines were for consumers to follow in their daily dietary choices. This was a product of the individualism that predominated in Anglo-American countries.

Box 4.3
Dietary Guidelines for Whole Populations –
Two Examples

For Americans

1. Eat a variety of foods
2. Maintain healthy weight.
3. Choose a diet low in fat, saturated fat and cholesterol.
4. Choose a diet with plenty of vegetables, fruits and grain products.
5. Use sugars only in moderation.
6. Use salt and sodium only in moderation.
7. If you drink alcoholic beverages, do so in moderation.

Source: US Department of Agriculture & US Department of Health and Human Services (1990) *You and Your Health: Dietary Guidelines for Americans,* 3rd edn, Home and Garden Bulletin no 232

For Australians

1. Enjoy a wide variety of nutritious foods.
2. Eat plenty of breads and cereals (preferably wholegrain), vegetables (including legumes) and fruits.
3. Eat a diet low in fat and, in particular, low in saturated fat.
4. Maintain a healthy body weight by balancing physical activity and food intake.
5. If you drink alcohol, limit your intake.
6. Eat only a moderate amount of sugars and foods containing added sugars.
7. Choose low-salt foods and use salt sparingly.
8. Encourage and support breast feeding.
9. Eat foods containing calcium. This is particularly important for girls and women.
10. Eat foods containing iron. This is particularly important for girls, women, vegetarians and athletes.

Source: Australian Government Publishing Service (1992) Canberra

In fact, guidelines and dietary goals were and are intended as guidelines for *nations* to follow. The food supply as well as individual consumers can adhere to guidelines. The 1990s have seen the rise of corporate nutrition policies as companies try to meet the demand for better – for example, lower fat, salt, sugar, etc – foods.

The potential effects of industry's adoption of dietary guidelines were spelled out in 1983 in a seminal paper by Fallows and Wheelock. They looked at fat and showed how fat was introduced into the British food supply. Even quite small changes, such as the substitution of fat reduced milks, could have major effects on the average person's fat intake, because products like milk and fat spreads are used so frequently.

While members of industry, government and the public have been increasingly enthusiastic about dietary guidelines, some nutritionists have raised doubts. Less than 1 per cent of Australians actually adhere to all of the guidelines. In fact, to do so requires a fairly high energy intake – higher than that of many people – to get enough of the various micronutrients. This would not have been a problem when people carried out a lot of physical work, but nowadays it is very difficult to expend much energy unless you are a sportsperson. To make the guidelines implementable, we will need structural changes

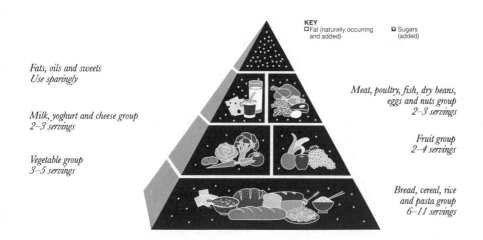

KEY
☐ Fat (naturally occurring and added)   ◨ Sugars (added)

Fats, oils and sweets
Use sparingly

Milk, yoghurt and cheese group
2–3 servings

Vegetable group
3–5 servings

Meat, poultry, fish, dry beans,
eggs and nuts group
2–3 servings

Fruit group
2–4 servings

Bread, cereal, rice
and pasta group
6–11 servings

Note: This pyramid is an update of the basic four food groups circle used in nutrition education in the USA since the 1950s. The pyramid was launched in 1992 and divides foods into five groups. It aims to convey three key elements of a healthy diet: proportion, or the relative amount of food to choose from each major food group; variety – ie, eating a selection of different foods from each major group each day; and moderation in eating fats, oils and sweets.
Source: USDA

Figure 4.3 USDA Food Guide Pyramid.

which enable us to do more physical activity – for example, more safe cycle routes.

Other concerns have also been raised, especially by social scientists. Most damaging are suggestions that the guidelines ban all that is pleasurable in ordinary people's food (salt, fat and sugar) and that they are merely social control of the majority by small groups of powerful (medical) people. They make people unnecessarily anxious without the means to do much about it. Similarly, they are unfair to women because the burden of implementing dietary guidelines in households usually falls to them and they usually do not have the power to enforce such nutritionally correct behaviour. In Australia, for example, it has been estimated that around one-third of women feel guilty about the food they purchase.

These social criticisms have a ring of truth. Perhaps they highlight the authoritarian ways in which governments and pressure groups sometimes act. On the other hand, there is a lot of evidence in support of the general thrust of the guidelines, which could be implemented more democratically (eg, via food-buying co-operatives and through more relevant information at the point of purchase). Many people are trying

to reduce their fat (and salt and sugar) intakes and to increase their fruit, vegetable and cereal intakes, perhaps because they accept that this reduces risks to their health.

Where next for nutrition?

Individualistic nutrition policies are now firmly part of the establishment in Western countries and are becoming so in newly industrialized countries, such as those in South-East Asia which have begun to experience the 'diseases of affluence'. Doctors, drug companies, food companies and health promoters all agree about the need to prevent these diseases. But many diseases of affluence are multifactorial and food intake is only one ingredient in the causal mix. Whatever the precise cause, there has been a marked drop in the prevalence of premature deaths caused by heart disease in some countries. In Australia and in the USA, heart disease deaths have been halved in the past two decades. However, cancers are a major cause of death, increasingly among women.

This simple fact is quite historic. Heart

disease blossomed in the twentieth century and now seems to be declining in several countries (although it is increasing in others) and the money for research is shifting towards cancer. During the past five years there has been a revival of interest in the micronutrients – substances like vitamins A, C and E – which play a protective role in heart disease. They also seem especially important as moppers-up of free radicals which can start off carcinogenesis (cancer formation). Indeed, vitamin E has been seized on by some popular reports as the latest wonder food ingredient. So in a sense we are returning to some of the preoccupations of the earlier part of the century. In order to understand many of the disease processes not previously associated with the immune system, such as cancer, heart disease and arthritis, the importance of the immune system is receiving more attention.

Another change has been the wider scope of nutrition. There is a realization that many co-factors are necessary for nutrients to function. Many of them, such as flavonoids, are found in foods. There has been a global examination of old folk beliefs in the utility of particular foods for maintaining or restoring the functions of particular body organs. For example, ginger has long been used to combat nausea and recent laboratory work confirms that it contains compounds which combat motion sickness. Folklore is being refined and tested in the laboratory so that foods (so-called functional foods) can be made with proven medicinal properties for an increasingly ageing but affluent population. Will nutrition end thus – being used by global companies to increase their profits from affluent populations? Will it provide the fabled 'elixir of life'?

Much of the nutrition research of the past 30 years has centred on the diseases of the affluent. Relatively few studies have had women or blue-collar workers as their subjects. But how much does it matter for the world's community if people live an average of 80 days longer because they have lower cholesterol levels, given the other problems that need solutions? It raises policy questions about the people for whom research is done, whether there are more pressing problems to be solved

and whether the money invested could be spent more wisely now. It would be ironic, to say the least, to solve the riddles of cancer and heart disease while polluting the ecosystems that make life possible. Thus, of the criteria for a good food system that we outlined in Chapter 1, sustainability and equity are fundamental to bring about the goals of health for all, espoused by the WHO.

Individuals and the community

When we think of health, we often think of illnesses which can be measured by their mortality rates (how many people per hundred thousand die from a disease) and their morbidity rates (the number of people per hundred thousand whose normal functioning is impaired by the disease). Health, however, is a positive state of feeling good about oneself and being able to function normally. WHO's definition is: 'Health is a state of complete physical, mental and social well being and not merely the absence of disease or infirmity.'

However, it is hard for individuals to be healthy if their communities are unhealthy. For example, water supplies, sanitation and harmonious non-violent relations between people are essential bases for individual health. Individual health depends on several factors, including the following:

1. biological or genetic factors – for example, a genetic predisposition to heart disease;
2. the environment – for example, if sewerage systems are absent disease may flourish;
3. behavioural and social factors, so-called lifestyle factors – for example, smoking increases the risk of disease;
4. the political-economic environment: if governments neglect the health consequences of their policies they may harm the health of some citizens – for example, tight financial policies may cause unemployment and so worsen the health of some in the community.

The foods we eat can impinge on community and individual health in each of these factors. For example, some people may be especially sensitive to particular food constituents (eg, salt or food additives (1). Food safety systems may fail or not be in place, causing outbreaks of food poisoning (2). Some people may increase their risk of ill health by consuming too much or too little of certain nutrients (3). Finally, governments may act to impede the availability of a healthy food supply or, indeed, they may act to promote an unhealthy food supply (eg, agricultural policies which subsidize the production of high fat foods) (4).

Dietary patterns alone, however, do not determine healthiness or disease. Other behavioural and environmental factors (accounting for up to 75 per cent of disease causation) include cigarette smoking, poor sanitation and sedentary lifestyles. But chemical and biological hazards can also be present in foods.

Food safety

Although the range of foods we eat may affect our individual health in the long term, food safety discussions usually focus on the more immediate effects that arise from consuming foods contaminated with some undesirable biological or chemical agent. This can lead to food poisoning, by far the majority of which is due to biological contamination. Food can be contaminated at any point – from production on the farm or in the sea to processing in the factory, in shops or in the kitchen. As the food chain has lengthened and the scale of processes has grown larger, now if something goes wrong, many more people may be affected. For example, if a large batch of tinned corned beef is not properly canned, many people may be poisoned by botulism.

Central to both the long-term and immediate hazards of foods, however, is the issue of risk and risk control. Who decides what risks should be taken? It is one thing to decide voluntarily to take a risk, but quite another if the risk is forced on you by the actions of another person or a company (eg, one which makes or sells a product containing pesticide residues or harmful bacteria).

Biological threats to health

Micro-organisms aid digestion and help to produce foods such as yoghurt, cheese, bread and wine, but some turn good food into a health hazard (see Table 4.1) Bacteria like salmonella and vibrio cholerae, viruses like those causing hepatitis A, and parasites like *Entamoeba* and *Trichinella spiralis* can find their way into people's bodies through the food they eat. After harvest, food can be contaminated by various poisons left by microscopic fungi or moulds – mycotoxins. The most infamous of these is aflatoxin which grows on groundnuts. Exposure to this has been linked to the high incidence of liver cancer in some African and South East Asian countries.

Like foods, the bugs can be traded, with some types of salmonella arriving in North America and Europe in imports of contaminated animal feedstuffs from tropical and sub-tropical areas. Ensuring that our food is biologically safe to eat again involves both individual and communal responsibilities, from good food handling practices, such as keeping fresh and cooked foods separate, to social provision for clean water and sanitation facilities which can all help to prevent microbiological contamination.

Food preparers and processors, at home or in the factory, try to eliminate the risks associated with these organisms. In mass production and the distribution of food this is a major task for manufacturers and quarantine services. It is a battle far from won, however, as the number of reported cases of food-borne diseases is increasing world-wide. In Venezuela, for example, from 1976 to 1987, the number of reported cases of food-borne diseases more than tripled, up to 90 per 100,000. In the USA, there are an estimated 6.5 million cases of food poisoning a year. Around 10 per cent of actual cases are believed to be reported in the industrialized countries, but in developing countries it is only about 1 per cent of cases.

Table 4.1 Unwelcome food bugs and their effects

Disease and organism that causes it	Source of illness	Symptoms
Bacteria		
Botulism Botulinum toxin (produced by *Clostridium botulinum* bacteria)	Spores of these bacteria are widespread but the bacteria only produce toxin in an anaerobic (oxygenless) environment of little acidity. Found in a considerable variety of canned foods such as corn, green beans, soups, beets, asparagus, mushrooms, tuna and liver pâté. Also in luncheon meats, ham, sausage, stuffed eggplant (aubergine), lobster and smoked and salted fish.	Onset generally 4–36 hours after eating. Neurotoxic symptoms include double vision, inability to swallow, speech difficulty and progressive paralysis of the respiratory system. Get medical help immediately – botulism can be fatal.
Campylobacteriosis *Campylobacter jejuni*	Bacteria on poultry, cattle, and sheep can contaminate meat and milk of the animal. Chief food sources are raw poultry, meat and unpasteurized milk.	Onset generally 2–5 days after eating. Diarrhoea, abdominal cramps, fever and sometimes bloody stools. Lasts 7–10 days.
Listeriosis *Listeria monocytogenes*	Found in soft cheese, unpasteurized milk, some seafood products, frozen cooked crab meat, cooked shrimp and cooked surimi (imitation shellfish). The *Listeria* bacteria resist heat, salt, nitrite and acidity better than many micro-organisms and survive and grow well at low temperatures.	Onset from 7–30 days after eating, but mostly reported 48–72 hours after eating contaminated food. Fever, headache, nausea and vomiting. Primarily affects pregnant women and their foetuses, newborns, the elderly, people with cancer and those with impaired immune systems. Can cause foetal and infant death.
Perfringens food poisoning *Clostridium perfringens*	In most instances caused by failure to keep food hot. A few organisms are often present after cooking and multiply to toxic levels during cool down and storage of prepared foods. Most commonly affects meats and meat products. These organisms grow better than other bacteria around 66–72°C, so gravies and stuffing must be kept above 78°C.	Onset generally 8–12 hours after eating. Abdominal pain and diarrhoea, and sometimes nausea and vomiting. Symptoms last a day or less and are usually mild. Can be more serious in older and debilitated people.

Table 4.1 continued

Disease and organism that causes it	Source of illness	Symptoms
Salmonellosis *Salmonella bacteria*	Raw meats, poultry, milk and dairy products, shrimp, frogs' legs, yeast, coconut, pasta and chocolate are most frequently involved.	Onset generally 6–48 hours after eating. Nausea, abdominal cramps, diarrhoea, fever and headache. All age groups are susceptible, but symptoms are most severe for infants, the elderly and infirm.
Shigellosis (bacillary dysentry) *Shigella* bacteria	Found in milk and dairy products, poultry and potato salad. Food becomes contaminated when a human carrier does not wash hands and then handles liquid or moist food that is not thoroughly cooked afterwards. Organisms multiply in food left at room temperature	Onset 1–7 days after eating. Abdominal cramps, diarrhoea, fever, sometimes vomiting, and blood, pus or mucus in stools.
Staphylococcal food poisoning Staphylococcal enterotoxin (produced by *Staphylococcus aureus* bacteria)	Toxin is produced when food contaminated with bacteria is left too long at room temp. Meats, poultry, egg products, tuna, potato and macaroni salads and cream-filled pastries are good environments for these bacteria to produce toxin.	Onset generally 30 minutes to 8 hours after eating. Diarrhoea, vomiting, nausea, abdominal pain, cramps and prostration. Lasts 24–48 hours. Rarely fatal.
Vibrio infection *Vibrio vulnificus*	The bacteria live in coastal waters and can infect humans either through open wounds or through consumption of contaminated seafood. The bacteria are most numerous in warm weather.	Onset abrupt. Chills, fever and/or prostration. At high risk are people with liver conditions, low gastric (stomach) acid and weakened immune systems.
Protozoa Ameobiasis *Entamoeba histolytica*	These are present in the intestinal tract of humans and are expelled in faeces. Polluted water and vegetables grown in polluted soil spread infection.	Onset 3–10 days after exposure. Severe crampy pain, tenderness over the colon or liver, loose stools, recurrent diarrhoea, loss of weight, fatigue and sometimes anaemia.
Giardiasis *Gardia lamblia*	Most frequently associated with consumption of contaminated water. May be transmitted by uncooked foods that become contaminated while growing, or after cooking by infected food handlers. Cool, moist conditions favour the organism's survival.	Onset 1–3 days. Sudden onset of water stools, abdominal cramps, anorexia, nausea and vomiting. Especially infects children, hikers travellers and institutionalized patients.

Table 4.1 continued

Disease and organism that causes it	Source of illness	Symptoms
Viruses		
Hepatitis A virus	Molluscs (oysters, clams, mussels, scallops, and cockles) become carriers when their beds are polluted by untreated sewage. Raw shellfish are especially potent carriers, although cooking does not always kill the virus.	Onset begins with malaise, appetite loss, nausea, vomiting, and fever. After 3-10 days patient develops jaundice with darkened urine. Severe cases can cause liver damage and death.

Source: Hecht, Annabel 'The Unwelcome Dinner Guest: Preventing Food-borne Illness', *FDA Consumer*, Jan–Feb 1991, pp22–25

Chemical threats to health

Some chemicals in food pose complex but real threats to health (see Box 4.4). The same chemical might pose quite different risks, according to whether a person is a producer (such as a farmer using a pesticide) or a consumer of a product with only a residue of the substance present. Similarly, a substance may be quite safe for humans but toxic for plant and animal life, thus harming the ecosystem which produces human food (for example, DDT which can destroy bird species but, in relatively small amounts, leave humans unharmed).

The vast majority of the toxic chemicals in food are entirely natural. They appear to do us little if any harm because we are exposed during our lifetimes to very low amounts of them. A serving of rhubarb, for example, contains about 1/300th of a fatal dose of oxalic acid.

Dosage (the amount of a chemical substance ingested per unit of body weight) and frequency of exposure are the key concepts in understanding chemical toxicology. Many compounds which are healthy or innocuous at low dosages can be highly toxic at high dosages. For example, at low dosages vitamin A is necessary for health, but at high dosages it causes liver damage and can be fatal. The maxim that 'If a little is good, more must be better' is not true.

The main reason we find toxins in foods is that plant species make toxins as a defence against predator animals and pests. Another cause is cooking. When meat is roasted, for example, toxic chemicals are produced. Agricultural and food processing methods may also introduce chemicals such as pesticide residues and mineral oils, but usually in very small amounts. Many of these compounds, whether natural or synthetic, cause cancer when they are consumed in large, frequent doses. The effects of prolonged exposure (over 20, 30 or more years) to toxic residues is little understood, nor are the interactions between the many different residues that may be found in our diet.

Some compounds found in food may damage genes through the release of chemicals called free radicals, perhaps the most common and damaging of which is ionic oxygen. Fortunately, mammalian bodies have elaborate defence mechanisms against free radicals. Vitamins such as A, C and E which are present in many foods – notably fruit and vegetables, especially green leafy and orange vegetables – can 'mop up' free radicals by combining with them so that they do not damage our genetic material. These and other protective substances are present in much higher amounts than the toxic ones. Health studies of different populations around the world have shown that people who eat large amounts of fruit and vegetables have fewer

Box 4.4
Toxicity Testing

Toxicity testing is used to try to minimize the risk to people from specific chemicals in food. Chemicals are usually tested on bacterial cells or on animals to see whether they induce carcinogenesis (cancer-causing processes); however, recent work on DNA damage has called into serious question present methods. The bacterial cell is a poor model of a mammalian cell. It is now possible to test the effects of a chemical compound directly on human cells either through cloning of cells from skin tissue, for example, or by examining the effects of normal dietary exposure to a compound through examination of cells taken from different parts of the body.

Chemicals are also tested by giving huge doses of particular compounds to animals to see whether they induce carcinogenesis or other forms of illness and causes of death. Under the US Food and Drug Administration's Delaney Clause, any compound shown to cause cancer in animals cannot be used in the manufacture of human food in the USA. It can sometimes seem that almost everything causes cancer. Indeed, many substances can cause cancer (in experimental laboratory rodents) but only at huge dosages, usually at thousands of times the levels humans are exposed to during the whole of their lifetimes.

Other attempts are being made to gauge the risks that various chemicals in food pose. One such is the Human Exposure Rodent Potency (HERP) index. This assesses the risk of getting cancer from chemicals in food. Basically, the index compares the dosages at which carcinogenesis was induced in rodent experiments with the lifetime exposure humans are likely to have to that compound. The risks are very small for most substances, and some fairly toxic synthetic compounds, such as DDT, pose minimal risks when exposure occurs through the human food supply. This work emphasizes the importance of dosage and frequency of ingestion (lifetime exposure). It provides a perspective which enables us to attend to serious rather than insignificant risks.

There are many circumstances, however, when HERP indices are not appropriate indicators of possible damage to health – for example, when large amounts of a compound might be ingested during particularly sensitive growth periods. Solanin, which is found in green potatoes, is believed to cause congenital abnormalities if eaten in large amounts during the early stages of pregnancy when foetal cell division is most prevalent. Even more common is the use of alcohol during the early stages of pregnancy which can cause serious problems such as mental retardation, small size and underdevelopment of the mid face in the foetus – so-called foetal alcohol syndrome.

Environmental pollution, especially the pollution of river systems by heavy metals and their subsequent concentration through the food chain up to their consumption by humans, may also cause damage. Minimata disease, named after a Japanese industrial area where people regularly ate fish which had been contaminated by heavy metals such as mercury, was among the first epidemics of human heavy metal contamination to be recognized. The Dutch dietary guidelines warn people of the dangers of eating fish from contaminated rivers and estuaries.

HERP indices also do not provide warning about those food compounds such as some preservatives and food colourings which may lead to food sensitivity reactions (so-called 'food allergies') in the form of skin eruptions, bleeding, or disturbed mood and behaviour, among other signs and symptoms. Tartrazine, for example, causes serious skin irritation and inflammation in some susceptible people. It was to allow people to avoid the ingestion of such compounds that the European Community invented its ingredient numbering system – the 'E numbers'. This includes compounds which are considered to be safe for general use as food additives, although some of these compounds may cause adverse reactions in small numbers of people.

Source: Gold, L S, Sloane, T H, Stern, B R, Manley N B and Ames B N (1992) 'Rodent carcinogens: setting priorities' *Science*, vol 258, pp261–65

cancers than those who eat lesser amounts. We are likely to learn much more about these complex, nutritional genetic defence systems in the coming decade.

Our Minds, Food and Behaviour

By and large the nutritional quality and safety of food do not determine what we eat. Our sensory perceptions of food, our feelings about it – what we might loosely term 'psychological factors' – play a major part. These perceptions determine what gets into our mouths repeatedly. They stem from both the individual human personality and the culture and society of which the individual forms a part. What, for example, is your reaction to eating dogs, horses, snails, insects, frogs, pork, beef or any animal product? These are just a few of the foods that people in some societies accept and eat enthusiastically but others reject.

Sensory control of food intake

Three senses matter most in eating – sight, taste and smell.

First, sight. We tend to reject foods that appear 'strange' or unsafe – for example, if there is a fly in the soup! Foods should appear as they are supposed to be – for example, green meat will cause revulsion. The appearance of food contributes to our enjoyment. A drink which is orange in colour but which tastes salty and not like orange juice will usually be experienced as unpleasant. A drink served from a brown beer bottle will taste 'beery' even if it is the best wine! Expectations confirmed by a food's appearance strongly affect its perceived pleasantness.

Secondly, smell. We can distinguish hundreds of different aromas. These are mixtures of molecules carried along in the air and breathed in through the nose. What gives aromas distinctive qualities is not fully understood, but is connected with the shape of the molecules. Our language for aromas, especially for pleasant aromas, is poor probably because our sense of smell developed long before language. The processing of sensory information about aromas is done by those parts of the brain which are poorly connected with the cerebral cortex and other parts associated with consciousness. Aromas play a central role in enabling humans and animals to avoid toxic or unsafe substances.

Thirdly, taste or gustation. Marketers and food retailers often say that food acceptance is all a matter of taste. If people do not like the taste, they will not eat it. Westerners generally recognize four tastes – sour, sweet, salt and bitter – which are associated with specialized receptors on different parts of the tongue and palate. The Japanese and some other cultures recognize a fifth 'savoury' taste called umami which is carried by monosodium glutamate (MSG) and related chemicals in seaweed, tomatoes and other edibles. It is a key component of soy sauce.

Our other senses are involved in food consumption, such as the ability to sense texture and temperature and to hear the sounds of eating and drinking. However, for most foods these play lesser roles than sight, smell and taste.

The sensory signals sent to the brain as we eat and drink are integrated into perceptions. Although we can sense many features of foods, our perceptions of them are organized along three central dimensions: character or quality, pleasure and intensity. Perceived saltiness and perceived sweetness are examples of the character or quality of food (eg, bacon and chocolate respectively). Carbonated fizzy drinks, spirits and chilli have strong intensity or impact.

The all-important pleasure which foods and beverages give depends mainly upon their taste and, to a lesser degree, their aroma. The pleasure that comes from eating particular foods can be predicted accurately because there are strong relationships between the concentrations of chemicals (eg, sugars) in foods and the perception of pleasure.

Since the middle of the nineteenth century, the science of psychophysics has studied the relationships between physical stimuli (eg, the volume of sounds) and the subjective perceptions of these stimuli (eg, the loudness of sounds). When applied to foods, this science is sometimes called sensorimetrics – the measurement of sensation or, alternatively, sensory evaluation. If we trace the relationship between the concentration of sugars in a drink and its perceived sweetness, there comes a point at which for any individual (or population of individuals) the sweetness perception (and thus the sugar concentration) is just right. Any more sugar and the drink will be too sweet, any less and it will not be enough. Bob McBride, an Australian sensory psychologist, has rather aptly called this the 'bliss point'. A particular food may have a whole series of psychophysical curves according to its ingredients and sensory characteristics – for example, its sugar, salt, fat, fibre, water content, temperature and aroma properties. Bliss points can be measured along each of these perceptual curves.

These mind-chemical interfaces are extremely important for food acceptance. However, people can alter their bliss points and so change their taste preferences, as at least two studies – one in Australia and one in Finland – have shown. Finns who changed from drinking full fat milk to zero fat milk found it unpleasant at first, but if they had a good reason for making the change (like avoiding heart disease) and if they persevered for some weeks, they got used to the taste and enjoyed it. They later found that whole milk was unpleasant or nauseating.

Psychological influences on food intake

The way we think about food is complex and surprisingly little is known about the psychological factors involved. For example, the ways in which children's food habits and preferences develop from infancy to adolescence remain largely unexplored. Our food behaviours are also influenced by other people (see Box 4.5). The psychological influences can be grouped loosely into several categories: cognitive influences, personality and motivational influences and learning.

Cognitive influences

'Cognition' concerns thought and thinking. Our beliefs, attitudes and knowledge about food, eating and drinking can affect what we decide to eat. A woman who believes that lack of calcium might cause her to suffer from osteoporosis ('weakened bones') may try to eat more foods which contain calcium. Essentially, beliefs and attitudes to foods and food behaviours are related to the personal significance which foods have for individuals. For some, foods have great and varied significance – they think about food a lot. For others, foods are merely a means to a state of pleasure or of non-hunger.

Many attitude-belief-behaviour models have been proposed to account for human decision-making. The main principle behind all of these attitudinal models is similar to the psychophysical 'bliss point' – that is, most people seek to maximize the benefits from their actions and minimize their 'losses'. So if you cut down on chips, you may hope to achieve a desired outcome (eg, weight loss or a longer life) which you might trade off against the loss of immediate pleasure. In this sense, these are rational models. They assume people are 'sensible' in thinking about the consequences of their behaviour.

However, most of our eating is so habitual that often we are unaware of its consequences, especially its long-term consequences. Sometimes we can become aware of possible negative consequences – for example, when a close friend dies of a heart attack or when we see a particularly graphic health-promotion message. Then we may decide to alter our food intake, but usually we will have to contend with other influences on our food consumption which are not wholly under our conscious control (like the sensory factors described previously).

Box 4.5
Food Behaviours

A chicken which has filled its crop with corn will start eating again when it sees another chicken eating. The effects we humans have on each other's eating and drinking habits can be similar. Such an effect is an example of behavioural contagion – we observe someone else eating and copy them (usually because they make us feel hungry).

When a spouse defers to the wishes of his or her partner to eat fruit rather than chips, or when people eat according to social norms or rules, it is called behavioural conformity. They act according to some rule or expectation. The influence of important people in our daily lives ('significant others' such as parents, spouses, children, friends and relations) on food consumption behaviours is relatively under-explored. Research in the 1940s showed that the food habits of families were controlled to a great degree by the 'homemaker'. In the United States in the 1940s, most women were 'homemakers' and controlled the choice and purchase of food and the ways in which it was prepared, served and consumed.

Nowadays, this gatekeeper role seems to have weakened in many countries, particularly as much food consumption occurs outside the home. The 'woman as gatekeeper' viewpoint also underestimates the influence of the husband in traditional families over food selection, and perhaps assumes too much power over the family food supply on the part of the woman.

Nevertheless, there is little doubt that members of households have a strong influence over each other's eating habits. Among adolescents and young adults the influence of peer groups is very strong, especially over food choice (among young women) and alcohol consumption (among young men).

Eating and drinking consists of many movements and actions and quite a few sounds. People stare at food, sniff it, play with it on their plates, move it to their mouths, shove it in in small or large heaps, chew it, swallow it and usually they intersperse all of these actions with talk. These 'microbehaviours', such as the speed of movement, the size of helpings, the speed of chewing, and the rate and quality of the conversation, can be measured. All of them may be influenced by factors such as the number of hours since food was last eaten (hunger), the type of food presented, the time of day, personality and gender, among others.

Generally, apart from when they are faced by starvation, people have some choice over what they eat. Even when faced by starvation, however, individuals may refuse to eat rats, insects or human flesh, or other nourishing but 'disgusting' foods. Just why individuals select the foods they consume is a major aspect of food consumption behaviours of especial interest for marketers

The actual food choices made by individuals may or may not be determined consciously. They will most probably be influenced strongly by the person who prepares the food and by the preferences of the 'head of the household'. Food choices appear to be largely habitual. The same limited number of meal menus will be consumed over many years by most people.

People develop fairly entrenched food preferences. These are important both for food consumers and for food manufacturers, retailers and caterers. Food preferences (and their opposites, food aversions) can be behavioural or attitudinal. Behavioural preferences refer to the actual selection of one food from a specific set – for example, 'I prefer chicken to other meats'. Attitudinal or hedonic preferences reflect the way we feel about particular foods – for example, when we say 'I like ice cream a lot' or 'I hate onions'. We can express our degree of liking for a food without comparing it directly with any other.

The difference is important to market researchers and others who wish to predict food consumption. Behavioural food preferences can be hard to measure but they are the best single predictor of a person's probable food choice. Expressions of liking or disliking for foods are useful predictors of food acceptance and are relatively easy to measure. For example, put a cross on the line below to express how much you like or dislike cabbage.

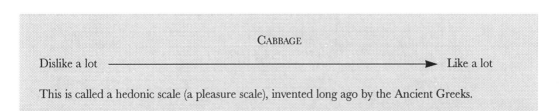

CABBAGE

Dislike a lot ──────────────────────────────▶ Like a lot

This is called a hedonic scale (a pleasure scale), invented long ago by the Ancient Greeks.

Personality and motivational influences

Personality studies examine the regularities in people's behaviour across social situations. For example, we might expect an introverted person to be relatively shy in most social situations, not just one or two. During the 1950s and 1960s, the relationships between food preferences and personality were investigated, but few clear findings emerged. However, some work linked introversion-extroversion with taste preferences and with the effects of caffeine consumption. Extroverts appear to crave stimulation – they tend to like intense tastes, like lemon juice, more than introverts.

It now appears that there are five major personality dimensions. A recent Australian study has shown that personality factors such as introversion-extroversion and 'tough-mindedness' may be major predictors of the habitual consumption of nutrients like fats and sugars. For example, men with high 'neuroticism' scores (a measure of emotional insecurity) reported eating more red meats and less fruit than those with low 'neuroticism' scores. Future studies may well demonstrate that particular food intake behaviours are related to such factors.

Our values – the guiding principles in our lives – may affect food intake. Some of us are more concerned with the effects of food on our appearance than on our health, or on being seen eating socially desirable or expensive foods. These concerns reflect people's different values. Some people, for example, may place a high value on animal life and so avoid animal products.

Values are really sets of motivational goals which may serve individual or group interests (see Box 4.6). Members of different social groups and cultures are socialized into these value orientations to different, characteristic degrees. For example, the high emphasis on achievement in American society contrasts with the emphasis on tradition and order in Confucian societies like Taiwan and China.

Values are essentially our basic assumptions about what is good. They help us to assess actions and activities according to certain sets of desirable outcomes. Conflict between different actors in the food system may sometimes be due to clashes between personal values. For example, in the recent past, marketers have emphasized individualistic values more than, say, health professionals or ecologists who tend to hold more 'holistic' values. Recognizing what the values of different actors or stakeholders in the food system are may enable working compromises to be reached to the satisfaction of all parties. Recent works shows that consumption of foods such as hamburgers and soft drinks is greater among people with hedonistic values and lower among those with universalistic values.

Which value types do you particularly identify with and how do they compare with the values espoused in common television food advertisements?

The ways in which people think about themselves, and the ways they would like to appear to others, affect their approach to food. Some people, for example, believe that their health and appearance are largely under their own control. Studies suggest that they alter their food intakes more than people who believe that these things are a matter of luck or fate. Others believe that they generally accomplish things that they decide to do; less fortunate people have far less confidence in their own ability (ie, they have less self efficacy or control over their dietary habits).

We all think of ourselves in particular ways

Box 4.6
Ten Motivational Types of Values

Following a review of people's values in several countries, Schwartz identified ten 'universal values' that help to guide behaviour. These are listed below along with common examples:

1. *Self direction* – independent thought and action – for example, freedom, creativity, independence, self respect.
2. *Stimulation* – the need for variety and stimulation to maintain an optimal level of activation – for example, the desire for an exciting and varied life.
3. *Hedonism* – pleasure or sensuous gratification for oneself – for example, pleasure, enjoyment of life.
4. *Achievement* – personal success through the demonstration of competence according to social standards – for example, being ambitious, influential, capable, successful, intelligent.
5. *Power* – emphasis on the attainment or preservation of a dominant position within a general social system – for example, social power, wealth, authority, social recognition.
6. *Security* – the motivational goal of this type of value is safety, harmony and stability of society, relationships and of the self. It derives from both group and individual requirements – for example, family security, national security, social order, health, cleanliness.
7. *Conformity* – the goal of this value type is the restraint of actions, inclinations and impulses likely to upset or harm others and violate social expectations or norms – for example, obedience, self-discipline, politeness, honouring of parents and elders.
8. *Tradition* – traditional modes of behaviour become symbols of the social group's solidarity, expressions of its unique worth,and presumed guarantors of its survival – for example, respect for tradition, being devout, humble and moderate.
9. *Benevolence* – concern for the preservation and enhancement of the welfare of people with whom one is in frequent personal contact – for example, being helpful, responsible, forgiving, honest and loyal, valuing mature love and true friendship.
10. *Universalism* – understanding, appreciation, tolerance and protection for the welfare of all people and for nature – that is, a broader focus than benevolence values, for example equality, unity of nature, wisdom, social justice, broad-mindedness, protection of the environment, a world at peace.

Source: Schwartz, S H (1992) 'Universals in the content and structure of values: theoretical advances and empirical test in 20 countries', *Advances in Experimental and Social Psychology*, vol 25, pp1–63

and like to measure up to these, to meet our 'psychogenic needs'. Several sets of these needs have been proposed, the most famous of which is Maslow's needs hierarchy which has many similarities to the value types discussed above. These needs include those for self recognition or prestige, affiliation (being accepted by others), achievement, dominance over other people, and the need to reduce guilt or shame. The person concerned is often not aware of his or her needs or motives, but these motives do appear to have some impact on food-related behaviours.

One example comes from a study of American working mothers. They were asked to list five nutritional priorities in their feeding of their children. Next, they were asked for two reasons for having each of these priorities and they were then asked to give their reasons for these explanations. This revealed a great deal of guilt on their part. When they were children their mothers had stayed home to look after them and they did not do this. They thought that the main way they could relieve this guilt was to ensure that their children's food was of the highest nutritional quality. They felt that this enabled them to be 'good' mothers.

Market researchers, as we discuss in Chapter 8, are particularly interested in how our motivations and needs influence our food-buying behaviours, as well as in the ways social behaviours change over time.

Learning

Although new-born human babies tend to prefer sweet tasting liquids to unsweetened liquids – breast milk is sweet – most of our taste and food preferences are acquired after birth. At first, babies try to ingest almost any object, although they prefer sweet tastes and reject bitter tastes. Infanthood is one long personal experiment in tasting objects in the environment, at the end of which we know what is a food and what is not.

Within a few years of birth most children have narrowed down the list of food or other objects they will eat. They do this primarily by developing aversions to foods and by developing a certain reticence about trying new, unfamiliar foods. Such aversions are learned rather than innate (although bitter foods are generally disliked in the absence of cultural learning processes).

People and animals appear to acquire taste aversions rapidly. This rapid form of learning can occur when a new taste (eg, of a new food) is accompanied within a day or so by gastro-intestinal or other upsets. For example, if a child tries a new vegetable and then happens to get a tummy upset, that new taste is paired with a set of noxious autonomic responses. The child's feeling of nausea is paired with the taste of the new food. She or he is likely to avoid that food in the future.

Other processes are also likely to play a part in acquiring aversions, such as the linking of disgust to foods. Children are taught that certain objects, such as urine and faeces, are disgusting and to be avoided. Many animal species develop similar disgust reactions. Humans, however, can attach disgust reactions to a variety of objects, even pleasant objects like fruit juice. Paul Rozin, an eminent food psychologist, has done experiments to illustrate this process. In one experiment he dipped a dead cockroach (a 'disgusting' object) into a soft drink which had previously been evaluated by the subjects. Thereafter, the subjects found the drink less pleasant, even when it was poured from a fresh bottle into a fresh glass. The image of the soft drink had been conditioned in the subjects' minds with the image of the cockroach.

Our Culture, Social Status and Food

Our cultural background and social position also affect our food intake as cultural and social norms become internalized in individual behaviour. These combine with our individual physiological and psychological needs to produce certain food habits. Cultural and social influences overlap, but they are not really synonymous. There can be differences between social groups within a culture, such as differences between age or social class groups, but similarities remain across all the groups within a culture. So Chinese foods can be judged according to how 'hot' or 'cold' (yin and yang) the foods are, Mexican foods have a

distinct hot 'chilli' flavour, and American foods a sweet flavour.

Some cultures have given rise to major gastronomies and cuisines, such as Thai or French or Turkish food. It is not that all the foods are unique to particular cultures; after all, many distinct cuisines use basic ingredients like fish and rice, although some things considered suitable for human consumption by one culture are considered unsuitable by others (eg, termites, dog and horse). It is the ways in which the foods are flavoured, prepared, cooked and served which are distinctive.

Cuisines are usually unique to a particular

culture. For example, French regional foods have unique tastes and unique methods of preparation. Children are usually socialized into knowing the difference between 'excellent' and 'mediocre' meals – a set of expectations is created during socialization as to what 'good' food should be like. Westerners often think that this relates only to organoleptic properties (such as taste, smell and texture), but Koreans, for example, hold the view that a 'good' food not only tastes good and is hygienically prepared, but must also be prepared in a sincere, well-intentioned way by the cook.

Foods can be used to express relationships between people. For example, many cultures allow the most powerful individuals to have the first and best choice of foods, and all of them use foods to celebrate important occasions such as holy days, weddings and initiations.

Food consumption is a common way in which humans regulate or control their bodies, often within the context of power relationships. St Catherine of Siena, for example, prevented herself from being forced into an unwanted marriage through a regimen of religious dietary restriction (like modern-day anorexia). She informed her parents that the suffering she put herself through was for their spiritual redemption! This path of self-denial led her to head a religious order and eventually to become a papal adviser. Officially, she had no power or authority, but dietary self denial was instrumental in her exercise of influence over others. The point has not been lost on recent politically inspired hunger strikers.

Food embodies several cultural paradoxes, including the following:

➤ food is the source of life, yet life must end to gain it (the meat-eater's dilemma);
➤ food gives life but can also end it – unsafe foods can kill us;
➤ food is a source of great pleasure and also a potential source of suffering (eg, when it causes food poisoning), so we need to balance the pleasure we get from it with the need to restrain our appetites.

Different cultures approach these paradoxes

in different ways. For example, Hebrew culture was wary of new foods or foods from unclean and hybrid animals. The Mosaic Law forbade many types of flesh as unclean; even the flesh of animals which were killed by unclean (unwhole) men (eg, eunuchs), was forbidden. Kosher food laws are attempts to ensure that food is fit to eat in the sight of God and man.

Present-day Western cultures emphasize the harm that food may bring to both men and women. For example, too much fat gives men heart attacks and makes women fat and unattractive. Quite the reverse occurs in many cultures, where fat women are viewed as 'normal' and attractive – hence the existence of fattening houses for young women in some parts of Africa. Western women are caught in a bind where much of the available food is highly calorific and fattening, yet they are under great pressure to attain thin body shapes which are quite abnormal in terms of both biology and most other cultures.

Vegetarianism reminds us of the first paradox above. Many people see animals as innocent beings, yet they eat them. They resolve the paradox in many ways. For example, English has one word for the live animal (cow, pig) and another for the dead meat food (beef, pork). Most people in the culture know the relationship between the two forms, but this semantic distancing technique keeps the violent aspects of meat eating out of most people's minds. Vegetarians do not allow themselves this device and resolve the paradox by accepting the sanctity of animal life.

We tend to think that people in other countries are bound by special cultural rules, but that we are not. What we do is plain common sense. Of course, that is how everyone feels! Western scientists and food technologists often think that nutrition is a highly commendable science and that foods are not much more than nutrients and taste-containing molecules. But to a French peasant, say, manufactured food is not very wholesome and certainly does not taste as good as traditional food. And to someone from a South East Asian culture, the idea of eating manufactured food may raise some concerns about its status in terms of yin and yang properties.

Foods are very much the product of cultural ideas about society: who should eat it, about the body and spirit, and which foods are good for which functions. In the West especially, where spirituality has been driven back by materialism, foods no longer have spiritual properties, whereas in more spiritual cultures they certainly do.

The uses of food listed in Box 4.7 do not include any spiritual uses and few mood-enhancing ones, but they do involve numerous activities found in many cultures. The ten value types listed in Box 4.6 summarize many aspects of human relationships. Food can be involved in all of them in some way. Our culture is intimately involved in the food habits we adopt.

Social influences on food intake

Societies are made up of individuals, groups and categories of people which interact with each other. Usually, the attributes which form the bases for social categories are: gender, social class, life stage and age, and ethnic and cultural differences.

Gender

Most men require more food than women, partly because of their greater size. In some countries, some men require greater energy intakes than women because they do more physical labour. However, in many, especially developing, countries around the world, women do more physical work than men and so may require relatively greater food intakes.

The food preferences of men and women often differ. In most cultures, women tend to consume more fruit than men, but men often consume more meat. There is some evidence that in some Western cultures men make greater use of salt and high energy fat and sugar foods. All of them have high taste impact. Most women appear to eat more foods which are regarded as healthy.

These differences may be related to differences in social status between the sexes, with tastier and more scarce foods being reserved for men. This may be appropriate for those social groups in which men expend more energy in physical work than women (eg, coal-mining communities). The energy needs of pregnant and lactating women, however, are considerable and are likely to be greater than those of the average man.

It is also likely that men's lack of responsibility for food preparation is associated with less awareness of the health consequences of food intake and with the greater opportunity to eat high fat convenience foods at work. They may also care less than women about the effects of their food intake on their appearance. Most studies show few, if any, differences between the taste perceptions of the sexes, so these do not explain the observed differences.

Gender roles dominate the food system. This is seen most clearly in industrial food production and in the preparation of food at home. The top positions in most companies and organizations are usually taken by men. Although girls tend to do better in all subjects at school, in the world of paid work women have been largely confined to supporting roles such as clerks and production-line workers. Women tend to be paid less than men and more of them are part time workers in the USA, the EU, Australia, Japan and elsewhere (see also Chapter 5).

In the home, women are usually the active decision makers and they do most of the work including the cooking, shopping, cleaning and child care. Furthermore, they are likely to be responsible for the adequate performance of these tasks. Usually, men are not held responsible for them and do not do them. In many societies, men are given first choice of food and are allowed to take the best foods for themselves. In some traditional Islamic societies, women prepare most of the food but eat separately from the men. In many agricultural societies, women prepare the domestic food and they also do the bulk of the manual farm work.

Box 4.7
Some Common American Uses of Food in the 1970s

Food		Use	
Pie	Yogurt	When I have to lose weight	With coffee
Potato salad	Roast beef	When I want something	When riding in a car
Milk	Spaghetti	easy to chew	To eat with my fingers
Tomatoes	Fried eggs	To eat with a spoon	For men
Chicken	Tossed salad	For teenagers	On cold days
Coffee	Orange juice	For dessert	When I am visiting someone
Jell-O	Ice cream	When I want something	In a salad
American	Baked beans	inexpensive	When I am really hungry
cheese	Rice	When I want something	A spicy food
Shrimp	Carrots	nutritious	For children
Tea	Bagels	When I do not have much	Served cold
Chili	Broccoli	time to eat	To eat with a fork
Vegetable	Cottage	For breakfast	Easy to digest
soup	cheese	For special holidays	With cocktails
Liver	Peanut butter	In the summer	When I want something I
Fish	French fries	In a sandwich	really like
Soft drinks	Ham	For guests	Just by itself
Meat loaf	Peas	When I want a little variety	When I want something easy
TV dinners	Strawberries	in my meals	to prepare
Watermelon	Potato chips	For a snack lunch	On a picnic
Steak	Tuna	At parties	Something you broil
Wine	Pizza	For a between meal snack	Something you try not to run
Dry cereal	Frankfurters	With friends	out of
Cake	Pickles	For lunch	
Dip	Onions	When I want to feel	
Chop suey	Parsley	creative	
Apples	Candy bars	When I am not very	
Chitterlings	Bacon	hungry	
Bread	Tacos	When I am unhappy	
Hamburger		As a main dish	
		For dinner	
		When I am not feeling well	
		When watching TV	
		When eating out in a	
		restaurant	
		When I want something	
		light	

Source: Schutz, H G Rucher, M H and Russel, G F (1975) 'Food and Food-Use Classification Systems', *Food Technology*, March, p52

One result of this maldivision of domestic labour is that women tend to care more, know more, and to be more active about food. In food matters women are usually the active sex. This may be partly the result of the different ways in which boys and girls are brought up. In many cultures, most girls expect that they will be responsible for the welfare and care of other people; most boys, on the other hand, have been raised to be more career oriented. However, even within the home, most men have great influence. In many families women are expected to cater to the tastes of their men folk. Men rarely cook main meals, but they frequently veto meal choices. Women are more often told what they should feed their families, particularly their children, but they are *not* told what to eat for themselves.

Of course, not all relationships are so unbalanced. Over the past 30 years there has been a considerable readjustment of the relations between the sexes in many countries and we now talk of neo-traditional and non-traditional families. In non-traditional relationships, men share domestic chores and take responsibility for seeing that they are done. The neo-traditional form of relationship may look traditional in that men are the main breadwinners and women do most of the food preparation, but this arrangement has been negotiated between equals in order to suit the prevailing economic factors. This seems to be quite common in some West European societies.

Outside the home, male dominance of the food system is still complete. Men make the decisions about investment, about price fixing, about new technologies. However, it is women who have to deal with the effects of these 'macro' decisions, such as how to feed a family on unemployment benefit.

One of the biggest changes to occur in most societies since the Second World War has been the steady increase in the number of married women who rejoin the workforce after having children. The result has been that many women now have more responsibilities. They perform work tasks outside the home in addition to their 'traditional' domestic work. Apart from a small minority, most men do not share the tasks of growing, acquiring and cooking food. One way of easing the burden has been through the introduction of convenience foods and labour-saving equipment, such as blenders and microwaves. This enables women to aspire to be the 'perfect housewife' while working full-time outside the home, but it does nothing to reduce the inequity in the 'gender status quo'.

Marketing campaigns, technological investment, shopping hours and product development are all centred on modern concepts of gender. Most are centred on the woman as consumer. They assume a degree of gender inequity almost unconsciously. Is such a system really equitable? Is the best way to address working mothers' work overload really through the purchase of microwaves and ready made meals?

Marriage and cohabitation involves the disruption of previous food patterns and the establishment of new ones. Married men tend to have healthier food intakes than single men. Conversely, married women tend to have less healthy diets than single women. Some foods are eaten more by couples than by single people. One of the few studies that has examined the effects of cohabitation was a study of the food habits of engaged couples in Sydney, Australia. This showed that soon after marriage the men tended to adopt more of the food habits of their wives than vice versa. However, as the years passed, the men reverted to their previous habits and their wives adopted some of them. In general, men's eating habits in several Western countries seem to be characterized by greater amounts of condiment, salt, sugar and fat intakes, and lower fruit intake.

Age

Our food needs do vary with age, but how these are fulfilled depends on sociocultural factors and to some extent on power relations between people, the relationship between parents and children being a crucial one.

The best food for babies is human breast

milk. But breast-feeding has not always been viewed as desirable. It is less than 200 years since the rich, educated elites of Europe scorned the idea of a woman feeding her own baby – wet nurses were in common supply in Dickensian England.

Infant feeding is subject to many pressures – hospital practices that may discourage breast feeding, social pressures to conform to what peer groups or role models do, economic pressures resulting in mothers having to work soon after birth in conditions where breast feeding is difficult, and psychological pressures compounded by advertising which may promote artificial alternatives to breast milk (milk formulae), and special weaning and infant foods as being a better way to show you are a good mother (see Box 4.8).

In many cultures mothers may use whitish-looking foods for weaning infants off

breast milk. Indeed, even coffee whitener has been used instead of breast or cow's milk. Young children may not be fed 'solid foods' because they are perceived to be bad for them, or they may not be fed meat because it is too 'powerful'. Some foods and beverages clearly are used to mark the transition from one life stage to another – for example, alcohol is reserved mainly for adults because it is believed to be bad for children.

Food in modern Western societies often becomes an area where parents and children battle for power over what the child eats. Surveys of 5 to 15 year old children suggest that their food patterns are similar in variety to those of adults, although food aversions (especially to sharp or bitter foods) may be common.

Adolescence is characterized by the adaptation to adult roles (the pursuit of self-identity) resulting in high self-consciousness and by

Box 4.8
Breast Milk is Best Milk!

Good nutrition starts with breast milk – it is so nutritionally complete that an infant normally needs no other food for the first 4–6 months. It is hygienic and inexpensive. It immunizes infants against common infections and it reduces the risk of breast and ovarian cancer in the mother.

But it is under threat from city living, women working more, hospitals that discourage breast-feeding and advertising to persuade mothers that bottle feeding is more modern and sophisticated. Yet the WHO estimates that over a million infants' lives could be saved each year if they were fed just breast milk for their first 4–6 months. Together with UNICEF, they have launched a 'baby-friendly' hospital initiative to ensure that babies born in hospitals get off to the best possible start.

Hospitals that follow the ten simple steps below can put up a baby-friendly sign:

1. Have a written breast-feeding policy which is routinely communicated to all health staff.
2. Train all health staff in skills to implement this policy.
3. Inform all pregnant women about the benefits and management of breast-feeding.
4. Help mothers initiate breast-feeding within half an hour of birth.
5. Show mothers how to breast-feed and how to maintain lactation even if they should be separated from their infants.
6. Give new-born babies no food or drink other than breast milk, unless medically indicated.
7. Practise rooming-in (allow mothers and infants to remain together) 24 hours a day.
8. Encourage breast-feeding on demand.
9. Give no artificial teats or pacifiers (also called dummies or soothers) to breast-feeding infants.
10. Foster the establishment of breast-feeding support groups and refer mothers to them on discharge from the hospital or clinic.

Source: UNICEF (1991) *The state of the world's children 1992* Oxford University Press, Oxford

physical growth spurts. In general, growth spurts occur about two years earlier in girls than in boys (around 10–11 years and 12–14 years respectively). These growth spurts are associated with huge demands for energy and associated nutrients. During a growth spurt, an adolescent may require as much as 10,000 calories per day (about three to four times as much as an adult man). This can result in almost continuous eating in an attempt to meet energy needs. If this involves the development of food patterns which include lots of fat and sugar which may have no obvious ill-effect during adolescence, it might lead to preferences for these foods into later life when energy outputs are less, contributing to obesity and heart disease.

Adolescents experiment with food patterns. Meals are skipped, and snacks and eating in the company of peers are more common. In Western societies, slimming diets are common among early teenagers of both sexes, although they are about twice as common among girls. As adolescence progresses, more girls and fewer boys experiment with weight-reduction diets. Serious eating disorders first appear, mainly among girls, during this period.

Many other food habits are related to age. A key difference between industrialized Western society and other societies is the low status associated with age in the former. In traditional societies, age is often associated with the acquisition of valuable knowledge and wisdom. In Anglo-American society, a person's money-earning power and social participation may be sharply reduced at retirement. This is a double affliction because in many countries people can now expect to live for far longer than any of their predecessors – for example, until 80–90 years of age. Since most families in industrialized countries are small and unextended, many years can be spent in relative solitude and loneliness, especially by widowed women.

In West European cultures, food preparation has to be readjusted to the needs of one or two people before or shortly after retirement. Some older people become less interested in food shopping and preparation when they no longer have a family to cook for.

Some see this as a form of freedom, others as the loss of an important social role. Single men have the poorest food patterns with very little variety. Widowed and single women have greater variety.

Social class

Social class is almost universally related to health as well as to social prestige, relative economic power and access to education. The poor generally have worse health and die earlier than the rich, and they often have worse diets. People with similar incomes may spend money differently on food, as studies of households headed by manual and professional workers in Britain have shown. Even in reputedly egalitarian countries like Australia, the top 10 per cent of income earners own far more than the bottom 10 per cent. In contrast, Japan has a more even, though far from equal, distribution of wealth. Apart from Japan and a few other countries, the last decade has seen a major decline in the equality of income distributions in most countries. In Britain and elsewhere, the rich have become richer and the poor have become poorer. This has been associated with parallel declines in the relative health status of the social classes. In Britain, this has been shown in a series of reports in community health journals since the landmark Black Report showed strong links between class and health. The diets of poor people appear to have been relatively unaffected by recent nutritional improvements in affluent countries.

Social class matters because it affects what people eat – what they choose and what they can afford. Many working-class consumers do not have the knowledge, skills, time or social contacts to use the food system to their maximum benefit. For example, they may not know how to complain about poor food products or have access to legal resources to voice effective complaints. When people are poor, food often has to take a back seat to the need to pay the rent, clothe the family and pay for transport to work. When people become unemployed, food expenditure can be squeezed more easily than fixed payments such as those for housing, which may lead not only to malnutrition (and possible deficiency diseases) but also to stigmatization of children if their parents cannot buy the advertised 'normal' foods their peers consume.

Ethnic and cultural differences

Different ethnic groups socialize their children in different ways, especially in the ways in which food is produced and consumed. Feasts and holidays are marked by special meals and foods, which may be prepared, served and eaten by different people. For example, when Australians prepare barbecues for gatherings of families and friends, cooking the meat – the main feature of barbecues – is done mainly by men. Most cultures mark their social occasions by special foods. Often, special foods are reserved for consumption by special categories of people – for example, meat may be served first to the men and then to women and children.

Ethnic minorities are found in many countries and national societies are becoming multicultural. Foods are often used by minorities to retain the autonomy of their own cultures. Members of these minorities are often less wealthy than members of the host culture, are often discriminated against and may perform poorly paid jobs.

Ethnic minorities may face major problems as consumers. Their accustomed foods may not be available at reasonable prices. They may resort to importing food from home (as in the case of Cook Islanders in New Zealand and Italians in Australia), thus breaching quarantine regulations. They also face the problem of adapting to new foods. This is often easier for the men who tend to work more outside the home than for the women who may be socially isolated (eg, Bengalis in Britain where there is a problem of vitamin deficiency malnutrition in women). A key problem for migrants is unfamiliarity with the host culture's language and shopping arrangements. Supermarkets, for example, can be incomprehensible and frightening for migrants who may not be able immediately to distinguish food products from animal and non-food products.

Although there is something of a fuzzy boundary between culture and society, one of the key trends today is the development of an overarching consumer culture which is spanning societies and cultures world-wide.

Food habits

Most people tend to eat the same foods with the same people in the same places at the same time of day, week after week. We even use our hands and/or cutlery in the same habitual ways. We commonly mark the time of day by meals – breakfast, lunch, tea, supper. These habits are so fixed that people often believe they are derived from our physiology – that is, that humans need so many meals a day. It is only when we look at cultures different from our own that we realize that this view derives from our culture. For example, in middle class Britain breakfast used to be a large, high protein meal (eggs, sausages, bacon, toast, tea and much more), whereas in southern Spain and much of Latin America, there was no early morning 'breakfast'; at most, it was a cup of coffee and a piece of bread.

Even within the same culture, not everyone eats the same foods in the same ways. 'Food patterns' describe the different ways in which large sections of the population consume foods. Food patterns can vary between people in many ways. Some food patterns can be decidedly boring and others can be unhealthy. Some people try to make their habits as healthy as possible, others are more concerned with the pleasure of eating food or with the effects of food intake on their appearance and social and sexual acceptance.

What we eat

People eat different amounts of foods from different groups of foods. However, saying just what the groups are is problematic. Your view of essential food groups might not be the same as someone else's. An American survey of lay people's groupings of foods showed that they differ from nutrition-based food groups (which themselves vary from country to country in the number of recommended groups). Lay people may also group foods in ways other than their apparent botanic or zoological similarity – for example, foods may be simply classified into 'slimming' foods or 'fattening' foods. Some foods are seen as healthy foods, others as less healthy (see Figure 4.4).

The types of foods which may be served to form a meal are pretty firmly laid down by local culture. For example, the main meal of traditional British working class families had a tight structure in terms of liquids and solids. First, liquid was served – soup; then came solids – the main course (meat, potato, vegetable and gravy); this was followed by a sweet, semi-liquid 'dessert' (eg, sponge pudding and custard); followed by a liquid – tea. Among Australian children, the rules of food combination are quite clear: savoury foods are served with savoury foods and sweet foods with sweet foods and never on the same plate with each other.

The amounts eaten vary, too. Some people have large meals, others have small meals. Although we might expect that the larger the meal the less the frequency of food intake, this is not generally the case. People who eat large amounts of food tend to eat individual foods more often, too! Hormonal as well as central nervous system influences affect the duration of bouts of eating as well as the time between meals.

Variety and food repertoire are key nutritional concepts. Most people are bored by the

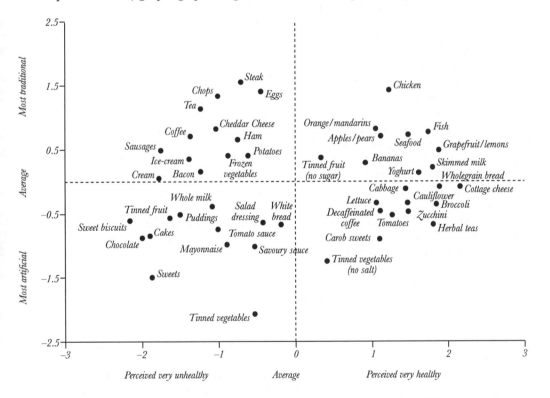

Note: Foods on the horizontal axis are represented along a dimension of perceived healthiness; the further to the right the healthier the foods were seen to be. On the vertical axis 'traditional' foods (top) are contrasted with 'artificial' foods (bottom).

Source: Worsley, A and Worsley, A J (1991) 'What should you eat? Sociodemographic differences in food consumption norms', *Appetite*, vol 16, pp239–47

Figure 4.4 South Australian women's views of the desirability of 48 common foods

same food, prepared in the same way, day after day. A variety of foods and tastes help to maintain our appetite – variety is the spice of life! By variety, however, nutritionists do not mean the need for varying alimentary sensations, but the desirability of choosing a wide variety of foods from the nutritionally defined food groups (in affluent Western countries, five groups are now generally specified – cereal grain foods, fruit, vegetables, meat and meat substitutes, milk and milk products). Nutritionally, we are better off if we eat from all of these groupings, and also if we choose a wide variety of foods from within each group. So, nutritionally speaking, we should eat many kinds of fruit, not just one. Such variety ensures that we get all the nutrients we need and that we do not get too much of one nutrient (or of a food contaminant). Variety is nutritional insurance!

In Western societies, many people have a relatively small food repertoire from which they make their choices, unlike in some traditional hunter-gatherer societies. For example, whereas today's Europeans eat about a dozen different fruits and vegetables, eighteenth-century Australian Aboriginals could, at different times of the year, select from about 120 foods from this grouping. One aim of nutrition education is to expand people's taste experiences beyond those of a few staples and snack foods.

Seasonality in food-consumption patterns is greatly reduced in Western countries where retailers buy foods grown in a variety of climates, produced many hundreds or thousands of kilometres from their shops, so fresh fruits, for example, are available in midwinter. Modern storage facilities such as low temperature and carbon dioxide storage of fruits, which maintain their firmness and attractive appearance (if not their taste), also enable retailers to overcome seasonality.

When we eat

Eating meals at precisely set times (eg, lunch at 12.30pm) depends on having the means to tell the time as well as the placing of a high premium on time-keeping. Some people eat until they feel satisfied, others eat until the plate is empty. One psychological theory ('externality' theory) suggested that some people's eating habits are governed more by external cues (like the clock) than by internal sensations of hunger. Many obese people were thought to be 'externalizers' who were unable to monitor their internal sensation accurately and so ate when others ate or when the clock said so, or even when they felt upset, confusing emotional disturbance with hunger. It appears, however, that non-obese individuals are affected by these factors just as much as obese people.

Many Westerners believe they should eat three times per day, with the main meal eaten either at midday or during the early evening, supplemented with two minor meals. Among the poor of the world, however, one or two meals per day is likely to be counted as fortunate. However, little is known about the prevalence of such regularity. Regular meal taking may be related to conservative personal habits, and more flexible arrangements have become popular during the past 20 years as part of the world wide trend to greater social informality. In the USA, for example, some affluent people have abandoned large meals and more or less continually eat or 'graze' their way through the day. This abolition of regular meals is not novel. Historically, native Fijians, for example, had no word for 'meal'; they ate huge amounts at irregular intervals at the completion of long traverses across country.

Dieting

Dieting, or voluntary restriction of food intake, is a common activity, especially for Western women. In any one year, around 80 per cent of American women try some form of weight-reduction diet. Why are slimming diets so prevalent? Two explanations appear plausible. First, affluent food supplies contain many processed foods which concentrate high amounts of energy into servings (mainly in the form of fats, sugars and carbohydrates). However, people use less energy in their normal living – work is less physical and people have more labour-saving devices (eg, cars). As a result, more people become obese or fat.

This alone need not cause people to take up slimming diets. The social norms of Western culture also play a part. Two social forces are definitely 'anti-obesity'. One is the so-called 'cult of thinness' in Western societies, where thinness is equated with beauty, and women especially are encouraged to lose weight by restricting their food intake. Such 'crash diets' are usually easier to follow than becoming more active – that is, increasing energy expenditures, or making gradual changes to lower energy intake through the consumption of lower energy foods such as cereals, fruit and vegetables.

This 'beauty myth' is deeply embedded in consumerist society with its emphasis on instrumentalism – that is, people often engage in activities to bring about desired goals but not primarily because they are good in themselves. Thus people jog, swim or diet, not because they are good to do in themselves, but because they believe that they will lead to other rewards such as social approval. When it becomes obsessive, dieting can be life-threatening and can certainly reduce the quality of life (see Box 4.9).

The second force is the creation of a visual image of what people should look like, through cinema and TV, which makes people focus closely on appearance. Film and video reveal all sorts of physical imperfections repeatedly to wide audiences, but also use relatively thin people in most advertising. A less important factor in prompting dieting is obesity, which puts people at some risk of premature death from a variety of diseases – for example, heart disease and diabetes.

Of course, slimming diets are just one kind of dietary restraint. Others are medically prescribed diets such as high complex carbohydrate diets for diabetics, low-fat diets for people at high risk of heart disease, or gluten-free diets for coeliac disease sufferers. In addition, there are vegetarian diets which avoid the use of animal products to varying

---

Box 4.9
Anorexia, Bulimia and Diet Cycling

Extreme forms of weight control are quite common. One to two per cent of adolescent and young adult women in European and North American countries suffer from anorexia nervosa – an excessive fear of being fat. The sufferers, including some men, eat very little food and usually avoid foods with fat or sugar. Their weight may be so reduced that normal body functioning may be seriously impaired and some die.

Another form of pathological weight control is bulimia nervosa. This appears to be a consequence of 'normal' dieting in which food intake is restrained for long periods at the end of which 'binge eating' occurs – eating of huge amounts of food more or less continuously for several hours. This may be terminated briefly by self-induced vomiting, which may be followed by another bout of binge eating. In this condition the victim maintains a normal body weight but achieves this in an extreme, dangerous manner.

A more recently recognized pathological phenomenon is 'diet cycling' in which people go on and off restrictive diets for many years. This cycling may reduce resting metabolic rates – the energy used each minute by the body to maintain itself when resting. This makes it increasingly difficult for the dieter to lose weight through food restrictions.

The causes of these conditions are controversial. Many psychiatrists opt for biological explanations. In contrast, some sufferers and feminists focus more on societal forces and unequal power relationships between men and women which impair the development of personal identities among girls and women.

degrees. Finally, there are many fad diets which people may take up for brief periods to avoid or gain some supposed condition – for example, fruit-only diets to 'cleanse' the body of 'poisonous chemicals'.

Dietary supplements and fortification

Dietary supplements are not foods, but they are made from some of the constituents of foods – for example, vitamins, minerals and substances like wheatgerm and evening primrose oil. They are either taken voluntarily by individuals or distributed on a population-wide basis by adding to certain highly refined foods such as white flour and breakfast cereals. The latter is known as fortification (ie, the addition of vitamins in amounts that are not present in the whole-food precursors of the product such as the addition of vitamin C to sweet confectionery) or restoration (ie, restoring vitamins in breakfast cereals to the levels found in whole grains).

There is some debate about the wisdom of fortifying foods with relatively high levels of these and other nutrients, because it may encourage under-consumption of groups of foods such as vegetables, fruits and cereal grains, which may also contain unknown but valuable anti-disease agents. Polyphenols in green tea, for example, appear to have potent anti-cardiovascular disease effects.

Individuals take dietary supplements as 'medications' either regularly or at infrequent intervals to maintain or improve their health. During the 1980s, about one-third of American men and almost half of American women took at least one nutritional supplement at least once a week, although this prevalence appears to be declining. Similar usage rates are found today in the USA and in other countries such as Canada and Australia.

In high doses, some supplements (especially fat-soluble vitamins A, D, E and K and iron) are dangerous. However, recent scientific investigations support the view that vitamins A, C and E may have anti-oxidant effects which may help to prevent the development of some cancers as well have beneficial roles to

play in the prevention of other diseases such as heart disease. The effects appear to operate only at low 'physiological' concentrations and forms, not at the high doses and forms found in many vitamin pills. In the 1990s, certain foods themselves – functional foods – are being promoted as having specific health-giving properties (see Box 4.10).

The widespread use of dietary supplements raises the question of whether it is a criticism of the quality of the commercial food supply reflecting low consumer confidence or a lack of confidence in the supplementer's ability to eat health-giving foods. Their use also can be seen as part of the modern cult of health – that it is extremely important to achieve maximal health and well-being. Nutritional surveys suggest that people who take supplements tend to eat more nutritious foods than non-supplementers

Dietary practices

Food patterns concern more than the kinds of foods (and beverages) consumed. They include the ways foods are cooked, prepared and served. For example, a person might eat ten slices of bread every day. This might be part of a healthy food pattern, but if the bread is eaten with lashings of butter rather than without fat or polyunsaturated fat, then it may be part of an unhealthy food pattern. About a quarter of all the fat consumed by British people, for example, is in the form of butter and margarine spreads.

Other food practices affect health, including the ways in which meats and vegetables are prepared and cooked. If they are fried in fats (especially lard, butter and other saturated fats) they may be far less healthy than if they are steamed, baked or microwaved. When it comes to health, it is not just what a food contains that matters but the amount consumed. Some cheeses, for example, contain around 40 per cent fat by weight. They are high fat foods, but if you eat only a small serving once or twice a week, they would not be a major source of fat for you. However, consumption at every meal would probably be unhealthy.

Box 4.10
Functional Foods, Nutriceuticals and Novel Foods

In many countries, folklore suggests that certain foods have medicinal properties – for example, carrots are good for eyesight, bananas and arrowroot relieve diarrhoea, fish is good for the brain, and so on. In the past, scientists generally dismissed such claims, but now opinions seem to be changing fast, thanks to the immense amount of research in the physiological roles of nutrients and other food constituents. Originally, this was done by health researchers but now, owing to the lure of niche marketing, food companies are spending a lot of money on what are called 'functional foods'. In Japan, the world leader in this innovation, sales of functional foods grew from nothing in 1987 to around US$2 billion in 1992.

Put simply, a functional food is a food which offers a health benefit. The Japanese Ministry of Health and Welfare defines a functional food as:

➤ a food derived from naturally occurring substances (not a capsule or a powder);
➤ a food consumed as part of the daily diet;
➤ a food with a particular function when ingested, serving to regulate a particular body process.

Examples include:

➤ bread on sale in Norway which has tiny granules of omega 3 polyunsaturated fatty acids which are believed to help prevent heart disease;
➤ salad dressing impregnated with eicosapentaenoic acid, an anti-cholesterol agent;
➤ oat bran breakfast cereals which are claimed to help to reduce serum cholesterol levels.

Food marketers have been quick to latch on to functional foods. John Cahill of CWFS Advertising in Australia claims that there may be as many as five categories of functional foods:

1. Foods that help the body's defence mechanism (by enhancing the immune system).
2. Foods that help prevent or control disease (eg, foods specifically designed to prevent cancer or to prevent or control diabetes).
3. Foods that aid recovery from disease (such as cholesterol lowering foods).
4. Foods that regulate body rhythms (eg, foods which act on the digestive function or absorption).
5. Foods that suppress ageing.

Much of this is just hypothetical. A great deal of laboratory research and many clinical trials would be needed to substantiate some claims. The US National Cancer Institute, with the best of motives, is spending around US$20 million on developing cancer preventative foods ('designer foods') which will incorporate naturally occurring compounds (from fruits, licorice and vegetables) into foods which will have cancer preventative properties. Would it be simpler, however, if people ate more fruit and vegetables to help prevent cancer?

Functional foods still face a major hurdle – the legislation in many countries which prevents the making of health claims on product labels. Originally, these prohibitions were introduced to prevent quackery – unscrupulous marketers making false claims ('eat this and you will live longer!'). There are determined efforts to undo these rules on the grounds that scientific evidence can now be made available to substantiate claims. Recent Japanese labelling regulations allow such health claims to be made. It may well turn out to be a hornets' nest. Consumers will have even more decisions to make – when is a food a food and when is it a medicine?

A nutriceutical is not a functional, or any other kind of food, but an ingredient (in a food) which has therapeutic applications – for example, vitamins, minerals and antioxidants like some of the polyphenols found in green tea.

Novel foods can mean just about anything, but usually the term refers to a new form of food which may or may not have health properties. For example, new super large tomatoes or tomatoes with four chambers instead of three which are supposed to be crispier than the older varieties. Bioengineering is often used to produce such novel foods.

Source: Cahill, John (1991) CWFS Advertising, Sydney

Summary

In summary, cultural and social influences on our food 'needs' are usually internalized within us so that we are largely unaware of them. Each of us uses food to satisfy a variety of needs, as we discuss more in the final chapter. How we try to meet these needs depends on other economic and ecological factors, such as the availability of different foods and our relative wealth or poverty. All the factors discussed in this chapter shape, constrain and determine the decisions we make about food. These decisions are greatly influenced by developments throughout the food system, by how the various economic actors within it are operating and by the tools they use to try to influence our decision making. In Part 2, we look at what each set of actors is doing and how they are working to shape the food system. Although there is variety within each particular group of actors, it is helpful to view each group as a whole and to note some of the conflicts of interest within the group and between groups.

Part 2

From Farm to Mouth – The Key Actors

In today's food system, individuals and small enterprises like farms face ever larger enterprises whose bargaining power is immensely greater than their own. Power and control over food is passing to these institutions which operate, quite expectedly, in their own interests. In most rich countries trends in the food economy are similar, even if their magnitude varies. The key trend is concentration – that is the economic jargon for small firms being swallowed up by larger ones and economic clout collecting in fewer larger firms. This has produced a market structure in which a few very large firms tend to dominate different market sectors. Food manufacturers led the way but, partly as a response to this manufacturing power, there have also been dramatic changes in wholesale and retail food distribution, which in some countries has shifted the balance of power between manufacturers and retailers. Other parts of the food system, from input supply industries making fertilizers, feedstuffs, machinery and chemicals, to importers and distributors, have shown similar tendencies to concentration. This process affects bargaining relationships, union organization, wages and social relations in the food system.

In Part 2 of the book we look at how the key actors – farmers, workers, traders, processors and manufacturers, wholesalers and retailers, caterers and consumers – work with the background basics discussed in Part 1. We start with farmers who produce the raw materials, move on to the workers in the food system, and discuss traders who ship the raw materials to where they are transformed into food products for distribution through wholesalers and retailers. We look too at the growing role of caterers in preparing food ready for consumers to eat and the many different factors affecting consumers. It is not a simple, neat linear system, but one with many different interactions and influences.

5

Farmers, Workers and Traders

Before the late 1700's there was probably no settled community in which at least nine-tenths of the population were not directly engaged in tillage. Rulers and priests, craftsmen and merchants, scholars and artists, were a tiny minority of mankind standing on the shoulders of the peasants. Under such circumstances any lasting change in climate, soil fertility, technology, or the other conditions affecting agriculture would necessarily modify the whole of society: population, wealth, political relationships, leisure, and cultural expression.

White jr, Lynne (1962) 'The Agricultural Revolution of the Early Middle Ages', in *Medieval Technology and Social Change*, Oxford University Press, Chapter 2, by permission of Oxford University Press

Until a few generations ago, most people lived in rural areas and had some connection with food production. Farmers were the lead decision makers about food – people had to buy and eat what they produced, which depended largely on their ability to farm the land and breed suitable varieties of plants and animals. What they produced was lightly processed – by milling, drying, salting, and the like – if at all, before use in the home or small-scale bakers and other enterprises in towns and villages.

Today, most people in industrialized countries, and in a growing number of developing countries, are remote from their food supplies. The farmers' lead role as decision-maker has gone. Farmers respond to others in the system who set the pace for change and the rules under which the farmers operate. These rules are set by national politicians and by the bargaining power of the different actors in the food system.

A major shift in these power relations came with industrialization and the new patterns of production it brought. Farm workers became fewer as activities were mechanized; new actors, such as food processors and manufacturers, became important employers. As food consumption patterns ceased to rely on local production, traders who were able to ship foods world-wide also came to play a growing role. It is these three groups – farmers, workers and traders – that we deal with in this chapter. In looking at them as separate groups in this and subsequent chapters, however, we must remember that the divisions are not so clear-cut, with some traders, for example, now also involved in food processing.

Farmers

farm units tend to buy inputs (and) sell output(s) to business units that operate in more concentrated markets, not common in the agricultural sector. Therefore the agricultural sector is vulnerable to shocks and pressures emanating both from the adjacent sectors, from outside the food economy and from the inherent instability of agricultural production itself

OECD (1981) *Food Policy*, OECD, Paris

Farmers are not a homogeneous group of people – they range from small-scale to large-scale operators, from tenant farmers to owners, from peasants to lords; they employ farm managers, workers, or family members; some work full-time, others part-time; some raise just one or two crops or animals, others raise many in a mixed farm or even an organic farm. Yet, despite this diversity, several basic factors and trends affect them.

One is that farmers are relatively small players in the food system today. This limits their power to determine what they do. They have to balance their use of the three classical factors of production in agriculture – land, labour and capital – and make as rational decisions as they can, according to prevailing circumstances. The decisions are made in response to a mix of economic circumstances, commercial pressures from their suppliers and purchasers and national policies. These have a major impact on how the three factors of production are used, especially labour. We could also add the use of energy, information and technology as vital factors of production, although to make use of them adequately requires capital and skilled labour.

An endangered species

Perhaps the most striking trend is that farmers and farm workers are becoming something of an endangered species. In industrialized countries, the percentage of the labour force in agriculture is a small part of the population, ranging from a low in Britain of about 2 per cent, just under 3 per cent in the USA to an average of 5.3 per cent in the OECD countries, although Greece has nearly 25 per cent. In the developing countries as a whole, about 60 per cent work in agriculture, but this ranges from highs around 90 per cent in parts of Africa and Asia to lows in the 10–20 per cent range in Latin America.

World-wide, the trend is towards fewer people working on the land, with larger holdings (see Figure 5.1). Even in the UK, which has a tiny percentage of its population working

in agriculture – just 1.5 per cent in England – numbers are expected to fall by a further 100,000 by the end of the 1990s to around 560,000. Falls are likely to be much greater in the European Union (EU) and in developing countries with huge implications for rural society, the countryside, and towns and cities.

Mechanization and economic pressures are the cause of the decline in numbers, and government policies have a lot to do with both. Many of the functions carried out by farmers in the past have passed on to other parts of the food system. Traction power, for example, is no longer provided by locally bred animals but by factory-produced machines. Cultivation practices requiring few purchased inputs but a great deal of labour have been replaced by practices requiring specialized machinery and chemical treatments produced in factories and bought-in seeds.

For those left on the land, there is constant pressure to increase labour productivity, by focusing on specific crops and products and through mechanization. Field cultivation is mechanized using tractors, drills and combine harvesters, and pesticides for weeding instead of hoeing. Animal production is mechanized through the use of intensive poultry systems, milking parlours and so on. These changes also transform the husbandry skills needed in farming. Farmers able to deal with complex

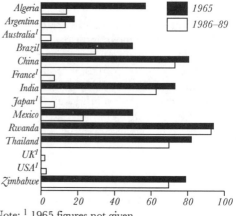

Note: [1] 1965 figures not given

Source: UNDP (1992) *Human Development Report 1992* Oxford University Press, Tables 16 and 36

Figure 5.1 Percentage of labour force in agriculture

ecosystems using recycling and feedback mechanisms are replaced by those who are able to manage more linear processes. These use prophylactic chemical treatments to contain problems that such production systems otherwise might incur, such as infection in large herds or diseases in crops.

Capital users

As farmers use less labour, farming becomes more capital intensive. Increasingly, a farmer needs more money to be successful. Access to capital and the terms on which it is available becomes crucial for farmers' survival and determines much of what they do.

The financial system in any country or area strongly influences access to money – through the tax structure, banking system and credit rules. In the UK, for example, post-war tax incentives encouraged investment in machinery and new buildings as they could be offset against tax. This has resulted in some of the world's most highly mechanized farming operations. Under the EU's Common Agricultural Policy (CAP) (see Box 5.1), the market is guaranteed for many farm products through the intervention prices, which are set above the cost of production; this, coupled with land-drainage grants and other land-development grants, encourages expansion into land that would otherwise be uneconomic for cereal growing. This land often produces poor quality cereals for which there is no real market. This is sold into intervention, put into stores and helps to produce surpluses.

As farmers find themselves having to make more money from their land, they may invest money in buying or renting more land or hiring labour, or they may buy products which make the land or labour or both more productive. In subsistence farming, very little outside capital, if any, is used. The family on the land provide the labour, they produce for themselves, and they may barter or exchange with neighbours for goods or for additional help when they need it. To invest money in land requires capital, which would have to be borrowed, probably at very high interest rates. For many who do not own the land they subsist from, with nothing to secure the loan on, credit is unobtainable.

Investing to increase returns from farming usually places farmers in debt, with banks or other money-lenders. While they make take on debts at one rate of interest and with one set of prevailing prices for their products, later they may find that they cannot service these debts (pay the interest on them) if prices fall or interest rates rise. In the 1980s many farmers in both rich and poor countries were squeezed out of business because they were unable to service their debts. In the USA in the mid-1980s, for example, over 200,000 farms – 11 per cent of all farms – went bankrupt as crop prices and land values fell. Often, to try to service the debts, farmers attempt to produce more from their land, resulting in the use of more inputs to maximize yields. If this, too, fails to provide sufficient returns, they have to sell the land and leave farming.

Land – control and benefits

Land is a farmer's primary asset. As we saw in Chapter 2, it is the biological basis for agriculture. But land ownership is also an established basis for power. It has underpinned the considerable political influence that rural landowners have had in the rich, industrialized world. As an asset, its value may determine how much can be borrowed to invest in increasing farm production. Who controls the land and how they do it affects how the land is used and to whom the benefits from its use accrue. This is why land distribution and ownership has been such a powerful political weapon.

The idea of owning land, however, has developed in particular societies and civilizations. Although people can make things – for example, bricks – from material found in or on land, land is part of the planet itself, part of a common heritage. Many peoples, for example Australian Aborigines and North American Indians, had a sense of the primacy

Box 5.1
The Common Agricultural Policy

The policy does not measure up well on either efficiency or cost criteria: it is extremely expensive, highly bureaucratic and prone to fraud. In terms of equity it is a highly regressive policy for both consumers and producers – that is, it artificially inflates prices in a commodity which takes up a particularly high proportion of the budget of low-income families and regions, while giving most help to the larger, richer producers. The system completely lacks transparency: it is largely incomprehensible to all but a select few, and strategic moves involving large sums of money are often made behind closed doors, with consumers being scarcely represented at the vital moments of decision.

National Consumer Council (1988) *Consumers and the Common Agricultural Policy* HMSO, London, p4, reproduced with the permission of the Controller of Her Majesty's Stationery Office

If the European Union's Common Agricultural Policy (CAP) is so bad, you might wonder how it has survived at all. The CAP was the lynch-pin on which the original European Community of six was founded and yet it became its most widely criticized policy – inside and outside the EU. CAP grew out of the post-war concern of the founding members – Germany, France, Italy, Luxemburg, Belgium and The Netherlands, but especially the first two – to ensure food security through maintaining food self-sufficiency and to support small farmers.

Thus it began with a variety of policy objectives: strategic, over-food security; agricultural, in promoting self-sufficiency; and social, in maintaining rural population. Cheap food was not one of the objectives. However, the CAP relied to an overwhelming extent on one policy instrument to achieve these different aims – pricing. For the CAP is based on the maintenance of a sufficiently high price at the farm gate for the producer to maintain small farmers in business. It encourages production and has provided almost guaranteed markets for many farm products until recently.

It has survived partly because in much of Europe, but not Britain, a remnant peasantry with smallholdings remains; the peasantry is part of a national identity, is politically important and has benefited from the policy, but not as much as larger farmers and middlemen storing the surpluses.

Consumers pay a higher price for food as domestic production is protected by a range of barriers from overseas competition where this would disadvantage community producers. Consumers are increasingly unhappy with this policy.

Despite the original Community's enlargement to 12 (up to 15 from 1995), including Britain which previously imported low-cost food and made deficiency payments to farmers to make up their income, the CAP basically remains wedded to that policy.

The effect of the high-price policy, coupled with a guaranteed market where the EU buys up surplus production at an intervention price (below the expected market price but above the cost of production), has been:

➤ to encourage high levels of production surplus to European requirement;
➤ to encourage the intensification of production with technology developed to squeeze higher and higher productivity from both workers and land;
➤ to benefit the rich most, as the larger farmers benefit most;
➤ to produce a system of regulations of Byzantine complexity making them more open to fraud;
➤ to enable the dumping at subsidized prices of surpluses on the world market with often disastrous consequences for farmers in poor countries.

In the early 1990s, when the vast expense of the policy – it took 35 billion ECU out of a total EC budget of 65.5 billion ECU in 1993 – and the surpluses it produced were an acute embarrassment to ministers, some controversial reforms were agreed. However, these reforms will still not change radically the policy. At the core of the reforms is a drop of about one third in cereal prices towards world market levels – from a target price of 155 ECU per tonne to 110 ECU per tonne in the 1996–97 marketing year. Farmers with less than 20 hectares and those who set aside 15 per cent of their arable land will get compensatory payments. These will equal the difference between the 1991–92 target price and the new, lower target prices – calculated on the basis of the average regional yield in the three years to 1990–91, so that any yields above this level will not get any compensation.

The European Commission expects productivity will remain the same but others, such as industry analyst Brian Gardner of EPA Associates, believes that 'the set aside effect will be very much less than expected and that productivity will be [increasing] 1.5 per cent per year or more'. The new regime also requires a substantial bureaucracy, with detailed claims of precisely what is grown where to be submitted by farmers. Although the Commission expects the reform to cut costs in the long term, many doubt whether it will and the Commission's own figures project a budget increase in the first few years. Much more radical reform of CAP is likely, especially as the EU admits more members.

---

of land – as lived off, or with, but which was not or could not be owned. Land carried an obligation to be passed on in as good or better condition to the next generation.

Today, individual, private property rights imply rights to the owners which are largely unfettered by obligations. The forms of ownership and land tenure influence the way that land is used or exploited. One example is inheritance rules – is land passed on to the first-born male or equally to all the family? The latter leads to continuous sub-division of land into smaller and smaller parcels from which it may be hard to make a living, as happened in Ireland. Some people leave, those who remain take over, working the land as sharecroppers or tenants, but that is divided still further when they die. Where the first-born inherit, great estates can be built up and passed on from generation to generation.

Common lands have been appropriated from native peoples and peasants and greatly reduced in area or enclosed, as discussed in Chapter 3. If the landowner is not the farmer, the benefits from good land management may not go to those working it. This in turn may affect how the land is worked. In Britain, land in the Middle Ages was under the control of a feudal lord but was worked by his serfs, who also had access to common lands for grazing animals; this common land was later enclosed for private use.

Just who owns what land in Britain today can be hard to discover, but Richard Norton Taylor in his book *Whose Land is it Anyway?* states that about 1 per cent of the population owned 52 per cent of the land in the late 1970s, with about 1300 individuals owning one third of the country. Of the agricultural land, itself about four-fifths of the total, just under two-thirds was owned by owner-occupiers and just over a third by landlords. Average farm size had grown from 63 acres in 1908 to 278 acres in 1980. Today, professional family farmers, often tenants of large landowners, supported by wage labourers, tend the land. These farm workers, however, are disappearing and so, too, are many family farmers. Institutions such as insurance and pension funds have bought land and gone into farming for investment purposes. Small enterprises find it hard to survive and are least able to bear the risks of farming.

Risk takers and risk minimizers

Farmers are at the riskiest end of the food system. Farm production depends on a wide range of factors, many of which are beyond a farmer's control, in particular the climate. Good weather can produce large harvests for which there are insufficient markets, and depress prices so much that it may not be worth taking the crop out of the ground. Poor weather can reduce harvests, produce huge crop shortfalls and high prices, and encourage more farmers to plant a particular crop for the following year, which is likely to produce another glut. Such gluts and surpluses are endemic to farming and increase the instability and insecurity of the food supply; this is highly undesirable for society as a whole and a major reason for government intervention in agriculture.

Economic policy decisions, often taken far away or in the interest of more politically powerful groups, also affect farmers, influencing what they grow, what they earn and whether they can make a living from their land. Diseases can wipe out crops or animals, or war and civil conflicts make sowing, harvesting, or other operations impossible.

With so many factors increasing risk, it is not surprising that in the past and today poor farmers favour land and animal management strategies which minimize risk but tend to require detailed, individual knowledge of complex farming systems. In developing countries, for example, farmers may plant mixed landraces of crops rather than a pure variety, so that some crops will grow even if one or other of the many hazards, such as drought or disease, strikes.

Yet farmers must adapt to changing times. Innovation, however, also means taking risk. Usually, larger farmers can afford to take risks. They can afford to put in capital, often borrowed and secured against their substantial assets, to buy more land, draining it or buying machinery and energy to replace labour to enable one person to do what five or ten did before. Smaller farmers may be unable to do this or they may need to come together in some form of co-operative before they make any capital investment so that the risk is spread.

Ensuring a stable food supply and markets for farm products underlies the major political goals of the rich industrialized nations. In Europe, post Second World War, this desire for food security helped farmers to gain far greater influence on policies than their numbers would

suggest. Indeed, they and their American counterparts have secured support policies of various kinds which have made overproduction the key problem for rich world farming. Farmers' and landowners' organizations have had much influence in government ministries and departments of agriculture. Some see this influence as helping to produce policies that have benefited larger farmers most.

Now, you might suppose that a national union of farmers would speak up for farmers of all kinds, and particularly for the small farmers who form the large majority. In reality, no one has any significant influence in the NFU [National Farmers Union] unless he is a member of its council. To serve on the council, a member must leave his farm to travel down to London most weeks in the year and there sit on committees and perform other duties which are likely to take two or three days a week. Small farmers cannot do that; they are too busy sitting on their tractor or milking their cows...

So the NFU Council has gathered to itself men less aware of the day-to-day realities than most farmers, and less sympathetic to them. Being the larger farmers – having 1,000 acres or more is not unusual for an NFU Council member – they have, almost by definition, been the beneficiaries of the system which has amalgamated tens of thousands of small farms. These are the men who have enjoyed the lion's share of the grants, subsidies and tax allowances; they have worked the system; and they have prospered. Prosperity has blurred their vision.

Body, Richard (1991) *Our Food, Our Land – Why contemporary farming practices must change* Rider, London, p139

As farming has become more industrialized, more and more food-production tasks are done in factories and processing units and fewer are carried out on farms. Industrial production aims to reduce degrees of uncertainty – for example, those caused by climate or natural variations, by controlling the environment in which processes occur. In agriculture, this leads to intensifying production using more inputs on the farm, such as increased fuel, fertilizers or pesticides, or in the technology of the suppliers who provide seeds and breeding animals.

Commodity producers – price takers not makers

Modern farmers grow primary commodities for sale – from grains to vegetables to beef cattle, from coffee to cotton, from poppies to tobacco. These commodities may be eaten, drunk, turned into clothes, or used as drugs. Only subsistence farmers, now largely confined to the poorer countries and often the poorer parts of them, produce for themselves using their own resources. Of course, some farmers grow their own crops to feed to their animals and then sell the animals. Agricultural trade is a significant factor in the world's trading system and the support that various governments give to agriculture was a major stumbling block to the agreement of changes to the world's trading regime. These supports have produced very complex trade links (see Box 5.2).

The terms of trade for primary commodities have declined over the long term, despite occasional fluctuations. This deterioration in terms of trade between rural and urban-industrial products is one reason that farmers find their margins squeezed, and why their ability to make a living from the same area of land is decreasing with time. As primary commodity producers, they find that the same amount of a commodity like wheat, sheep, or soya will buy less of the industrial products they need as time goes on. This happens both to individual farmers and to whole countries that depend upon agricultural commodities for their export earnings.

For at least a decade, the prices of agricultural commodities have tended to fall on international markets while those of manufactures have tended to rise. These contrary movements have resulted in a decline in the net barter terms of trade between agricultural commodity exports and the imports of manufactures and crude petroleum. In 1992 the decline was 2 percent. A comparison of the three years 1990–92 with the years 1979–81 shows a decline of 30 percent, that is an average annual rate of 3 percent. The decline was close to 40 percent for the agricultural commodity exports of the developing countries and 20 percent for those of the developed countries. Some countries have achieved gains in productivity sufficient to outweigh the decline in real prices (the barter terms of trade) but, for many, the decline has reduced earnings per hectare of land cultivated (the single factor terms of trade). Furthermore, the global decline in prices has been so large that it has generally offset the expansion of production, thus actually reducing overall earnings (the income terms of trade).

FAO (1993) *The state of food and agriculture 1993*
FAO Agriculture Series No 26, Rome, p51

Table 5.1 shows examples of the degree to which increases in crop yield and production have been outweighed by declines in the barter terms of trade. Countries exporting coffee have generally been big losers and other commodities showing large losses have been cocoa, sugar, rice and maize. Prices are also subject to great fluctuations, owing to the effects of climate on production, and so returns change from year to year.

This decline in the terms of trade between farm producers and the industrial producers of goods and services has increasingly disadvantaged the farmer. Farmers have had to increase output to maintain their livelihood. Many farmers borrowed money at variable interest rate loans when interest rates were low. When interest rates rose, they found themselves on a treadmill which required greater productivity from their factors of production just for them to stand still. Many farmers end up being forced off the land. Larger enterprises, however, have been able to benefit and investors looking for maximized financial returns have also entered farming and may manage the factors of production for short-

Box 5.2
CAP, GATT and Agricultural Trade

Trade in agricultural commodities is of vital interest world-wide. For net exporting countries, it offers a major source of foreign exchange earnings or an opportunity to dispose of surplus stock, while for net importers the terms of trade greatly affect their food security. Until the proposals agreed in the General Agreement on Tariffs and Trade (GATT) Uruguay Round, in December 1993, agriculture could be regulated and subsidized by countries as they wished. However, following this GATT agreement trade in agriculture will be largely deregulated.

Disputes between the world's two biggest agricultural trading blocks, the USA and EU, delayed the Round many times. It took the US and the EU (then called the EC) seven years to reach a compromise deal, after almost getting into a trade war. The deal was finalized at Blair House in Washington in November 1992. It included:

➤ limits on the area the EU devotes to oilseeds to 4.615 million hectares;
➤ limits on subsidized exports to 21 per cent of the average level from 1986 to 1989, to be reduced over six years from 1994;
➤ US acceptance of the EU's compensation to farmers who set aside 15 per cent of their arable land for price cuts in internal CAP reforms.

Even so, some farmers, from the American mid-West to the dairy producers in Europe, were far from happy.

Hans Bender, director of the Danish Dairy Board, called it a 'General Agreement to Terrorise Trade'. He believes that the negotiators focused on grain and bulk products like oilseeds and left the dairy sector – which faces immediate cuts of almost half in cheese exports in the first year of implementing the outline agreement – with a lousy deal.

Even sections of the food industry were unhappy, according to Alec McClumpha, of the UK Food and Drink Federation and former general manager of Nestlé UK's overseas trading division. For them, the problems include the focus on agricultural bulk exports, putting controls on volumes of exports, and treating added value and consumer-ready products in the same way as cereals, sugar, meat and milk.

The EU is the only major exporter of value added products, he points out, exporting two-thirds of the 25 billion ECU of EU agricultural exports as added value products of some kind. The proposed changes could hit these exports.

Changes in Europe could also have a major impact on developing countries. With falling cereal prices in Europe changing the relative cost of animal feed, Thailand could be hit as a supplier of manioc (tapioca) for feed, according to Professor Stefan Tangermann, of the Institute of Agricultural Economics at the University of Göttingen in Germany. With tapioca and other such marginal feedstuffs, 'the producing countries have to feel the pain for one or two years before they reduce production', says cereal trader Raymond Rogers of Finagrain, a part of Continental Grain.

Most exporting countries – for example, Argentina – are expected to benefit from reduced cereal subsidies in Europe. Such reductions would relieve the downward price pressure on the world markets, with the GATT agreement expected to raise world cereal prices by 10–15 per cent, more for milk and beef and even more for sugar.

But while agricultural exporting countries may gain, importing countries will lose out without some assistance. High food prices should also encourage such countries to give a higher priority to improving their own agriculture according to Simon Harris, Chairman of the International Trade Policy Working Party of the UK Food and Drink Federation, and Director of Corporate Affairs for British Sugar.

Source: Revised from a *Gemini News Service* feature, London, 12 March 1993

Table 5.1 Percentage changes in yield, production and terms of trade for selected food commodities for all countries, 1979–81 to 1990–92

Commodity	Change in yield	Change in production		Change in terms of trade	
			Barter	Single factor	Income
Coffee	3	14	−66	−65	−61
Cocoa	18	42	−66	−60	−52
Tea	27	36	−2	−8	−2
Sugar	10	27	−55	−50	−43
Soyabean	16	26	−36	−26	−19
Rice	29	32	−48	−33	−31
Wheat	36	29	−35	−12	−16
Maize	15	18	−35	−25	−23

Note: Barter terms of trade = export prices (of agricultural products) deflated by import price (of manufactured goods and crude petroleum); income terms of trade = export earnings deflated by import prices; single factor terms of trade = net barter terms of trade adjusted by changes in productivity (yields per hectare)
Source: FAO (1993) *The state of food and agriculture 1993* FAO Agriculture Series No 26, Rome, Table 4, p 52

term financial gain.

Smaller farmers are constantly squeezed in a cost/price wedge that promotes higher production using greater technological inputs, and encourages larger holdings or part-time farming. It is a squeeze that encourages farmers to be the first to maximize production from their resources. This is because the benefits from innovations tend to go to those introducing them first. As soon as many people adopt an innovation, overproduction tends to make prices fall. It also encourages farmers to seek support from society as a whole for their enterprises.

One double-edged risk management strategy has been the development of contract farming, whereby a farmer contracts to supply a food processor or retailer with a specific crop, with defined specifications at a specific time. This gives farmers a definite market but requires them to plant specific varieties and to treat them as laid down in a contract. If the eventual produce does not appear at the right time or meet size or other requirements, it may be rejected by the buyer and left on the farmer's hands. Such contract work, in effect, shifts the risk of climate, disease and so on more fully on to the farmer, while the buyer can seek other sources of produce (see also Chapter 6).

Farm subsidies

Despite the fall in numbers and relative increases in farm size, farmers are still predominantly small-scale producers and family businesses. There are many more farmers compared with the number of tractor, seed, chemical and other firms from which they buy (see Box 5.3) and the food manufacturers and supermarkets to which they sell. Economically, farmers and farm workers are often poorly organized, although politically farmers are usually better organized and have much greater clout in the industrialized countries than their numbers suggest. The power derives from various quarters, depending on the country. In the UK, it stems in part from the power and influence of the landed classes, in the USA from the political clout of farm associations where agricultural commodity exports are a vital source of national revenue, and in France from the political importance of the peasantry.

In developing countries, however, despite their greater numbers, an urban bias in development policies often means that farmers have very limited political clout. Urban populations tend to influence centres of government more, and governments themselves see agriculture as relatively unimportant in the strive to develop, which is often seen as synonymous with industrialization.

The political influence of farmers and the strategic importance of food supplies has meant that the major industrialized nations have been careful to protect farming, if not all farmers, from free markets with their gluts and surpluses in a variety of ways since the Second World War. Whatever their rhetoric, few governments leave agriculture to market forces, while in the past the former centrally planned economies tried to plan production in a command economy. They all intervened, with varying degrees of success, to influence agriculture. In 1947, the British Government, for example, introduced the Agriculture Act which supported farmers in maximizing production

Box 5.3
Input Suppliers: a Farmer's Best Friend?

Several major groups of new players in agriculture have developed over the past few decades. They supply inputs to farmers. Before the industrialization of agriculture, farm supplies were mostly generated on the farm – seeds for the next season, crops to feed animals, dung for fertilizer, and draught animals, raised or bought, for power. Ploughs and small pieces of equipment were bought from local blacksmiths.

With the development of more capital-intensive, high-input agriculture, farmers increasingly relied on outside suppliers to provide their tools, buildings, fertilizers, seeds and feedstuffs. Most of these suppliers have grown to become huge conglomerates, often servicing a global market and taking a global view of their business. They can be divided into four groups:

➤ Agro-chemical industries provide fertilizers and biocides (toxic substances designed to kill animal or plant pests and called more specifically pesticides, herbicides, aphicides and fungicides).
➤ Seed producers and feed manufacturers provide seeds for planting and feedstuffs for intensively reared livestock and supplements for more extensively reared livestock.
➤ Biomedical companies provide drugs for prophylactic use in intensive farming, for vaccination and for curative purposes.
➤ Equipment and energy companies provide farm machinery and fuel to power it.

Although farming is quite clearly geographically based, input providers need not be and can distribute their products over wide areas. They often started as local companies but have become huge transnational enterprises, marketing their wares throughout the world. Often they use the names of companies they have absorbed in their growth to maintain some local identity.

Input suppliers have a strong interest in maintaining markets for inputs to intensive agriculture, from energy to hybrid seeds. A few large companies tend to be the dominant suppliers in each market, as the tables opposite illustrate. As pressures mount for changed practices to meet environmental concerns, these large input companies are diversifying or regrouping to ensure that they can remain important players. For example, seed-producing companies, until recently usually local or national concerns, are being bought by the major chemical industries, which are also moving into genetic engineering, with many acquisitions taking place in the 1990s. It is the interest of these suppliers, rather than farming, environmental or other interests, that is driving the move into biotechnology.

Box 5.3 *continued*

Top ten pesticide suppliers in 1988

	Pesticide sales (US$ billion)	Percentage of global market
Ciba-Geigy (Switzerland)	2.14	10.70
Bayer (Germany)	2.07	10.37
ICI (UK)	1.96	9.8
Rhone-Poulenc (France)	1.63	8.17
Du Pont (USA)	1.44	7.19
Dow Elanco (USA)	1.42	7.11
Monsanto (USA)	1.38	6.89
Hoechst (Germany)	1.02	5.12
BASF (Germany)	1.00	5.00
Shell (Netherlands, UK)	0.94	4.69
Total	*5.00*	*75.02*

Top ten seed corporations in 1988

	Seed sales (US$ millions)	Percentage of global market
Pioneer Hi-Bred (USA)	735	4.90
Sandoz (Switzerland)	507	3.38
Limagrain (France)	370	2.46
Upjohn (USA)	280	1.87
Aritrois (France)	257	1.71
ICI (UK)	250	1.67
Cargill (USA)	230	1.53
Shell (Netherlands, UK)	200	1.33
Dekalb-Pfizer (USA)	174	1.16
Ciba-Geigy (Switzerland)	150	1.00
Total	*3098*	*20.65*

Source: Hobbelink, Henk (1991) *Biotechnology and the Future of World Agriculture* Zed Books, London, p44 and p46, Tables 4.4 and 4.6

through subsidies for fuel, fertilizers, drainage, hedgerow removal and credit and used deficiency payments to top up their income if returns from sales fell below certain levels.

In the EU, farmers work within the framework of the widely criticized Common Agricultural Policy (CAP) which gives preference to agricultural commodities produced in the EU, provides price supports to farm products and attempts to produce an equal market across the member states (see Box 5.1). In the USA, where agriculture is a major export earner and there is a substantial farm lobby, the Government intervenes by offering direct support to farmers' incomes, through a mixture of price supports via loan schemes which effectively act as a floor price, public stock management, production restraints through

area reductions and land conservation programmes.

The amount of money involved is huge.
Total transfers to agricultural producers in the
OECD countries grew from US$299 billion
in 1990 to US$335 billion in 1993, according
to an OECD Secretariat estimate. The
OECD uses two main measures of how agricultural policy assists producers. The
Producer Subsidy Equivalent (PSE), measures
the value resulting from agricultural policy of
monetary transfers to farmers from consumers of agricultural products and from
taxpayers – market price, direct income, indirect income and other forms of support like
extension and research services. The
Consumer Subsidy Equivalent (CSE) is the
implicit tax paid by consumers as a result of
market price support.

The support levels vary greatly among
member countries (see Figure 5.2). Much of
this money does not reach the farmers, but
promotes storage, disposal and other policies
and goes into the pockets of various middlemen. The complexity and large sums of
money to be made encourages fraud. In the
EU, officials believe that the scale of fraud far
exceeds the nearly 270 million ECU reported
for 1992.

Some, such as British MP and farmer
Richard Body, argue that the various subsidies and schemes promoted by governments
for production at all cost have far broader
effects than encouraging fraud. One result is
an artificial increase in the cost of farming
land to far above its real farming value. This
has encouraged speculative investment which
has pushed up land costs and driven many
farmers out of business. This leads to a greater
concentration of the ownership of land.
Another view is that high prices or guaranteed markets for farm products allow
manufacturers of inputs to price their products more highly than would otherwise be the
case, in the knowledge that farmers' agricultural costs are being subsidized.

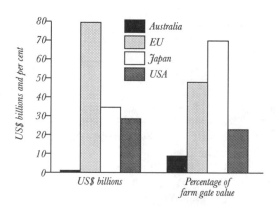

Note: The figures are provisional. The percentage of
PSEs indicate the total value of transfers as a
proportion of production (valued at domestic prices),
adjusted to include direct payments and to exclude
levies and feed adjustments

Source: OECD (1994) *Agricultural Policies, Market and Trade –
Monitoring and Outlook 1994*, OECD, Paris

*Figure 5.2 Producer subsidy equivalents in US$ (and as a
percentage of the value of domestic production), in selected
areas in 1993*

Farmers today

Farmers are becoming outworkers in an industrialized food production system in which
human labour is usually replaced by machines
and processes are carried out in controlled,
centralized units where possible. Overall, it
produces farming methods that allow one person to produce far more than ever before.
First, it can make working conditions easier,
but high levels of inputs are used. Secondly, as
land has become more expensive to develop –
the best land being farmed first – and as
demand has grown, increasing yields from the
land has been a priority to give a good return
on capital invested. Research and development has focused on these ends. The addition
of external inputs such as artificial fertilizers
and pesticides, and the use of new crop and
animal varieties selected for production capacity, means that farmers now use more capital
per unit area and produce more per person
employed than ever before in the rich countries. Whether this makes for efficiency,
however, is debatable (see Box 5.4). Other
strategies, like companion planting and inter-
cropping to maximize yields and to reduce

pest damage or mixed land-race planting, have not received the same degree of research and development as they are both more labour-intensive and less likely to generate privately marketable profitable products.

The increasingly scientific basis of farming and the need to balance a whole range of issues, policies and factors together to survive have transformed the management skills and information needed to farm successfully. At the same time, the development of machinery and other inputs to increase yields, as well as the growth of large-scale units, has in some ways deskilled the job of farming. It has moved away from a close understanding of the natural environment in which it operated, to an attempt increasingly to control the environ-ment. This is perhaps taken to its extreme in the development of poultry 'factories'. Here, 100,000 birds can be kept indoors in cages with movement restricted and feed, water and prophylactic drugs delivered to them. Similar developments have occurred with pigs and calves. These trends have been aimed in part at reducing the risks that farmers face from natural fluctuations in their environment caused by weather, climate, disease, etc. Such practices are producing resistance among consumers concerned about animal welfare.

Now, a farmer needs to know about equipment maintenance, accounting, marketing, use of chemicals, as well as long-standing soil and animal management skills. The diversity of crops and animals to which these skills

---

Box 5.4
So Who is Efficient?

Modern farmers are often said to be very efficient. But in what sense and which farmers? Critics of industrial farming argue that it is not enough to measure efficiency in terms of lowest cost of production or productivity per unit of labour or capital employed. They want to look more broadly at the economics involved, at the efficiency of resource usage and environmental sustainability:

Conventional wisdom would have us believe that big units are more efficient than small ones. On very small units, if the labour input is fairly costed, unit costs do tend to be higher. Set against this, however, land worked with intensive labour input (for example, organic horticultural holdings), as opposed to intensive fossil fuel input, often make very efficient use of natural resources, whilst also producing a high output.

In the case of small- and medium-sized farms, versus big farms, the economics of production have become hopelessly skewed by subsidies. In economic theory, farmers produce for a market, and compete with each other to supply that market: the most efficient thrive. In Europe and in the US, taxpayer subsidies are linked to the volume of output: the biggest and most intensive farmers pick up the lion's share of the subsidies, and they thrive. This should not be interpreted as economic efficiency. In fact, the current situation, where the EC for instance overproduces grain, the larger the volume of grain a farm produces, the more it costs the EC in storage and export subsidies. Taken to its logical absurdity (which it now has been) it becomes cheaper to pay farmers to produce nothing (set-aside) than to buy their grain. Within this economic framework, small, less-intensive farmers should be seen as being more efficient as they produce fewer unwanted surpluses.

In effect, it is almost impossible to tell who is producing more efficiently. Past subsidies for fuel, for drainage, for irrigation, for research into high input/high output agriculture, not to mention current subsidies, produce a completely distorted picture. All that can be said with any certainty is that current policies continue to favour large, intensive farmers.

Source: Clunies-Ross, Tracy and Hildyard, Nicholas (1992) *The politics of industrial agriculture* Earthscan, London, pp16–17

are applied, however, has generally declined as monocultural practices, in which one or two kinds of crops or animals are farmed, have developed.

Modern agriculture has essentially become industrialized. Farmers rely increasingly on scientific and technical inputs developed by those who do not farm. Scientific research on all aspects of farming supplements or replaces the experience-based knowledge that guided farmers until recently. The crucial question is who now decides the future direction of farming.

Bit players in a changing world

The net result of all these factors in a changing, increasingly global market economy is a massive experiment with the food system – the introduction of a technologically dependent, high-input type of farming which produces an increasing range of man-modified organisms over which farmers seek ever greater control. As the focus of individual enterprises has narrowed and high-tech inputs have increased, the economic and ecological consequences of such methods are being questioned. The industrialized world faces massive overproduction of foodstuffs for which there is no market without subsidies. This leaves an underproductive, input-starved agriculture for the vast majority of humankind, which is too poor to put effective demand on its farmers to produce food. In addition, the farmers themselves lack the capital and tools to improve their production sustainably .

For some, the time has come for the role of farmers in industrialized societies to be rethought. Gordon Dickson, Professor of Agriculture and Head of Department at Newcastle University for many years, would be happy to see the number of farmers fall, provided that those who remain are supported by sensible policies to encourage efficient food production and rural land use and that the land is not turned into a rural museum perpetuating an uneconomic structure.

Agricultural and rural social objectives are confused in today's policies. Dickson believes that too many people want to preserve the seventeenth-to nineteenth-century features of the landscape, instead of using today's knowledge and technology to create the new and enhanced landscapes of the future. He wants change, not stagnation. It comes down to a matter of livelihood. On the hills, especially the more remote ones, he sees the case for change – to grouse moors from sheep, for example, in Northumberland where a farmer can quadruple his income by making such a switch.

He sees the Common Agricultural Policy as a ludicrous burden on the population, despite which agricultural incomes are at a 50-year low because most of the money goes into handling surpluses and the services industries, with the latter pricing up their products because they know that the end price is supported.

Dickson abhors the objective of the EU to maintain all farmers on the land. 'Why,' he asks 'should agriculture not be treated like other businesses' – for example, steel, with the least efficient going out of business? With the fixed market in Europe and advancing technology, which increases productivity by about 2–3 per cent per year, he argues that 2–3 per cent of land per year could be released. He believes that there should be fewer people in farming with larger, economically sized units.

Not everyone sees the industrialized approach as the way forward for agriculture. A growing concern for the long-term sustainability of present methods has led to an increasing interest in alternative methods, often lumped together under the term 'organic farming'. In the UK, the most famous organic farmer is the Prince of Wales who has helped to draw attention to the organic movement, which does not use artificial inputs such as fertilizers, pesticides and prophylactic drugs for animals. While some farmers seek alternative methods for philosophical reasons, alternative agriculture can also make commercial sense (see Box 5.5). There are many obstacles to developing alternative agriculture but interest is growing into research and development of more ecologically sound, sustainable farming systems. In Australia in 1991, for example, the Australian Consumers Association and horticultural groups agreed on a charter to reduce pesticide use.

Farmers are also no longer being seen as basically food producers; they are managers of a countryside in which urban dwellers also

---

Box 5.5
Alternative Agriculture

After the US National Research Council reviewed alternative agriculture in the mid-1980s, which included 11 detailed case studies, they concluded the following:

➤ Farmers who adopt alternative farming systems often have productive and profitable operations, even though these farms usually function with relatively little help from commodity income and price support programs or extension.

➤ Alternative farming practices are not a well-defined set of practices or management techniques. Rather they are a range of technological and management options used on farms striving to reduce costs, protect health and environmental quality, and enhance beneficial biological interactions and natural processes.

➤ Well-managed alternative farming systems nearly always use less synthetic chemical pesticides, fertilizers, and antibiotics per unit of production than comparable conventional farms. Reduced use of these inputs lowers production costs and lessens agriculture's potential for adverse environmental and health effects without necessarily decreasing – and in some cases increasing – per acre crop yields and the productivity of livestock management systems.

➤ Alternative farming practices typically require more information, trained labor, time, and management skills per unit of production than conventional farming.

➤ Many federal policies discourage adoption of alternative practices and systems by economically penalizing those who adopt rotations, apply certain soil conservation systems, or attempt to reduce pesticide applications. Federal programs often tolerate and sometimes encourage unrealistically high yield goals, inefficient fertilizer and pesticide use, and unsustainable use of land and water. Many farmers in these programs manage their farms to maximize present and future program benefits, sometimes at the expense of environmental quality.

➤ Fertilizers and pesticides are often applied at rates that cannot be justified economically without consideration of present or future farm program payments.

➤ Federal grading standards, or standards adopted under federal marketing orders, often discourage alternative pest control practices for fruits and vegetables by imposing cosmetic and insect-part criteria that have little if any relation to nutritional quality. Meat and dairy grading standards continue to provide economic incentives for high-fat content, even though considerable evidence supports the relationship between high consumption of fats and chronic diseases, particularly heart disease.

➤ Current federal pesticide regulatory policy applies a stricter standard to new pesticides and pest control technologies than to currently used older pesticides approved before 1972. This policy exists in spite of the fact that a small number of currently used pesticides appears to present the vast majority of health and environmental risks associated with pesticides. This policy inhibits the marketing of biologically based or genetically engineered products and safer pesticides that may enhance opportunities for alternative agricultural production systems.

➤ The results and design of basic, discipline-oriented research programs often are not sufficiently integrated into practical interdisciplinary efforts to understand agricultural systems and solve some major agricultural problems

> Research and extension program funds to study, develop, and promote alternative farming practices are inadequate. It is unrealistic to expect more rapid progress in developing alternative practices to farmers without increased funding.

> There is inadequate scientific knowledge of economic, environmental, and social costs and thresholds for pest damage, soil erosion, water contamination, and other environmental consequences of agricultural practices. Such knowledge is needed to inform farm managers of the tradeoffs between on-farm practices and off-farm consequences.

> Research at private and public institutions should give higher priority to development and use of biological and genetic resources to reduce the use of chemicals, particularly those that threaten human health and the environment.

Source: Reprinted with permission from *Alternative Agriculture*, copyright 1989 by the National Academy of Sciences. Courtesy of the National Academy Press, Washington, DC, pp8-16

have a stake for recreation and environmental sustainability. Their political clout in securing subsidies may also be slipping. In Europe, the days of massive support to farmers from European taxpayers and consumers may be numbered. The Agriculture Commissioner of the EU pointed out to a farming conference in 1993 that agriculture then accounted for less than 9 per cent of the EU's exports and 2.6 per cent of its GDP. 'Should agriculture be seen as an obstacle to a GATT deal,' he asked, 'can the sector expect to maintain its present level of support long-term, especially at a time when the numbers unemployed are almost double those in farming?' In the event, agriculture did not stop a GATT deal (see Chapter 9), but the sentiment is likely to surface again.

Even if the number of farmers continues to fall and farm sizes to increase, farmers will still operate relatively small enterprises compared with the companies with which they deal. Compared with these companies, individuals and even groups of farmers have relatively little bargaining power. They are becoming more peripheral to the decisions about how food is produced.

Workers

While farmers and farm workers are becoming an endangered species, people working in the food system are not. The food system as a whole is a major employer but as it becomes more industrialized the places where people are employed change. Fewer people work on farms and more in supplying primary producers with inputs and transporting and transforming the outputs. At least one in seven of all British workers, for example, work with food, not counting those who supply the industries directly in the food chain (see Table 5.2). In 1992, over 2.8 million people worked in the British food and drinks industry – agriculture (farming), manufacturing (processing) and distribution (retailing, wholesalers and catering) – according to official figures, out of a total work-force in employment of 21.4 million people.

Similar trends are found in other rich countries. In the USA, for example, employment in what is termed the food and fibre economy fell from 24.5 million people in 1947 (41 per cent of the total workforce) to 21 million people in 1985 (18 per cent of the total). The major shift was in those working off the farm, which grew from around 60 per cent in 1947 to 90 per cent of those in food and fibre in 1985. Over the same period, the size of the work-force involved in farming fell from about 10 million to 2.5 million people.

In Australia, out of a population of 17 million in the mid- to late-1980s, some 380,000 people were employed in agriculture and services to agriculture. About half that number, 172,000,

Table 5.2 Employees in the British food industry in September 1992

	1981 (thousands)	1992 (thousands)	Percentage of women
Agriculture, forestry and fishing [a]	354.6	288.8	27
Food, drink and tobacco	658.9	500.3	41
Distribution and retail:			
wholesale	251.2	224.8	33
retail	572.0	600.1	67
Hotel and catering[b]	900.7	1194.4	65
Total in waged employment	*2737.4*	*2808.4*	

[a] More than double this number of people work in farming when owner occupiers, partners and directors in farming are included.
[b] The figures include cafes and pubs, but not public service catering. In 1991, the British Hospitality Association put the numbers in the hotel and catering sector, including public service and contract catering, at 2,135,000.
Source: HMSO *Employment Gazette*, London, February 1993

were employed in food and beverage manufacturing industries, a further 317,000 in food service industry operations, such as catering establishments, and over 430,000 in food retailers, grocers, confectioners and tobacconists.

Wages and conditions

For workers in any part of the food system, however, wages and conditions tend to be poorer than average. Manufacturing and processing are usually the best paid part. In the mid-1980s in the UK, for example, taking average earnings for a basic working week, excluding overtime, in all industries and services as 100, in the food and drink industries sector (the statistics also include timber, rubber, plastic and paper) the average was 93, in agriculture 65 and in hotel and catering 61.

The averages hide variations between men and women. According to UK Government figures, average weekly pay for men in manual work in 1985 in the food, drink and tobacco sector for an average 45.1 hours per week was just over one and a half times greater than for women for an average 38.7 hour week. The main reason for this difference seems to be that

women tend to occupy the lowest grades. They work less core hours and overtime than men. Often they do not qualify for shift or night-work premium. Average hourly earnings for women were 73 per cent of the average earnings of men in 1985.

Within those sectors, some people do even worse. The figures for hotel and catering (61 per cent of average earnings), for example, included managerial wages. In fact:

➤ half of all waitresses earned less than half the weekly average in 1985;
➤ half of all women kitchen hands earned less than 52 per cent of the weekly average;
➤ female chefs and cooks fared better, at almost 58 per cent of the weekly average.

Many workers received even less than these amounts. Much of the hotel and catering industry was covered by the legal minimum rates of pay set by Wages Councils, although these were abolished in the 1990s by the Government. In 1985, the legal minimum rate for a typical adult worker in the sectors covered by Wages Councils was around 45 per cent of the national average weekly earnings. Despite this, and despite the legal obligations on employers over minimum rates of pay, the Wages Inspectorate found that in nearly four out of every ten (37.3 per cent) licensed hotels and restaurants visited in 1985, the employer was failing to comply with the Wages Council rate.

Canteen workers were usually a few pence an hour better off than hotel workers, but were not covered by any statutory safeguards. The Industrial and Staff Canteens Wages Council was abolished in the mid-1970s, and since then no new national agreement has been reached to cover the 200,000 industrial canteen workers, mainly women. During this time, contractors have moved in and new technologies such as cook-chill have been introduced.

About a quarter of all UK workers are part-timers, but part-time women workers outnumber full-timers in areas such as hotels and catering, school meals and retail distribution (see Table 5.3). Male part-timers outnumber full-timers only in public houses, bars and clubs. Even in the food manufacturing industry, women workers are more likely to be part-time than men.

Unions

A major factor influencing workers' rights and the terms and conditions of employment is their ability to bargain with their employers. If done on an individual basis, this is pretty much a David and Goliath affair. When workers band together in trade unions and bargain together (ie, collectively), their power is increased and they can secure better terms and conditions (see Box 5.6). A growing trend in the UK and elsewhere, however, supported by Government legislation, has been to reduce the power of trade unions to negotiate on behalf of their members.

Even so, where unions do work together, they can affect the policies and practices of large companies. In the late 1980s and early 1990s, unions in Denmark caused a change of heart by an international fast food chain over recognition of the right of their workers to negotiate through trade unions. Despite a company policy of not dealing with trade unions, the company was forced to decide whether to move out of Denmark or to recognize some form of negotiating rights, as a result of action by milk, bread and other workers who stopped supplies to the company. In

Table 5.3 Full-time and part-time male and female employees in Great Britain in various food sectors, September 1992

	Male (Thousands)			Female (Thousands)		
	All	Part time	Percentage part-time	All	Part time	Percentage part time
Agriculture, forestry and fishing	210.0	30.3	14.4	78.8	29.4	37.3
Food, drink and tobacco	294.4	13.0	4.4	205.9	73.7	35.8
Distribution						
wholesale	149.8	9.9	6.6	75.0	26.5	35.3
retail	199.8	68.2	34.1	400.3	296.1	74.0
Hotels and catering	420.0	180.9	43.1	774.4	551.1	71.2

Source: HMSO *Employment Gazette*, London, February 1993

Box 5.6
Men and Women – Equal Treatment?

In many countries, legislation safeguards the rights of workers. One of the accepted functions of government has been to ensure a just and fair system in which firms and business operate. Providing equal opportunties – and sometimes ensuring access for groups traditionally excluded, as has been the case of affirmative action programmes in the USA – is deemed a socially desirable policy that has been introduced by law. In the UK, it is perhaps not surprising that one of the most important tribunals on equal opportunities started in the food sector.

Julie Hayward was a cook (not chef!) at Cammell-Laird shipyard in Liverpool in the 1980s. Aided by her union, the General, Municipal, Boilermakers and Allied Trades Union (GMB) and the Equal Opportunities Commission, she took a case against her employers to a tribunal claiming that she did not receive the same basic pay as men doing work of equal value.

Before 1984, many women manual workers were largely unaffected by the Equal Pay Act. The Act provided that women should have equal pay with men doing the 'same or broadly similar' work – for example, in local authorities men work as gardeners, refuse collectors, road-menders and caretakers; women are home helps, cleaners, cooks and care assistants. So for most women, it was not possible to claim equal pay with a man doing like work. After the 1984 amendment to the Equal Pay Act, women could claim equal pay for doing work which is different but of equal value to that of men. For the first time women doing 'women's work' could compare the value of their work with that of a man; thus a cleaner's job could be compared with a road-sweeper's job and that of a school meal attendant with a gardener.

The union enabled Julie Hayward to take her case through the Tribunal, Appeal Tribunal and eventually to the House of Lords, the highest court in the land. They ruled that while Julie's wage in total was similar to a man's of equal value, her basic wage was not and it should be.

Government legislation, however, can also hinder attempts to gain equal opportunities. For example, some local authorities in Britain were concerned about the Local Government Act 1987, which reduced their powers to promote equal opportunities for ethnic minorities by restricting authorities to act against unlawful discrimination, a much narrower role than promoting equal opportunities. The Act prevents authorities from imposing conditions of equal opportunities when contracting out services. This contract compliance could affect school meals services, employment monitoring and purchasing of food.

Source: Clutterbuck, C *Trade Unions and Food*, Fact Pack 1, unpublished

the end, the company joined the employers' association, which had agreements with unions over the terms and conditions of the people employed by its members, thus avoiding direct negotiations but accepting union-negotiated agreements.

In the UK, the degree of unionization in the food sector varies enormously. For example, 70 per cent of the workforce in hotels and catering is unorganized, yet over 90 per cent of the bakery workers in large plants are in trade unions. Most sectors in the food industry have some national agreements between employer groups and appropriate trade unions. Many individual company agreements also exist. However, several employers in the food sector, particularly in fast food, refuse to recognize trade unions for bargaining.

Industrial relations vary enormously. There may be quite close relations between a farmer and farm worker, or sometimes a more Victorian master-servant style relation in hotels, or more sophisticated industrial relations in large plant bakeries. Throughout the food system, stoppages are few. In the food, drink and tobacco sector in 1986 in the UK, 1214 days were lost on average, compared with 2,105 for all industries and services.

Food businesses, however, are no longer simply national and trade unions are responding to the internationalization of food. In Europe, for example, the European Committee of Foodworkers Unions has set up working parties to develop joint union policies on EU directives, collective bargaining, working practices and the position of women. It is also, along with its parent body, the International Union of Food and Allied Workers Associations (IUF), developing at least European level consultations with major transnational companies like BSN and Nestlé. The IUF has a small secretariat of less than 20 in Geneva which acts as a clearing house for information and as a co-ordinator of activities requested by member unions.

Workers' organizations are concerned that the trends throughout the system are towards the replacement of human labour by machinery, which does not bargain, and towards increasing the amount produced per person employed. Usually the most skilled processes, which are also the most costly in terms of labour, are mechanized first; the less skilled work, such as packing boxes, is mechanized next. Another trend is to replace full-time jobs with part-time ones. These tend to be less well paid and do not receive the social insurance and pension rights of full-time jobs.

The IUF recognized that in the EU the concept of a national market has become less meaningful after the creation of the EU single market in 1993. In a report on the development of a large, highly automated £60 million soft drinks plant in the North of England, the IUF highlighted concerns about the closure of older plants, cutting out older unions, reliance on systems that use non-returnable containers, and the effects on working patterns and conditions.

Despite these activities, however, workers have relatively little power over the direction and pace of change throughout the food system.

Traders

The continued use of intermediaries in the supply and distribution of imported food products suggests that they fulfil functions which both suppliers and consumers find useful. These functions are primarily linked to the difficulties for both original and final consumers in finding a partner for direct trading between themselves, and an aversion to assuming the risks involved in doing so, as well as a lack of expertise in the execution of relatively technical functions such as arranging freight and insurance.

Harris, Simon (1983) 'The Role of the Intermediary', in *The Food Industry: Economics and Policies*, Burns, Jim, McInerney, John and Swinback, Alan (eds), published in association with the Commonwealth Agricultural Bureaux by Heinemann, London, p153

Traders – importers and exporters, brokers and merchants – are the least visible group of actors between farm and mouth, yet these intermediaries are responsible for most of the world's food trade. It is a massive business in which commodities from developed countries dominate (see Box 5.7). Primary commodity exports make a major contribution to trade in many industrialized countries, with such trade in the USA, for example, accounting for 14 per cent of merchandise exports in 1991, or about US$58 billion, according to the World Bank. Relatively little is known about the activities of these intermediaries that move most of the food and agricultural commodities about the world, as Simon Harris noted in a review of their role in the UK in the early 1980s, despite importers handling about 40 per cent of the food consumed at that time. The reasons he outlined were that official statistics do not allow identification of a food importing sector; their trade associations, where they have them,

Box 5.7
Commodity Trade

'Before World War II,' as Dan Morgan noted in his seminal work on the grain trade, *Merchants of Grain*, 'the amount of grain that crossed borders, or oceans, seldom exceed 30 million tons a year. By 1975, this figure reached nearly 160 million tons, a growth only slightly less spectacular than the growth of the oil trade.' By the early 1990s, total exports of rice, wheat and coarse grains were around 200 million tons, according to FAO; this was about 10 per cent of world production.

The value of grain exports world-wide in 1991 was almost US$34 billion, but this was just a fraction of the total value of agricultural and fishery products traded that year which was US$367.1 billion – itself about 10 per cent of all merchandise exports that year (see table below). Some of these products are traded in packs, such as tinned salmon, ready for sale to consumers. Mostly, however, they are traded in bulk for further processing into food and beverages or for use as animal feedstuffs with around 5 per cent used as raw materials for industry such as textiles.

Exports of agriculture and fishery products by value and origin, 1991 (billion US$)

Export group	Developed countries	Developing countries
Beverages, sugar, bananas and pepper	6.3	19.8
Oil-seeds, fats and products	16.7	13.7
Cereals	27.9	6.0
Meat	37.8	7.1
Milk and milk products	20.1	0.5
Raw materials	8.3	7.8
Other commodities	13.9	3.6
Other agricultural products	107.1	32.2
Fishery products	21.2	17.3

Source: Food and Agriculture Organisation of the United Nations, (1993) *Commodity Review and Outlook 1992–93*, FAO, Rome, p12, Table 5

do not produce public information; they generally import commodities which will undergo further processing, or if they import consumer packs, such as tinned fruit or salmon, these generally have the retailer's name on them.

Fewer and bigger

One trend was clear, however: increased concentration throughout trading activities. By the early 1980s, for canned goods in Britain, Harris reported that 'most of the first hand distributors and agents have either been taken over, amalgamated, or gone out of business, with the result that a major processed food import into the UK, such as canned salmon, has today only six importer distributors of any consequence and more than 60 per cent of the trade is in the hands of just 2 companies.' Relatively few companies dominate the different sectors, from bananas to cereals to tinned salmon (see Figure 5.3).

Traders sit between farmers and the other actors in the system. Brokers buy and sell on behalf of a client for a commission, while full traders do so on their own account. Traders may also handle shipping for cash or barter trade

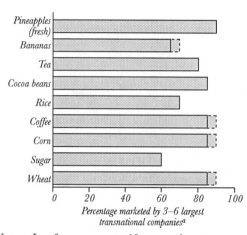

Note: a In a few cases, up to 15 companies account for the bulk of the market. Dotted areas indicate range.

Source: Brown, Michael Barratt (1993) *Fair Trade: Reform and Realities in the International Trading System* Zed Books, London, p 51 Table 5; leading traders in 1980 for a range of commodities are also listed in Table 10, p 71

Figure 5.3 Corporate control of global commodity trade, 1983

done on a bilateral basis between countries.

Interest in traders grew after 1972 when the then Soviet Government was able to make deals with grain traders that helped to precipitate the food crisis of the mid-1970s. The Soviets used the traders to buy up large quantities of grain to make up for a poor harvest. They bought both food and feed grains, thus avoiding the massive slaughter of livestock which were dependent on the feed grains. The result, though, was that they cornered the world market. Prices for the remaining grain stocks rocketed out of the reach of many poor food-importing countries, leaving them faced with famine.

By the end of the 1980s, six companies, five of them private, dominated the world's grain trade: Cargill and Continental (both USA), Dreyfus (France), André/Garnac (Switzerland), Bunge and Born (Brazil) and the newcomer Mitsui/Cook (Japan), the only publicly held company (see Box 5.8). In the USA, the world's largest grain exporter, the six companies accounted for 95 per cent of US corn and wheat exports.

Traders are less concerned with the basic price of commodities and foods than with the money to be made from buying, selling and moving them around. The essential point is that trade takes place, not the cost of the raw material traded. However, the traders' job is complicated and they deal with differing regulations, risk taking through use of the futures markets, and financing of imports from developing countries.

A complex business

Commodity trading is a big and complex business today. Agricultural commodities become available at specific times, which vary depending on the part of the world and the seasonal production cycle. The demand for these commodities, for either fresh or processed foods, runs right though the year. Post-harvest, prices tend to be lower as much is available. Later in the year, prices tend to rise as stocks are used up. Annual production varies, however, depending on the weather, and this affects the price.

World-wide trading can smooth out supplies, but stocks must be held and the traders hold many of them. The futures (or terminal) markets have developed as a way of spreading the risk involved in buying commodities for which there is no buyer at the time of purchase. This is called 'hedging'. In these futures markets commitments are made to buy, or sell, a specific amount of a specific commodity, which is defined to an agreed standard, at a specific time. Those involved do not need actually to have the commodity in their hands. It is this aspect of futures trading that also allows speculation – trading futures for products you do not have in the hope of buying or selling the contract at a profit before the time to fulfil it is up.

So, instead of buying a commodity for cash and hoping it will sell, at some price or other, traders can buy a commodity for cash and immediately sell a futures contract on it, to sell at an amount that covers the costs and vice versa. Thus, the futures markets allow buyers and sellers to manage their exposure to price fluctuations of the physical commodity by hedging, but they also encourage speculation.

Box 5.8
Cargill's Global Reach

Cargill began life in 1865 when William Cargill started buying up grain elevators in the US mid-West as the railroad pushed north into Minnesota. It grew to become one of the, if not the, largest grain traders. It is now one of the largest food multinationals in the world, operating from about 800 locations in 60 countries. It still specializes in commodity trading, but says of itself, 'Cargill links suppliers and producers through its skills, knowledge and international expertise in commodity trading, processing, handling and transportation; food and food ingredient processing; and risk management.'.

Its turnover was US$46.8 billion and profits, which it does not define further, were US$450 million in 1991. It likes to grow its managers within the company and its 66,000 employees work in five areas including:

➤ agriculture – animal feed and ingredients, fertilizer and seed;
➤ food processing – chocolate products, coffee, fruit juice blends and concentrates, hazel nuts, malt, milk products, rice, soya protein, sweeteners and vegetable oils;
➤ industry – petroleum, steel;
➤ meat – beef, chicken, pork and turkey;
➤ trading – grain, financial markets, ocean transportation, consultancy services.

It is a private company, still largely owned by descendants of the founders, and says it takes a longer-term view than most publicly owned companies. Cargill's stated aim is to double its size every five to seven years. In 1990, it owned country elevators in the USA, Canada, Argentina, Brazil, England, France and Thailand; it bought, stored, sold and processed grain, oilseeds, citrus fruits and other commodities, managing most of its international commodity trading from its Geneva offices. It chartered 30–40 million tons of cargo capacity annually and controlled 1.6 million tons deadweight of shipping. It provided farm inputs – seeds, feeds and fertilizers – did research in plant breeding, animal production and other areas relevant to its businesses, and processed poultry and beef. It provided commodity and financial futures brokerage services and management consultancy; its financial marketing department traded in financial instruments and foreign currencies for Cargill and other companies.

Source: The World of Cargill and other company information

The power to choose

From Cargill's standpoint, being the feed supplier, the banker, the buyer of the finished cattle, their butcher and their wholesaler, creates a tidy system that gives Cargill maximum control and return with the major risks – weather and animal health – being shouldered by others. It is also a very good way to market cheap grain, as long as you are not the farmer who grows it...

Cargill has consistently pursued a three-level policy based on maximizing returns available on the basis of low grain prices:

➤ *trading grain internationally;*
➤ *adding value to cheap grain by using it as the raw material for livestock feed manufacturing;*
➤ *adding further value to the cheap grain by turning it into meat.*

By positioning itself at each of these three levels, Cargill is able to shift its emphasis from one level to the other depending on current conditions and prices. It has also been policy for some time to secure its position at each level by pursuing a double-track policy of either buying out or driving out its competition while integrating its operations vertically. This has been evident in Cargill's meat operations around the world.

Kneen, Brewster (1990) *Trading Up: How Cargill, the World's Largest Grain Trading Company, is Changing Canadian Agriculture*, NC Press, Toronto, p67 and p100

Knowledge of the real world conditions for production and any other factors, such as probable currency fluctuations, political actions affecting production and so on, can give the edge over competitors in deciding what and when to buy for how much.

Michael Barrett Brown attributes traders' power to:

➤ 'the scale of their resources;
➤ the integration of their operation from grower right through to consumer;
➤ their access to finance;
➤ their ability to manipulate the market and increasingly
➤ the freedom they enjoy to switch operations from one source to another, and even to switch operations from plantations to outgrowing'.

This switching of suppliers according to price and conditions can leave primary producers, especially the developing countries, facing unstable demands for their produce. The biggest players are developing their own world-wide information systems using the latest technology so that they can buy and sell according to the most up-to-date knowledge of growing conditions world-wide.

Cargill, for example, writes Brewster Kneen,

prefers to make its money not in speculating, but in putting huge amounts of any given commodity through its system, where it can collect charges and fees for handling, storage and transport as well as making small margins on the trades themselves. To minimise the risk inherent in dealing in different currencies and other unforeseen risks likes wars and weather – events over which it has not *managed to gain control, that is – Cargill engages in what the trade calls hedging. Thus when Cargill buys a carload of real barley at price x, it sells a contract to deliver a like amount of barley at the same, or slightly higher, price at a specified future date. Whatever the fate of the real barley, Cargill knows that it can actually sell that amount of barley at a known price in the future. If, in the meantime, it can get a higher price for the real barley, it will sell it, perhaps at the same time redeeming (calling for delivery on) another futures contract that it had purchased earlier.*

For example, if Cargill decides to sell oats that it has in storage to Saudi Arabia because the price is right or the US Government is offering a subsidy, then at the same time it will buy a contract to take a delivery of a similar amount of oats (possibly in Canada) at a future date for a price that balances, more or less the sale just made.

Commodity traders, then, can offset a physical forward commitment to buy or sell a commodity by an equal and opposite transaction in the futures market. But who takes the risks involved? In short, the answer is speculators with capital who think they will make money on the deals by buying futures for less than they will be able to sell for at the end of the trading period.

Paper trading

Most futures trading in commodities is done by technical traders who do not and never will actually deal in physical commodities but will only speculate in the market. The volume of futures trading has grown rapidly; in the USA between 1970 and 1985 the volume of futures trading grew from 13.6 million lots to 158.7 million lots, but the actual volume of international trade in commodities over the same period only doubled. All this activity makes the futures markets quite volatile and can lead to great fluctuations in price. In addition, a physical amount of a commodity may be bought and sold many, many times.

Much of the trading in grain between US grain firms is paper trading. It is not done on the total cost of the grain through futures contracts but on contracts dealing in the 'basis' or premium. This is the difference in price, for a specific time and grade of commodity, between the futures price quoted at the Chicago Board of Trade (CBT) futures price and the fob (free on board) cost of the same commodity at the US Gulf. This fob price is less than the cif (carriage, insurance and freight) cost at Rotterdam, say, or any other place for delivery. These trading deals can be very complex.

On 5 March 1985, for example, the US Department of Agriculture (USDA) reported three different prices for US no. 3 yellow corn for April–May shipments: CBT, US$2.71 per bushel or US$107.42 per metric ton; fob US Gulf US$120 per metric ton; and cif Rotterdam $133 per metric ton.

Buyers who buy at cif prices forfeit all the logistics flexibility to the seller, who will try to fulfil the contract at cheapest cost and cash the buyer's letter of credit as soon as possible after the sale. If they have the expertise, information and decision-making capacity, buyers can control both the price and logistics of imports by buying fob and handling shipping themselves; this is what the major commodity traders do.

Box 5.9
A Soya Bean Fob-string

In 1976 the *M/V Meraklis* was chartered by Deutsche Conti in Hamburg to ship 25,000 tons of soy beans from the US Gulf to Europe. Deutsche Conti then nominated the ship to Louis Dreyfus, from whom it had a purchase contract. The string finally worked as follows:

1. Deutsche Conti(nental), FRG (nominated to)
2. Louis Dreyfus Corp, Paris
3. Italgrani, Italy
4. Continental Grain, New York
5. Cook Grain, Memphis, TN
6. Garnac Grain, New York
7. Cook Grain, Memphis, TN
8. United Grain Growers, Portland, OR
9. Garnac Grain, New York
10. André (Garnac), Switzerland
11. Finagrain (Continental), Switzerland
12. Tradax (Cargill), Switzerland
13. ICIC, Italy
14. Andre (Garnac), Switzerland
15. Alfred Teopfer, Germany
16. Finagrain (Continental), Switzerland
17. Continental Grain, New York
18. Sosimage (Louis Dreyfus), Rome
19. Finagrain (Continental), Switzerland
20. Desarollo, (Continental), Italy
21. Continental Grain, New York
22. Cook Grain, Memphis, TN
23. SIAT, Switzerland
24. Mantovani, Italy
25. Continental Grain, New York
26. Peavey Grain, Minneapolis, MN
27. Peavey's elevator, New Orleans, LA

Peavey's grain elevator loaded 25,000 tons of soybeans on the *M/V Meraklis* bound for Europe for the account of Deutsche Conti, who in turn may well have had a carriage and freight sale of soybeans to another German importer or end user.

Source: Pinniger, Simon (1987) 'Trade and Pricing Issues of International Grain Marketing', in *Agricultural Marketing Strategy and Pricing Policy*, Dieter Elz, ed, World Bank, Washington DC, pp44–55

In what may seem like a simple transaction – say, a European importer placing an order for a commodity from a US trader – a 'string' with dozens of contracts, all paper trades, may be involved (see Box 5.9).

Consider what happens when a European importer or US exporter with a cif contract charters a ship to load the corn cargo purchased for April–May shipment on 5 March. The buyer nominates the ship to the company selling the grain. The selling company may not own the grain or have facilities to load ships so it will then renominate the ship to another company, from which it has also purchased no. 3 yellow corn for April–May delivery. This renominating of the same ship from buyer to seller can go on time after time until it eventually reaches a grain seller with an elevator with yellow no. 3 actually in it and with a free berth to load the ship in April–May, who decides not to pass the nomination on to another grain seller but to nominate the ship to its own elevator for loading. All these traders between the buyer with the ship and eventual seller with a loading elevator are 'in the string' and make paper trades.

To facilitate all the paperwork and speed up the final physical transaction, all those in the string settle up the respective price differences between each buyer and seller when the ship is nominated and before the cargo is loaded.

Not all traders are happy with the price fluctuations on futures markets, let alone the producers. In 1990, the chairman of Archer Daniels Midland, then the thirteenth largest food-processing company in the world with US$7.75 billion sales and the second largest flour miller and soya-bean crusher in the USA, said his company was going to reduce its dependence on futures contracts. He was reported as saying 'The Chicago Board of Trade is a gambling institution and no longer contributes to orderly marketing of commodities.' The company planned to develop a world-wide computer network linking its 100 trade offices, 100 ships, 2000 barges, 7000 rail cars and 10,000 trucks to its headquarters in Illinois, so that it would only have to rely on the futures market to hedge against major fluctuations.

Although we have focused on these companies as traders, today the major players are far more than just traders, as the example of Cargill shows. They act right across the food system from farm to shop and see themselves on a par with other companies that are thought of more as food processors and manufacturers, and it is to this group of actors that we turn next, in Chapter 6.

6

Processors, Distributors and Caterers

The production of processed food and beverages amounts to some $1.5 trillion a year, making it one of the world's largest industries. Production in the OECD region is about $800 billion, in Eastern Europe and the [former] Soviet Union about $200 billion, and in the developing countries about $150 billion. Large multinational companies (MNCs) play a major role: the 100 largest OECD-based companies are estimated to account for about 20 per cent of global production of processed food and drink.

OECD (1992) Investing in food, OECD, Paris, p17

Food production is only one aspect of the food system. As urbanization and industrialization have spread, close links between farmer and consumer have disappeared. First, the food processing and manufacturing industries affected the food that people ate, and increasingly these industries determined the output from farms and the terms on which it would be bought. Today, farm products are much more highly processed and many new products have been developed for consumption by food manufac-turers. Distribution to increasingly urban end users became essential, at first through networks of wholesalers and small shops, then through large multiple retailers. Factory work, increasing wealth and time constraints brought caterers into the picture, transforming food into meals eaten out or taken home to be reheated. In this chapter, we look briefly at these three actors and discuss some of the trends that are helping to shape the future food system.

Food Processors and Manufacturers

Food processors often started as preservers of food, but now may manufacture foods from a basic set of ingredients, such as sugar, starch, fat and flavourings. They have moved or are moving from being craft-based industries producing goods in batches to the use of continuous industrial production lines with minimal amounts of skilled, or indeed any, labour. Some research in the rich, overfed world now goes into producing fun foods, minimum nutrient and even non-food ingestibles, such as imitation fat with little or no nutritional value.

Food processors, of course, are businesses first and foremost. They just happen to deal in products people eat instead of those they drive, watch, or listen to. They employ all the tools that are necessary to secure supplies of their raw materials and to sell their products. They have increasingly sought market segmentation

– breaking markets up into an ever-growing number of niches for different products – as well as new and more profitable products.

Until the 1980s, the processors grew, diversified and became the biggest players in many countries, particularly in Britain and the USA. Even in 1977 in Britain, there were about 5600 food and drink firms, but the top 30 companies accounted for 60 per cent of employment and value added in the industry. The top five firms in over 60 food product markets accounted for over 70 per cent of those markets. By 1988 according to OECD, in Western Europe as a whole, the sales of just eight firms – Unilever, Nestlé, BSN, Cadbury–Schweppes, ABC, United Biscuits, Hillsdown and San W. Berisdorf – made up 70 per cent of the US$250 billion food and drink market. By the early 1990s, in Europe, most product markets were quite concentrated, with the top three suppliers tending to dominate (see Table 6.1).

Getting bigger, keeping focused

The Anglo-American domination of large food firms is slowly beginning to change. In 1974, about 50 per cent of the largest and most concentrated firms (by turnover) were based in the USA with 20 per cent in the UK. The average size of firms by employment was largest in the UK, with 4.3 per cent of enterprises (with over 500 employees) employing 82.8 per cent of all employees, reported the OECD. By 1987, out of the 110 largest food and drink companies in the world, 43 were American and 18 British (see Box 6.1). However, firms in other OECD countries were trying to compete by a series of mergers and acquisitions.

As industrial concerns, firms want to minimize the risks they face, control the processes they use and the costs they incur. One way to achieve this has been to invest in technology and auto-

Table 6.1 Percentage share of top three suppliers in the major markets in Europe, 1991

	UK	*France*	*Germany*	*Italy*	*Spain*
Baby food	62	99	85	85	80
Biscuits	70	50	50	67	54
Bread, industrial	na (58)[a]	69	10	15	na (50)[b]
Breakfast cereals	64	83	66	84	89
Fruits, canned	25	na (30)[b]	25	na(30)[b]	na (55)[a]
Chocolate confectionery	79	51	51	50	49
Coffee	78	64	63	67	88
Frozen foods	35	33	53	79	49
Ice-cream	58	56	88	80	62
Jam	59	56	58	63	na
Milk	64	40	18	21	na
Mineral water	50	76	24	34	36
Pasta	27	58	57	52	na
Pastries, industrial	58	44	14	50	na
Pet food	87	79	85	78	71
Ready meals, frozen	53	75	na	84	40
Snacks	91	53	na (64)[a]	65	na (56)[a]
Sugar confectionery	41	58	na	43	53
Tea	60	89	67	66	na
Yoghurt	62	65	39	67	84

Notes:
[a] Figure for top two suppliers
[b] Figure for top supplier 1990–91
Source: Seymour Cook (1992) *The European Food Report – 1992*, Food Research International, London, p5

mate processes, reducing the need for skilled, expensive labour. Another way for food processors to minimize risks has been to shift some risks on to the farmers supplying their raw materials and to use traders to source their supplies.

Farmers found that increasingly they had to meet industrial requirements for supplies at the right time, and of the right size, colour and the like. Firms began to contract farmers to produce, imposing more conditions on purchases – not just conditions on price but also on growing methods, planting times, pesticide treatments and harvesting dates. Crop and animal breeders seek to produce plants and animals with uniform characteristics suitable for industrial processing, such as potatoes designed to make good crisps.

Many of the trends in food processing and manufacturing have gone further in the USA and Britain than in other countries. The key trend has been concentration. The number of firms supplying the different food products has reduced, their size has increased and market power has become concentrated in fewer and fewer hands, as Table 6.1 illustrates. 'The process of concentration', as Sir Adrian Cadbury, the famous chocolate name, noted in a speech at the launch of Food from Britain in 1983, 'has changed the nature of the manufacturing side of the food industry in two ways. In the first place, the large companies have broken out of their original product boundaries. They are no longer the bread, beer, milk, meat or confectionery companies they were only a relatively short time ago – they are food and drink companies. Secondly, their expansion across market boundaries has enabled them to develop their businesses overseas at a faster rate.'

He believed that the large international food companies had to focus on a limited number of branded products and market in areas where worthwhile volume sales could be built up. This they have been doing. Take United Biscuits, for example, which had a £3.12 billion turnover in 1992. Their self-declared prime corporate objective was to become a world leader in snackfoods and they restructured the company in the late 1980s towards that end. The company sees technol-

ogy – both food technology and information technology – as essential to its success. It spent £17.8 million on research and development in 1992.

Although it began as a UK biscuit maker, United Biscuits (UB) had five groups producing snackfoods in the broadest sense. These groups were developed after selling peripheral businesses and buying up others in Europe and the USA which strengthened the core business aim. Until early 1993, the groups were:

➤ McVitie's, the second largest biscuit business in Europe, with market leadership in the UK, The Netherlands, Hungary and the Nordic region;

➤ Keebler, the second largest producer of cookies and crackers in the USA which also produced snacks and was entering the quality frozen speciality foods area (in 1993, United Biscuits bought Bakeline which supplied 25 per cent of the market in own label cookies to US grocery retailers);

➤ KP, a UK and European brand in the salted peanuts and snack market which was going for more profitable, up-market products, both developing its own (see Chapter 8) and buying in niche suppliers, like Phileas Fogg, supplying speciality products to particular market segments;

➤ Terry's, including Terry's chocolate which made branded and own-label chocolates, Callard & Bowser sugar confectionery and Chocometz in France;

➤ Ross Young's, a frozen food manufacturer and wholesale catering supplier which began to use the McVitie's brand on some products.

The last two could still be considered rather peripheral and not leading brands. In March 1993, Terry's of York was sold to Philip Morris, thus expanding the Philip Morris chocolate operations. This left four main groups in UB and gave it the flexibility and resources to develop its world-wide snack food strategy. The company bought the snacks division of Coca-Cola Amatil in late 1992

Box 6.1
Three Food Giants

Unilever

And yet the name Unilever is relatively unknown among the public at large, reflecting its decentralised structure, which places great emphasis on decision making and brand marketing by its operating companies. It is these companies – such as Van den Berghs, Lipton, Lever Brothers and Elizabeth Arden – and with their familiar brands that the public identifies. Millions of people buy Blue Band, Flora, and Becel margarines, Cornetto ice cream, Brooke Bond and Lipton teas, Surf washing powder, Lux soap, Timotei shampoo, Signal toothpaste, Pond's skin creams, Chloé and Eternity perfumes without ever seeing the name Unilever.

Unilever World-wide, Unilever External Affairs, London, no date, p2

Just over half of Unilever's £27,863 million (US$41,795 million) turnover in 1991 came from its food businesses and 49 per cent of its almost £2427m operating profit, before adjustment for costs involved in restructuring the business. Nearly 55 per cent of sales were in Europe, 20 per cent in North America and the other 25 per cent over the rest of the world, where the company expects further growth.

A joint Anglo-Dutch multinational company, founded in 1930 by a UK soapmaker and Dutch margarine manufacturer, Unilever operated through about 500 companies in some 80 countries with around 295,000 employees world-wide in 1993. It made over £1 billion capital investments each year from 1989 to 1992 and has focused more firmly on its four core businesses in the 1990s: foods, detergents, personal products, such as skin care products, toothpastes and deodorants, and speciality chemicals, including starches, resins, flavours and food ingredients.

Three directors determine the strategic direction of the food businesses world-wide and co-ordinate them. Product teams report to the chairman of this Foods Executive and are responsible for implementing the global food strategies and supplying specialist marketing, research and technology know-how in the following specific areas:

➤ oil and dairy-based foods, like margarines, mayonnaises and cheeses (£4317 million sales in 1993);
➤ meals and meal components from frozen ready meals and frozen foods to soups, meat products and sauces (£3792 million sales);
➤ ice-cream, beverages and snacks, from cheap children's ice-creams to expensive 'impulse' buy adult bars and luxury ice-cream desserts, tea and coffee, and savoury peanut or meat snacks (£3312 million sales);
➤ professional markets like caterers and bakers with special packs and formulations of a wide range of products (£2930 million sales).

In 1991, agribusinesses operated in four areas – animal feeds, fish farming, plantations and plant breeding – and accounted for 4 per cent of turnover. By 1993, only the last two areas remained which were combined into a plantations and plant science group in 1992. This group, with trading activities, accounted for 6 per cent of turnover. Unilever spends about 2 per cent of its turnover on research and development – some £518 million in 1993 – and employs about 4000 scientists and technicians spread over six major laboratories in three continents – Europe, North America and Asia.

The company says its global research allows it to use the results of its scientific work across a range of product types. For example, an understanding of fats and oils technology can be used manufacturing soaps, skin creams, ice-creams and spreads.

'Unilever's strength', says the company's Research and Engineering Division in *Research and Engineering: Take an Idea... and Create the Reality*, (London, 1991) 'stems from its diversity of operations, global scale, and the prominence of its brands'. Indeed, the chairman, in his introduction to the company's 1993 performance, called brand equities 'the most valuable items in our stewardship' and saw 'the power of our brands as the engine of long-term growth'. During 1993, the company spent almost 12 per cent of its turnover (£3284million) on advertising and promotional investment.

Source: Unilever *1990 Annual Report, 1991 Annual Review and Annual Accounts* and other company materials

Nestlé

The Swiss food giant Nestlé celebrated its 125th anniversary in 1991 with sales exceeding 50 billion Swiss Francs (or US$36.3 billion) for the first time and a trading profit of over Sfr5 billion (US$3.6 billion). Sales rose by 8 per cent in 1992 to Sfr 54.5 billion (US$38.4 billion), with a trading profit of Sfr5.6 billion and a net profit of Sfr2.7 billion (US$1.9 billion). Over 90 per cent of its sales come from food and beverages, with nearly half the sales in Europe, a quarter in North America, about 10 per cent in Latin America and Asia, and the rest from Africa and Oceania.

The company employed just over 218,000 people world-wide and operated 482 factories in 69 countries at the end of 1992. It was investing heavily in expansion, especially in developing country markets. It spent 5.9 per cent of turnover (Sfr3.2 billion) on capital investment in 1992 – the fourth successive year it spent over Sfr2 billion on capital investment.

The company name appears on many products, but it also uses many other brand names. In 1992, it divided its food and drink products into four product groups:

➤ *Beverages.* Apart from instant coffee, the company also produces roast and ground coffee, chocolate and malt based drinks, fruit juices, wine and mineral water. Beverages accounted for 25 per cent of sales in 1992. Brands include Hills Bros (bought in 1988), Milo, Nescafe, Nesquik, and over 35 mineral waters such as Ashbourne, Buxton, Vittel and Perrier bought in 1992 to give Nestlé about 20 per cent of the global market (some 34 billion litres in 1991), 71 per cent of which was in Europe and 26 per cent in the USA.

➤ *Milk products and dietetics.* This includes powdered, sweetened, condensed and evaporated milks, non-dairy creamers, milk-based infant formulae, infant cereals, soya milk, breakfast cereals, cheese, desserts, yogurts and ice-cream. These products produced 27 per cent of 1992 sales. Brands include Carnation, Chambourcy and Coffee-mate. Nestlé bought Clarke Foods in 1992, number two in the UK ice cream market and owner of the Lyons Maid brand.

➤ *Chocolate and confectionery.* The company produces a wide range of solid chocolate and candy bars as well as sugar candy. Some were being marketed with Disney characters following an exclusive contract with Disney for use of their name. Confectionery accounted for 16 per cent of sales in 1992 and brands include Kit Kat (after the take over of Rowntree in 1988), Polo, Baci, Milkybar, Smash (Brazil and Chile) and Baby Ruth.

➤ *Prepared dishes and cooking aids.* This area was the largest contributor to 1992 sales (29 per cent) and includes culinary products (such as bouillons, soups, sauces, canned or dehydrated pre-prepared dishes), pasta, frozen foods, chilled pasta and sauces, and cold meats. Brands include Bavarois, Buitoni, Chambourcy, Crosse & Blackwell, Findus, Frisco, Herta, Maggi and Stouffer. Some food service products – semi-prepared foods and industrial bases, like tomato based sauces, frozen foods and drinks in packs – were geared to the catering trade under brands such as Chef, Santa Rica, Que bueno and Stouffer.

Nestlé also produces pet foods and pharmaceutical products, and has acquired interests in cosmetics. It spent about 1.2 per cent of sales on research and development in 1992, some Sfr669 million, through its Swiss-based scientific research centre and technology development centres in 11 countries. Although company sales doubled between 1980 and 1992, research and development (R&D) expenditure quadrupled. The company uses R&D to launch new products, improve existing products, adapt products to raw materials, local conditions and specific tastes in a wide variety of markets, and to introduce new, more efficient and less expensive manufacturing processes.

Source: Company annual reports, 1991 and 1992

Philip Morris Companies Inc

The 161,000 people working world-wide for Philip Morris in 1992 produced a US$59.1 billion turnover – 49 per cent from food, 43 per cent from tobacco (including Marlboro), 7 per cent from beer and 1 per cent from financial services and real estate. The company had a 14.6 per cent pre-tax profit margin, or US$8608 billion in 1992. It has greatly expanded its food interests since the mid-1980s. General Foods was bought in 1985, Kraft in 1988 and Jacobs Suchard in 1990.

General Foods and Kraft operated as Kraft General Foods (KGF) before the acquisition of Swiss-based Jacobs Suchard, parts of which were sold off if they did not fit in with the KGF structure. The company's reporting and management structure for food was reorganized in 1991 into North American food, which then accounted for US$20.2 billion sales, and international food, which accounted for US$7.9 billion sales. In 1992, these grew to US$20.3 billion and US$8.7 billion, respectively, generating an operating profit of US$2.8 billion. KGF's European businesses were renamed Kraft Jacobs Suchard early in 1994.

The North American products include Post cereals, Nabisco's Shredded Wheat, Kool-Aid drinks, Maxwell House coffee, Kraft and Philadelphia cheeses, Oscar Mayer and Louis Rich refrigerated meats and Jell-O desserts. KGF International's three core product groups are coffee, confectionery and cheese and include some of the above brands, plus others such as Jacobs, HAG and Maxim coffee, and Milka, Côte d'Or, Suchard and Toblerone chocolates.

Philip Morris wants 'to be the most successful consumer packaged goods company in the world' and wants Kraft to become the world's leading food company with the international operation matching the scale of the US operations. The company intends to increase its brands' volumes, shares and margins as it expands geographically by using its marketing and technological strengths. In 1992, it spent over US$3 billion buying and investing in businesses, mainly in food. It made 21 investments or acquisitions in 16 countries, including Freia Marabou in Norway, the company's fourth largest acquisition ever, and RJR Nabisco Cereals in the USA.

Other elements in the company's strategy to meet its goals include maintaining the highest quality of people through increased training and incentives, protecting and building its brand franchises, in which it invested about US$11 billion in world-wide marketing in 1992, and managing with a global perspective, creating cross-border management for core products such as cheese, and world-wide strategy councils for tobacco, coffee and spoonable dressings.

Source: Company annual reports, 1991 and 1992

(renamed The Smith's Snackfood Company), giving it market leadership in snack markets in Australia, Singapore, Malaysia and Papua New Guinea. These complemented existing activities in China, Hong Kong, the Philippines and Japan, and gave United Biscuits a presence in the world's largest snack food markets – North America, Europe and Asia Pacific.

The constant buying, selling and restructuring of companies in the food sector will continue through the 1990s as players try to consolidate their positions. In some cases, companies may well withdraw from areas they diversified into in the 1980s, if the diversification has not produced a major market-leading brand or if it has led them away from a core business.

Pressing suppliers

As food manufactures change, they also place new pressures on their suppliers. Ingredient suppliers to food manufacturers, for example, face a double challenge, as Dr Rainer Wild of Rudolf Wild GmbH & Co, a German fruit preparations and flavours manufacturer, told the European Food Industry Conference in Brussels in 1992. The first challenge is the internationalization of customers, and the second is the move towards globalization of the trade.

Internationalization does not necessarily imply standard global products. Although there may be global brands, there will be few global products which contain the same ingredients. This is due partly to differing tastes and partly to differing legislation governing, for example, which additives may be used. Companies may retain the product name world-wide, but the product formulation may vary. Fanta, for example, varies in sugar and fruit content for different markets.

Ingredient suppliers who process natural raw materials, like fruit, have to cover any fluctuation in harvest quality or quantity as their customers – the manufacturers – expect standardized products and fixed prices. The quality of a product is also taking on a new meaning, beyond the legal standards and product features, to include what Dr Wild called 'inner quality'. This means ensuring that the product is produced in a manner that is acceptable to the final consumer and does not contain any undesirable elements such as pesticide residues.

The main stimulus for innovations, however, does not come directly from the market, in Wild's experience. The majority of new developments is initiated by the industry, be it the use of a new raw material, a new production process, the presentation of a completely new product category, new packaging or other equipment.

Ingredient suppliers are no longer at the end of a food chain but are part of a more complex matrix, interacting with manufacturers and retailers, and indirectly with consumers. They supply competing manufacturers by ensuring that their business is kept confidential, although large companies may try to put pressure on their suppliers not to supply competitors.

Creating markets, extending brands

The food industry has to create and maintain markets among consumers whose basic food needs have been met. The problems are discussed most openly at industry conferences, in annual reports and in the trade journals, although the public may be addressed in different terms. An essential aim of food manufacturers is to increase the 'value-added' or 'added value' on their products to increase their profits (see Box 6.2).

Changes in attitudes and lifestyle also offer food manufacturers opportunities – for example, in concern about health and diet, and the psychology of affluence (particularly American); if people feel themselves to be better off, they reward themselves with higher cost foodstuffs. Other changes include the following:

Box 6.2
The Added Value Nut

The basic question for many food companies is how to turn a cheap commodity product, like wheat, potatoes or peanuts into a profitable item that people want to buy from them. With food consumption unlikely to grow in the rich country markets and cost reductions through labour cuts unlikely to provide long term profit increases since these were one-off, many firms see 'added value' as the key to profitability.

Mr J E Willet, senior vice president of Nabisco Brands Incorporated, described what this meant for peanuts at the 1983 Food Manufacturers Federation conference in London, entitled 'Food For Profit'. Nabisco was then ranked 52nd in Fortune's top 500 US industrial corporations with US$5.87 billion sales, over 55,000 employees and a net income of US$314.7 million in 1982. Its international operations were consolidated together and accounted for about US$2.6 billion of the turnover, covered 38 countries, with 36,000 employees in companies with turnovers from US$1–200 million.

the Planters product...is not a new product...it's 75 years old as a brand. It is a commodity product. You can buy shelled peanuts for, I think, about 25c a pound, and we sell our peanuts under the Planters brand, for about $3 a pound – that is not bad....

How do we prevent a product life (like) that from suffering from what I call hardening of the arteries...Well, along with all of the things you normally do in terms of your advertising, promotional packaging, etc., we have added new segments under that brand...first...an oil roast product....a second range of products called dry-roast products....in the United States, this product had a flavour to it that people liked and it had a perception – once again a perceived idea of lower calories. There are not, in fact, lower calories in it but people thought that it had lower calories. Consequently, a very strong development of that product.

We have a range of unsalted nuts and it appeals to the 39 million people who have high blood pressure... (we now) test market... Planters light-nuts. This time we have the promise – it does actually deliver 25 per cent fewer calories – and there will be a lot of people who do not normally eat nuts because there are 14 calories per nut, who will say, my gosh, I only eat 9 so now I think I will start buying – at least we hope so.

The results of all this activity have been first of all to maintain a brand alive and interesting....Perhaps more importantly, every one of those extensions had a better gross margin than the original product.

What does that allow us to do? Firstly, it allows us to advertise and spend a great deal of money keeping our franchise and brands strong. It allows us to put money into promotion and to make a better contribution to our profitability, which is what it is all about...

The crucial thing about added value is not the fact but the perception of it. A food might cost more and as long as it was perceived to be worth it, then greater profits could be made from the same or fewer ingredients. 'Lite' foods, which have something taken out, be it alcohol in the case of wine and beer or salt with almost any product, or status-related foods like Perrier, or apparently healthy foods, like yoghurt or granola bars offer opportunities for added value. As the song says, 'The more it costs, the more it's worth'.

➤ an ageing population with food and dietary restrictions and a smaller disposable income;

➤ a growing segment of the population remaining single, which has implications for convenience food;

➤ individual portions and disposable products for discretionary incomes – the money left after essential bills are paid;

➤ ethnic growth which influences food habits and through restaurants helps to create demand for exotic food products;

➤ working women with little time to cook who might want to reward themselves with special (expensive) products.

Different food manufacturers produce what are essentially very similar products in different product categories – from baked beans to breakfast cereals, from soups to fruit juices. To market them, heavily advertised brand names are used to differentiate between similar offerings in different product categories (see Annex 4 for an example of leading UK brands). However, in future only the very strong manufacturers with leading brands in particular product categories or very strong niches are likely to be able to command price premiums and space in the supermarkets. In Canada, recently, one multiple grocer launched a premium own-brand cola drink which took a 50 per cent share in-store from nothing in two years (but see Box 8.10 in Chapter 8).

For most food manufacturers in countries where just a few multiple retailers sell most of the food – that is, in concentrated retail markets, there is little value for manufacturers in offering a wide range of product categories such as tinned fruits, vegetables, soups, and so on, as retailers tend to buy from different suppliers by category. It may still be useful in markets like Spain and Italy, where a sales force of 3000 people provides direct coverage of 80,000–100,000 retail outlets. But in the UK, for example, in several product categories a sales force of 15-20 people calling on 50–100 key decision-makers

can serve 90 per cent of the relevant market, distributing their product to a small number of centralized distribution warehouses.

When manufacturers dealt with a very fragmented retail industry, their powerful consumer brands allowed them to set their terms. They had most knowledge of the consumer, through market research, and the capacity to develop products, production technology and point-of-sale representation. However, this is no longer the case. With this concentration of buying power, manufacturers have been forced to give discounts, become involved in in-store operation, produce and distribute according to retailers' specifications and so on, to keep their brands on the supermarket shelves.

Adapting to change

Although the food business is dominated by multinationals, they have tended to operate on a national domestic basis rather than integrate all their activities on a world-wide basis. Some, however, like Campbell's, are changing, at least regionally (see Box 6.3). Manufacturers need to capitalize on their brands and make people want them specifically. Those with strong brand images are trying to extend them by using the name on new products – for example, using a name associated with chocolate for dairy desserts, biscuits or cakes. Those who do not have leading brands often produce own-label products on a contract basis for supermarkets.

The various pressures in the food system have, however, led to a large number of mergers and acquisitions among food processors and manufacturers and, more recently, to a number of joint ventures as they try to regain their power relative to the retailers, and to ensure their survival and profitability. Competition is likely to become tougher for food manufacturers and for branded products, with the major European transnationals, such as Unilever and Nestlé, facing increasing competition from major US transnationals such as Philip Morris,

Box 6.3
The Euro Biscuit

The 1990s will see major changes in the EU following the development of the single market and the opening up of Eastern Europe. The single market is likely to bring about:

➤ a restructuring of Europe's food industry;
➤ the replacement of traditional forms of organization and management with new forms, such as networking organizations;
➤ the 'Americanization' of food consumption habits.

This is according to Edward Moerk, president of Campbell's Biscuit Delacre. He expects the number of biscuit-producing plants to fall from 700 to about 150, with production methods moving from artisanal to mass production and mass marketing.

Campbell's World-wide Biscuits and Bakery division is the fourth largest biscuit company in the world, with sales of around US$1.3 billion. The European side has undergone a radical transformation and is taking a determinedly pan-European approach, aiming to be the number one of an expected three pan-European brands in ten years. That means being a low-cost producer with high profitability.

To achieve this, the company sold off an Italian loss-making operation and reduced overheads by 15 per cent by cutting senior management positions in head office and in the field. They centralized all European production and sourcing under one person, consolidated quantities bought and developed a preferred supplier system which cut the number of suppliers by 50 per cent.

Next they consolidated their existing 5 brands into 1, with pan-European packaging turning 12 packaging types and 3 logos into 1 design for 100 products in 50 countries. They also moved their advertising from six agencies to one and picked one public relations company. Moerk believes that Europeans, too, often overestimate their differences and underestimate their similarities, and it is these similarities that mean that they can expand into a bigger European market. This is also helped by many young, educated, English speaking Europeans who are adding a global element to their culture through the pan-European communication media networks and satellite TV channels.

These major changes were made within 100 days and were part of a broader strategy for the whole company which diminishes the power of national managers used to producing for a national market. All the firms in the European operation are expected to work as an integrated network, with shared values and purpose but clear accountability, following a central strategy but with local execution.

Source: Food Industry and Food Trade: Cooperation or Confrontation? Impact on all levels of the food chain. *The European Food Industry Conference, Management Centre Europe*, Brussels, 29–30 June 1992

to undergo restructuring to achieve optimum efficiency levels. This will force cuts in manning levels and increased productivity. Job losses in the food industry are probable.

Concentration will continue, to enable manufacturers to establish a strong number one or number two position in each market segment. Then the brands will become powerful enough for them to be pulled on to the retailers' shelves by consumer demand. In the 1980s, Unilever, for example, has focused its activities on its core business – fast-moving consumer goods, food being a major part. In the eight years after 1981–82, the company went on a shopping spree, buying 246 concerns and selling 128 at a net cost of £3.3 billion, two-thirds of which was spent on food companies, half in Europe. At the same time the group invested £2.9 billion in its food companies. Expansion is also likely into new markets around the world.

Eastern Europe has millions of product-hungry but financially stretched consumers. Western food firms are buying up old state industries there, although it will be some time before any benefits to Western companies are felt. By the mid- to late-1990s, however, earlier investments in Eastern Europe should begin to pay off. Companies like Unilever will have substantial operations equipped with the latest machinery staffed by a skilled and relatively inexpensive labour force. This will happen just when pan-European food retailing networks are likely to be in place. This will give such manufacturers the ability to supply the lowest cost, branded or even private label products at a price that smaller, less efficient, regional manufacturers will be unable to match.

Asian dragons and tigers

A much bigger and more lucrative market is developing in East and South Asia and there is already major multinational interest in the area. Affluence is spreading from countries such as Japan, Taiwan, Hong Kong and Singapore to Indonesia, Malaysia and Thailand, although its distribution is far from even. The average age in the region, which has a population of about 1.9 billion people without the Indian sub-continent, is 23 years, compared with 38 in Europe and 37 in the USA and Mexico. Very high economic growth rates averaging 7 per cent imply a doubling of GDP every decade. The proportion of world trade occurring in the region is expected to grow from 19 per cent in 1990 to about 40 per cent by 2010.

Apart from Japan, Singapore, Hong Kong and Taiwan, incomes per head are low – around US$2000 per head or less. But GDP figures underestimate the purchasing power of populations for products and services. Because of socialist infrastructures which provide low rents, the People's Republic of China, for example, has the greatest purchasing power in the region, second only to the USA. According to the International Monetary Fund (IMF) calculations of purchasing power parities, in 1994 China's purchasing power was estimated to be US$2850 trillion, compared with Japan's US$2450 trillion and the USA's US$5800 trillion. By 2002 China's imports are estimated to be worth US$2 trillion, three and a half times those of Japan.

Already the OECD estimates that around 60 per cent of global trade is conducted by multinational corporations and the majority of these flows are intra-company trade. The major share of global Gross Domestic Product (GDP) is focused in three 'circles' of prosperity and much attention is going towards the third circle in Asia/Pacific (see Figure 6.1). Food companies within the region are also expanding beyond their national base and into the new markets – for example, Taiwan's President Foods is moving into China (see Box 6.4).

In the major markets of the USA and Europe, although farmers were affected by changes in their buyers first, that process has been repeated for the manufacturers and processors, as their buyers have ceased to be small shops and have become major retail chains, as we discuss next.

Box 6.4
President Foods China Strategy

President Foods is Taiwan's largest food and beverage company, with net sales of US$1.6 billion in 1993–94. It specializes in convenience forms of traditional Chinese foods such as instant noodles. This makes it well suited to manufacture these foods for the newly open markets of the Chinese mainland, which is 60 times the size of the 'local' Taiwanese market. This is how the company is setting out to become the dominant national brand in China.

To begin with, President is attempting to sell traditional Chinese foods at prices most families can afford. Its entry products are instant noodles, carbonated and non-carbonated drinks, flour and livestock feed – all everyday items rather than premium goods. It is also concentrating on a Chinese taste with red bean and soya drinks, shrimp rice flour and pork bone tea noodles. The company's initial targets have been urban areas with a per capita annual income of US$500. Most importantly, it has divided the mainland into eight marketing zones and has carefully gone about selecting a manufacturing hub in each one. Satellite factories are planned to follow in regional towns. Seven factories are already operating and two more have just been set up, bringing total investments to US$80 million. Eight more flour, noodle and beverage plants are just about to be, or soon will be, established for which another US$65 million has been earmarked.

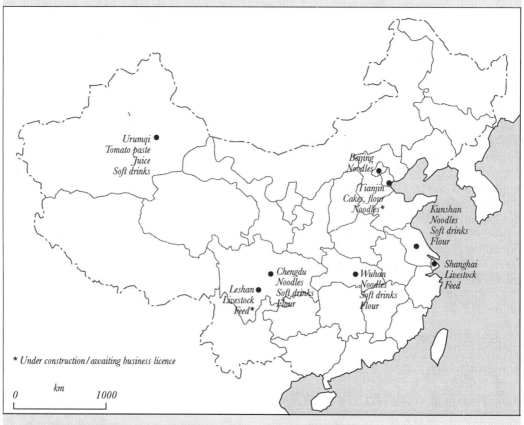

* Under construction / awaiting business licence

Source: Carrol Partners International (1994) 'A fourth quarter perspective' Carrol Partners International, Melbourne, p38

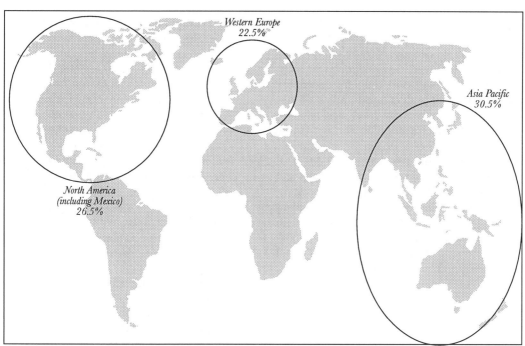

Source: Carrol Partners International, Melbourne, 1994

Figure 6.1 The three circles, share of the world GDP at end 1993 ('Purchasing power parity basis')

Wholesalers, Shops and Supermarkets

...there is no doubt that retail concentration is the biggest challenge that food manufacturers, both large and small have had to face since the Second World War. In the UK, where a company the size of Heinz may depend on only a handful of accounts for some 60 per cent of sales, this has been the case for some years. The same is now true in Scandinavia and Holland, and increasingly so in France and Germany...
The emergence of pan-European retail alliances will re-inforce the pace of retail concentration by allowing members to benefit from joint sourcing and purchasing.

Seymour-Cooke, (1992) *The European Food Report 1992* Food Research International, London, p14

Wholesale and retail distributors move foods to the point of sale. The trend with them has also been to ever larger businesses, with multiple retailers coming to dominate food distribution in many countries in the 1980s and many wholesalers being squeezed out. In Australia, the UK and the USA, for example, a few companies control the vast majority of the food moving into consumption, producing what economists call a highly concentrated food retailing sector. The use of different supermarket brands can make the number of retailers appear to be more diversified than it is, since one group may operate several different supermarket chains aimed at different segments of the market.

In Britain, for example, the proportion of independent grocers fell from just over half in 1961 to one-third of the total in 1978, while the multiple retailers, running expanding supermarket chains, grew from about a quarter to just over a half the total. In 1978, there were 41,500 grocery and large food retailing businesses with over 52,000 outlets, but even then just 58 large food retailers controlled more than 7000 outlets and handled 65 per cent of total turnover. By 1989, according to the Institute of Grocery Distribution (IGD) Research Services, a dozen or so multiple retailers – those with ten or more outlets – had increased the average size of their outlets and reduced the numbers to 4290, while the number of cooperative outlets had fallen to 2941 and the number of independent retailers, with from one to nine outlets, had fallen to 37,500. By 1993, just five multiple retailers handled 65 per cent of the retail food trade.

Retailers

Retailers are the closest most food buyers come to food producers. They used to be called shopkeepers or grocers, but these names conjure up a picture which is not in keeping with the scale and sophistication of retail selling. Retailers gather, display and sell the products which shoppers, depending on the money in their pockets, buy. When Mr Alistair Grant, group managing director of Argyll Foods PLC, addressed the Food Manufacturers Federation 'Food for Profit' conference in 1983, he presented what turned out to be a fairly accurate view of trends from the retailers' point of view. At that time, Argyll was the parent company of Allied Suppliers, Lo-cost Stores and Cordon Blue Freezer Centres, but by the 1990s it was the owner of Safeway, one of the major multiple retailers in Britain, with a turnover of over £5 billion and operating under the Safeway, Presto and Lo-Cost brands.

Grant saw a static or declining market in which economic power would swing towards retailers, with a concentration of the retail sector concentrating into a small number of major businesses which would engage in heavy consumer advertising. He expected each store to be treated as a profit centre and

retail management refocus more strongly upon optimizing sales and profits from individual products and product groups – as the scope for sales growth by new store development diminishes. The stores were also likely to become more similar, resembling food markets with long-life packaged groceries only providing a small part of the stores' range. Computer-based systems in the stores will enable managements to have better information about what is and is not selling and encourage the focus on goods with high gross profit margins. The stores are likely to be more influential in deciding what they take from manufacturers and how these products are managed and marketed.

He was mostly right, the major exception being his comments about new stores. These continued to be built up to the mid-1990s. At the heart of the changes, though, is the greater power of retailers to dictate terms to their suppliers, from processors to farmers who supply them with fresh produce.

The power of the multiples

The shift in power between retailers and manufacturers has been most pronounced over the past 30 years in Northern Europe – the UK, France, Germany and The Netherlands. It has been followed by a shift in what one industry analyst calls the share of the 'total food industry profit pie'. In the UK, for example, multiple grocers, which make about two to three times the operating profit of their European counterparts – around 7-8 per cent, compared with around 2 per cent in other EU states and 1 per cent in Australia – more than doubled their share of this pie, up from about 20 per cent in 1981 to over 40 per cent in 1989, at the expense of manufacturers and processors. The pie itself, however, has grown enormously, up from around £1 billion to £5 billion (see Figure 6.2)

This retail concentration has yet to be duplicated nationally in the USA, where it tends to be regional – in a sense it is rather like Europe – and has yet to occur at all in

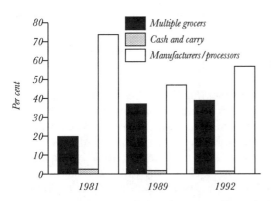

Source: 'The shift in balance of power between retailers and manufacturers', Andrew Wileman, OC & C Strategy consultants, paper presented at Food Industry and Food Trade: Cooperation or Confrontation? Impact on all levels of the food chain. The European Food Industry Conference Management Centre Europe, Brussels, 29–30 June 1992

Figure 6.2 UK food industry – share of the profit pie

Southern Europe. However, this shift in the balance of power is likely to occur throughout the national markets in Europe, but at different rates in various countries. In Spain and Portugal, for example, the wave of retail concentration that has happened in other EU countries has not yet appeared.

These changes are likely to be prompted in the EU following the creation of a single market since 1993, which allows companies to treat the whole EU as one market and operate without trade barriers within it. Retailers are likely to benefit through exercising pan-European buying power. Already multiple retailers can exercise a lot of clout in seeking discounts from manufacturers or imposing other conditions, since they can deny a manufacturer a substantial part of the market by denying them shelf space in their stores.

Some retailers which began as national firms may expand into other markets on their own, but others may co-operate in pan-European buying groups and alliances. The turnover of the retailers in the ten largest pan-European buying groups in 1991 was US$237 billion – well over one-third of the total European food market and heading for one half. Such groups could help to facilitate European distribution for manufacturers, but could also negotiate particularly good deals for their members.

Branded shopping

Retailers must ensure that customers come through their doors rather than go elsewhere. As food retailing has become dominated by a few major players, retailers have focused on themselves as the brand with which they want customers to identify. In Europe, the trend has developed furthest in the UK, where retailers such as Safeway, Sainsbury and Tesco market themselves as the brand image to which consumers should come to meet their food needs. (see Box 6.5) The stores stock a wide range of own-label products which are manufactured for them. Many of the own-label goods sell for prices similar to those of manufacturers' branded goods. In the early 1990s, around one-third of the products were own label, which was about twice the proportion in continental Europe.

Retailers can adopt a 'neutral' position on individual food products, which has enabled them to champion healthy eating and ecological causes – for example, by removing additives from their own brands or setting salmonella standards. As consumer concerns appear the retailers can respond by adding new niches to the product lines they stock. The development of the retailer as brand is reflected in advertising spending. The top spender on brand advertising in 1991 in the UK was the supermarket chain Tesco, which spent £24 million. The fast-food caterers McDonald's was fourth with £17 million and Sainsbury's supermarkets eighth with £15 million. Coca-Cola was well out of the top ten, spending only £8 million. Both of the supermarket chains have greatly increased their spending since the mid-1980s.

They also sell more 'value-added' – often own-label – products, and use new information technologies to improve profitability. Computer-controlled, laser checkouts are linked to automated ordering systems and carefully organized, central distribution systems to minimize the use of skilled labour, stocking levels and wastage. Lines that do not sell quickly enough are not stocked and own-label brands may push out the third or second

Box 6.5
Sainsbury's

Sainsbury's origins go back over 125 years to a single grocer's shop in London in 1869. In 1969 it had 89 supermarkets, but by 1994 it had 338 supermarkets, ten hypermarkets (Savacentres) and one club warehouse (Bulksava) in the UK, as well 87 supermarkets trading under the Shaw's name in New England, US. The average size of the stores opened in the last 15 years doubled to 34,900 square feet, with the larger stores stocking some 17,000 lines, half of which are own-label products. Over 300 suppliers in 60 countries supply fruit and vegetables for their more than 7.5 million customers per week. The company works with suppliers to produce new products – for example, with farmers and abattoirs to produce premium meat products.

Laser scanning at checkouts and a new high capacity data communications network are used to improve stock control and ordering throughout the supply chain, with over 850 suppliers in electronic contact with the company in 1994. It runs its own distribution division which schedules deliveries by computer; this has reduced the number of vehicles needed and miles covered by the fleet. Turnover has grown from £2.3 billion in 1983 to £9.2 billion in 1992, and UK operating margins have increased from 4.37 to 7.92 per cent in the same period. Pre-tax profits also grew from £110.7 million to £628 million.

Sainsbury's had over 11 per cent of all sales in food and drink in the UK in 1994 and employed over 100,000 people in the UK. It did not reach all parts of the country, however. It spent £617 million on new stores in 1992, opening 21 stores, six of which replaced existing ones. A further 22 new stores were planned for 1993. Despite cut-backs in expansion plans by its rivals in the mid-1990s, the company sees plenty of opportunity to expand further in the UK to regions not yet covered and to redevelop and enlarge its more than 100 stores with less than 20,000 square feet of sales area. It is also expanding and enlarging its Shaw's supermarkets in the USA as well as increasing the number of own-label products there.

Source: Sainsbury's Annual Reports and Accounts, 1992 and 1993

brands in each sector.

Retailers have become food marketers seeking to draw customers in to their specific branded premises, where they hope that the customers will meet their food needs rather than remaining distributors of manufacturers' produce. The market research required to test out own-label products, combined with the detailed sales analysis that is made possible from the electronic point-of-sale technologies, gives them powerful tools to monitor consumer behaviour and response to new products and marketing techniques. These market data are likely to underpin future developments, for both retailers and manufacturers.

Rethinking roles

Retailers, however, may not simply provide more of the same in the future. The equivalent of the whole of Holland visits Dutch multiple retailer Koninklijke Ahold stores every 17–18 days. Pierre Everaert, President and Chief Executive of the group, believes that with the delivery and information system they use to service their stores, they could take on other roles, from package delivery to issuing airline tickets – increasing, as he sees it, the power of the retailer to provide a service for the consumer. In the last few years in Holland, there has been a change in consumer shopping patterns, with customers coming more often but buying less each time, cutting out what was called secondary shopping in other smaller stores.

The major multiple retailers, which provide well-appointed supermarkets temptingly laid out for shoppers with over 15,000 lines, also face competition from discounters. The discounters sell a very limited range of goods which are often own label and much cheaper (see Box 6.6) Mr Everaert expects discounters to take about 30 per cent of the market in Europe, but this, he expects, will increase unemployment in the grocery sector. In America, for example, one discount store needs 200 people to do the work 600 did before. However, he still expects that 60–70 per cent of consumers will want to pay for service, as they do now for a ready cooked hamburger which costs over six times more than the basic product.

Economic recession, with job losses and tighter budgets for many people, contributes to the potential for discount stores and other types of cut-price suppliers such as warehouse clubs, which are spreading to the UK from the USA. These clubs supply in bulk a wide range of items to people who pay an annual membership fee. The growth of discounters has been rapid, with the UK market increasing from £1.9 billion in 1989 to £4.5 billion in 1992, about 10 per cent of the total grocery market. The IGD expects that share to double in the UK. Pressure to compete, however, will

Box 6.6
Aldi

The privately owned, German discount retailer Aldi, Albrecht Discount, is spreading rapidly throughout Europe and to the USA. The first Aldi opened in 1962 in Dortmund, but brothers Karl and Theo Albrecht had been involved in grocery retailing since before the war when they worked in their mother's shop. Aldi is divided into two groups in Germany – Nord and Sud – each with regional branches.

By 1988, one in three of the 6000 discount stores in West Germany was an Aldi store, but the company took almost DM16 billion of the total DM30 billion turnover of discounters. By 1991, following expansion into former East Germany, the estimated group turnover was DM25 billion. The company sells a limited number of lines, mainly good quality own-label products, at a discount – around 480 lines in Aldi Sud and 580 in Aldi Nord. Stores are simply laid out with goods put in cartons on the shelf; they have few staff, who are expected to be able to do most jobs. Check-out staff often have to memorize prices.

About three-quarters of German shoppers do some of their shopping at Aldi which was estimated to hold market shares ranging from 18 per cent for coffee to 40 per cent for canned vegetables in 1989. In 1990, Aldi had almost 10 per cent of the total German grocery market of DM248.27 billion. As its coverage of the market has saturated, it has gone into new ventures at home, such as sub-letting space in larger stores or moving into the food halls of department stores. It has also moved abroad.

By mid-1992, Aldi was operating a total of over 1000 stores in Denmark, Belgium, The Netherlands, France, Austria, the USA and the UK, and was expected to expand further into Eastern Europe, Spain and Italy. Aldi also had an indirect interest in the US market through Marcus Stiftung's 11 per cent stake in Albertson of Boise, Idaho, the sixth largest US supermarket operator which had a turnover of US$8.22 billion in 1990, over ten times that estimated for the Aldi stores. Theo Albrecht is chairman of Marcus Stiftung.

Aldi spends very little on advertising – 0.2 per cent of its turnover – and that is usually press advertising showing products and prices.

Source: The Institute of Grocery Distribution (1992) *Aldi, Germany*, IGD International Key Account Profile

mean that stores will continue to cut down on stocks and go more and more for last-minute stocking. The existing major multiples may also have to cut margins and curtail their expansion plans.

Competition will not come just from other retailers, but from changing food habits and increased food consumption outside the home. There are a limited number of meal occasions per person per year for supermarkets to target – about 1100 in a traditional three-meal-a-day pattern – but the stores have already lost the midday meal, which is no longer eaten at home, to caterers.

Even more major changes could be in the offing, believes Christine MacNulty, Chief Executive of Applied Futures Limited, who has spent 25 years studying consumers in the Western world. She believes that retailers themselves might find their role changing as the use of the interactive technology now becoming available in stores and homes could make shopping much less of a chore with tele-shopping being used for much routine shopping, especially for packaged food. This will become more likely as more and more people work at least part-time from home. In the UK, 15 per cent of people already work from home and 50 per cent are expected to work part-time by the end of the decade.

All this will influence retail formats and locations and offer the opportunities for competition from new sources – directly from manufacturers or from communication industries. It raises the question of who is the middleman. With burgeoning satellite and cable communications, coupled with interactive technology, radically different food retailing patterns could emerge, with some consumers interacting right down the food chain.

The independent retailers who are left, however, find that they have to work harder to remain in business – for example, they are opening longer hours to attract customers and entering into new links with their wholesalers. They are likely to cater more for those excluded, by virtue of their lack of transport, poverty, or infirmity, from access to the superstores of the multiple retailers or the bulk buying of warehouse clubs.

Wholesalers

In the past, wholesalers bought from a wide range of manufacturers in bulk and sold on to individual shops who could get their supplies from one source and then sell them to individual customers. As small shops have declined so too has the role of wholesalers. In the mid-1970s, food wholesalers were estimated to supply about half the purchases of food retailers. By 1991, wholesalers accounted for less than a quarter of total retail sales – just over £11 billion sales out of total retail sales of foodstuffs of £45 billion in the UK, according to the IGD – as manufacturers deliver direct to the increasingly centralized depots of the multiple retailers, from which they distribute to their own stores.

Wholesalers responded to this decline by a series of mergers that created fewer, larger firms, so that by 1980 the wholesale grocery trade in Britain was controlled by ten firms. They also developed links with the remaining independent retailers and helped to develop voluntary symbol groups, such as Spar and NISA. In these groups, individually owned shops trade under a common name and use their combined buying power to secure better prices from the manufacturers through their wholesalers. The wholesalers also provide some management assistance and develop some own-label products for the groups.

The nature of wholesaling also shifted away from delivery services to the cash and carry area and, more recently, to catering supplies (see Box 6.7). The numbers of wholesale depots in Britain fell by about a quarter between 1981 and 1989. The biggest loss was in those dealing in delivered trade (see Figure 6.3).

The linking of independent retailers and wholesalers to buying groups to increase their bargaining power has reduced still further the number of buying points with which manufacturers have to deal. In 1970, about 600 buying organizations accounted for about 70 per cent of the grocery market; in 1979, 285 accounted for 80 per cent and just 90 buying points accounted for 66 per cent of the grocery trade, putting great power into the hands of a few buyers for the multiple retailers and large wholesalers.

Box 6.7
Wholesale Change

Nurdin and Peacock (N&P) had 17 per cent of the British cash and carry market in 1991 – second after Booker with 26 per cent. Some 40 per cent of this market was still in the hands of small operators – which N&P saw as vulnerable, lacking financial and managerial resources to introduce new technology and develop their branches. They also felt that these small operators lacked buying power and own-label marketing expertise.

With a 1992 turnover of £1.44 billion, pre-tax profits of £30.2 million, over 5000 staff, and 46 branches nationwide, with one central distribution point at the end of 1992, N&P aimed to take over some of the market that is presently going to the small operators. It was expanding across the country to increase its market share. With 90 per cent of its turnover accounted for by just 10 per cent of the 60,000 lines carried in 1991, it decided to cut this range by half, keeping a core range of 10,000 lines and optional range of 20,000 lines for each branch to be stocked according to local preference and at the discretion of branch managers. Sales monitoring in 1992 allowed it to cut the nationally stocked core range to 8000 lines.

It also centralized its purchasing and distribution operations to improve its buying power and cut the costs of supplies. Computer scanning at each checkout produced itemized price-lists, showing guide selling prices and profit on return at those guide prices. It also moved to seven-day trading.

It was promoting its own symbol group – Happy Shopper – which had already been adopted by 1800 customers in 1992, and was developing its range of own-label products from an initial 26 lines worth £26 million at retail prices in 1986 to almost 700 lines worth £200 million in 1993, with 300 more lines to be added in 1994. These own-label lines have also become more profitable for the shops, with retailer's profits up from an average of 17.5 per cent in 1986 to 24 per cent in 1991. It also launched the first ever TV advertising campaign in Britain by a cash and carry group for its Happy Shopper brand in 1990.

The firm has also moved into the catering market using the Happy Chef label. Given the increase in the number of people eating out, it aimed to increase its share of sales from caterers – 20 per cent of total own-brand sales in the early 1990s and up by a third in the previous five years in the UK. At the end of 1993, its Happy Chef label had over 500 lines and £35 million in sales.

After studying the US membership club warehouse industry, N&P decided to expand into that in the UK in the mid-1990s and acquired four out-of-town sites for its first club warehouse operations. A new staff structure, staff training and performance-related pay, with monthly bonuses based on meeting target sales, were also part of the development package introduced in the early 1990s. In 1994, it went beyond food, converting its food cash and carries to trade and business warehouses and broadening its range to include office and business supplies.

Source: Company annual reports, 1991 and 1992 and other company materials

Co-operation between independent retailers and wholesalers also enables them to respond to the opportunities offered by the new European single market. One example in Britain is Nisa Today, which links independent retailers with annual turnovers greater than £2 million and wholesalers with annual turnovers over £3 million. The group has linked into a European purchasing organization – European Marketing Distribution – to increase its buying clout. It is also able to develop pan-European brands which can be sold through independent retailers across Europe, although the first products are washing powder and washing-up liquid, not food.

The development of warehouse clubs will

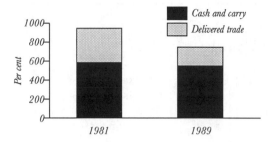

Source: IGD Research Services

Figure 6.3 Decline in wholesale depots in Britain between 1981 and 1989

a wide range of goods to members in large sizes and formats, including a small range of bulk packs of groceries, have taken a significant slice of the retail market, with US$27 billion sales in 1991, and are likely to spread into Europe where they will compete with existing retailers. In the UK, the biggest cash and carry wholesaler, Booker Cash and Carry, decided, initially at least, to stick with its existing approach in the UK while expanding into Spain and Portugal. Its competitor, Nurdin and Peacock, however, decided to move into the warehouse club side in 1994, in what, in effect, is wholesaler retailing.

also challenge traditional wholesalers and retailers. In the USA, warehouse clubs selling

Caterers

Happiness: a good bank account, a good cook, and a good digestion.

Attributed to Jean-Jacques Rousseau, 1712–78

Both food retailers and wholesalers have noticed that more and more food is consumed outside the home – at work or at school, while travelling, and in cafés, restaurants and fast-food outlets. For them, it represents lost business, but is a growing trend in rich world markets.

In traditional societies, people ate out in the homes of friends and relatives, in men's and women's meeting houses, and at ceremonies marking religious and other events, such as weddings. Hotels, alms houses and similar institutions arose when it became possible for strangers to travel for long distances across the territory of others. These institutions sell hospitality to strangers usually for days or weeks.

Today, fast-food outlets, cafés, restaurants, vending-machines and, to a lesser extent, transport catering on motorways, trains, ferries and airlines, increasingly supplement family food (see Figure 6.4). At varying costs, they all provide services which relieve or extend family food provision, usually for a few hours only. Few people can afford to live permanently off their services.

Although definitions differ, we can group catering into two main types:

➤ consumer catering where outlets are open to the public – the restaurants, cafes, pubs, fast-food joints;
➤ contract, industrial, cost centre, or subsidized catering with a closed clientele where a contractor or in-house caterer provides an often free or subsidized service.

In a sense, contract caterers are increasingly carrying catering into the home through the development of ready-prepared main dishes and meal elements that are now sold through the cook-chill and frozen cabinets of the supermarkets; this also helps the supermarkets to recapture some of the business otherwise lost to outside catering. In the UK, for example, in 1990, the UK Food and Drink Federation put retail spending on catering at £13 billion, with another £3.5 billion spent on subsidized meals.

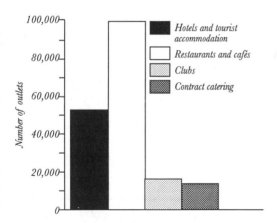

Note: Figures for hotels, restaurants and clubs relate to 1991 and are from the Hotel Catering Training Company; those for contract catering are from the British Hospitality Association and relate to 1993. The figures are only a general guide, as different sources give different numbers depending upon how they define different categories.

Figure 6.4 Number of catering outlets in the UK by category in early 1990s

This compared with an estimated £7 billion spent on buying food by these catering establishments. In 1980, only about one in 12 meals were eaten out in the UK, but this rose to one in seven in 1990, which still fell a long way short of the one in 2.5 in the USA. There, almost half of food dollars were spent on meals and snacks away from home in 1990, up to 46 per cent from 30 per cent in 1965 (see Table 6.2).

With the rise of mass affluence since 1947 in the USA, and since the 1960s and 1970s in much of Western Europe, Japan, Korea and Taiwan, disposable income, tourism and fast-food franchises grew rapidly. This has been driven by many factors, including the greater earning power of households, cheaper and safer air transportation and the increased availability of investment capital in these parts of the world. Taken together, they mean that many more people can eat out if they want to and have to when they travel. Hospitality and catering are huge 'service' industries which employ millions of people world-wide.

As affluence and tourism are spreading, the consumer catering that has become the hallmark of Western industrialized countries is

Table 6.2 American personal spending on food, 1960–90

Year	Disposable personal income [a]	Expenditures on food in billion US$ At home [b]	Percentage of income spent on food Away from home [c]	Total [d]
1960	360.5	50.6 (14.0)	12.6 (3.5)	63.1 (17.5)
1970	722.0	74.2 (10.3)	26.4 (3.7)	100.6 (13.9)
1990	4042.9	297.3 (7.4)	177.4 (4.4)	474.7 (11.7)

Notes:
[a] These figures include the income of the entire US population. If household incomes are used as the basis for assessing the percentage spent on food, it rises to 15 per cent because household income excludes the incomes of persons living in military barracks, prisons and institutions. Personal income also includes more items, such as payment by employers of Social Security, health insurance and retirement, which are not included in household income.
[b] Food bought from grocery stores and other retail outlets; includes food bought with food stamps and food produced and consumed on farms, but exludes Government-donated foods.
[c] Purchases of meals and snacks by families and individuals and food provided for employees; exludes food paid for by Government and business, such as donated foods to schools, meals in prisons and other institutions and expense-account meals.
[d] Totals may not add up owing to rounding.
Source: 'Percent of Income Spent On Food Holds Steady' in *Food Review*, UDSA-ERS, July-September 1992, vol 15 no 2, p11

spreading, too. The market research organization Euromonitor, for example, believes that South-East Asia is the largest and most attractive of these markets, especially Hong Kong, Singapore and Thailand. The spread of this type of large-firm consumer catering could threaten the many millions of people in developing countries who provide street food (see Box 6.8).

Average growth in food imports in East Asia is 13.8 per cent, compared with 6.7 per cent for the EU and 2.4 per cent for the USA. There appear to be two phases in food consumption growth-rates. First, the satisfaction of nutritional needs, and second, the influence of convenience, health and image concerns of relatively affluent consumers. An Australian firm has come up with a rule of thumb for these changes:

Box 6.8
Street Foods: Feeding the Developing World's Cities

Street foods – ready-to-eat foods and drinks prepared and/or sold by vendors in streets or similar public places – are a vast global business. In Malaysia, for example, street food sales by more than 100,000 vendors are estimated at US$2.2 billion a year. In Senegal, UNICEF estimated that in 1979 street foods employed 40–50,000 people, while modern agri-business and food industries gave jobs to only 6800 people.

This multi-billion dollar activity boasts no multinational corporations, yet it provides a livelihood for millions, and food for hundreds of millions, especially the poor.

Street food sellers produce a huge variety of foods from banana fritters in Bangkok to raw fish in lemon juice in Peru. Usually the dishes are based on nutritious, traditional foods, grown by local farmers. Women play the major role, although in some countries they may be behind the scenes rather than selling the goods.

The authorities have often ignored street food activity or treated it with hostility. Many regarded it as transitory. However, it remains, and with almost half the world's population expected to live in ever growing cities within a decade or so, it seems likely to stay.

Many poor families would be worse off if street food disappeared. For those living in shanty towns, with little or no cooking facilities, street foods provide cooked meals. Poorly paid workers, who have to travel long distances on crowded city buses, rely on street foods for lunch. But they want safe food. This is the main problem with street food.

Bad noodles killed 14 in Perak, Malaysia, in 1988, cholera was spread by contaminated ice in a sugar-cane drink in Pune City, India, in 1981, and poor street food hygiene is thought to be partly to blame for the major cholera epidemic in Peru in the early 1990s.

The challenge is to improve street food and the way it is produced. This requires action to overcome the basic problems facing street food producers. They need safe water and garbage disposal facilities as well as training in good food handling and preparation practices. Experience in Nigeria, Peru and Colombia shows that sellers willingly accept basic hygiene and sanitation training.

New legislation, food inspection and analysis services to check on street practice may also be needed.

Street-food consumers tend to have low- to medium-level incomes, relatively little education, and little knowledge of good hygiene. For them street foods are the right food in the right place at the right price.

Source: FAO (1992) *Food and Nutrition: Creating a Well-fed World* FAO, Rome

when per capita income is below $US4,000, a 10% rise in that income produces an 8% increase in the demand for farm products. Above that level of income, the impact declines quickly. Richer people spend more on food but the additional expenditure is largely on packaging and presentation

Hassall and Associates (1992) 'The medium and long term outlook for Australian Agriculture' Canberra

Diets are changing in East Asia. Ice-cream and Coca-Cola consumption has increased rapidly but, for example, rice consumption has dropped by 30 per cent in Taiwan, while wheat-based product consumption has grown by 30 per cent. There is a trend towards convenience-based products as the price of domestic labour rises with the advent of new industry-based jobs. However, products tend not to sell unless they are either exotic luxuries like McDonald's hamburgers or unless they suit local tastes and customs.

Many hands, many changes

Caterers are now the single largest employers in the food system – from highly skilled hotel chefs to production-line workers making sandwiches or cook-chill meals for supermarkets; from airline and factory caterers to local cafés and fast food outlets. In the UK, British Hospitality Association estimates put the number of those working in the catering and hotel sector at 2,191,000 in 1993.

There is an economic and technological revolution going on here, too. Large catering companies are developing which operate in many areas of catering (see Box 6.9). Technical changes, such as cook-chill and cook-freeze, in which dishes are prepared, then blast-cooled and stored chilled or frozen until reheated, and other centralized-production methods, are being introduced. These changes concentrate production facilities and require well-controlled storage and distribution networks. Such systems demand high technical

and safety standards for effective operation. In-house operations are also being replaced by contract caterers who provide services for a range of operations and companies. Caterers, not eaters, determine the ingredients, recipe dishes and cooking methods for markets based on a whole range of market information about consumer taste.

The fast food industry

Although drinking houses (pubs), bars, cafés and specialized food shops (for example, fish and chip shops) have provided relatively inexpensive, speedily served meals for centuries, a new trend emerged in the USA after the Second World War – the fast-food franchise. This consists of independently owned food shops, all using the same name, in which the same very limited range of standardized meals could be eaten on the premises or taken away.

Usually these chains provide one type of meal such as roast chicken, pizza or hamburgers. About half a dozen variations of the main meal are offered (eg, cheese pizza, bacon pizza, wholemeal pizza). Non-alcoholic beverages from a very limited range (coffee, cola, orange fizzy drink) may be available together with a few ice-cream desserts. Each item is bought separately (see Table 6.3).

This sector of the industry has been very successful (see Box 6.10). In 1991, Euromonitor reported that fast foods accounted for 19 per cent of the global consumer catering market, then worth over $730 billion and expected it to grow to 25 per cent of that market by 2000. They attributed the growth to continuation of trends such as spending less time on eating, having meals at non-traditional times and the Americanization of eating culture around the world. In a 1993 publication, Euromonitor reported that US fast-food sales reached US$78 billion, ten times the size of all the key European markets put together (see Figure 6.5).

Although the reasons for this financial success are complex, one key factor is product standardization. The product's sensory prop-

Box 6.9
Compass – Finding the Right Direction?

The Compass Group PLC had its origins in 1941 in the UK serving meals to munitions workers in the Second World War, but was formed by a management buy-out from Grand Metropolitan in 1987. By 1993, it had grown to a £497 million turnover company, working in both contract and consumer catering and healthcare. Catering accounted for most of the turnover, £436.1 million, and £33.7 million of the £41.5 million pre-tax operating profits. Some 20,000 staff in the UK were producing 3 million meals per week in about 3000 locations. Compass aims to be the leader in specialist catering and is developing a number of strongly branded catering concepts through a range of specialist companies which act as brand names.

Contract catering in the workplace, through Compass Services, was still the group's main activity, but it also operated in education, as Chartwells, with catering services in over 200 education establishments. It operated in sport and events catering as the up-market Letheby and Christopher, a company and brand name acquired in 1992, and was developing its executive dining services through a link with the Roux brothers to produce Roux Fine Dining.

The group sees great growth potential in all the catering markets it targets – for example, in healthcare and education where it sees a potential market worth over £3 billion in the UK alone. It set up a new division, Bateman Catering, to specialize in catering services for healthcare, although it has been servicing the Group's own hospitals and homes for some time. An unsuccessful attempt in 1992 to take over the Gardner Merchant part of Forte PLC Catering, one of the main competitors, cost it £2.3 million, but also gave impetus to the development of a branded base to its expansion plans.

Consumer catering was given a boost with the acquisition in 1992 of Travellers Fare, which operated at railway stations and airports with the Upper Crust and now superseded Casey Jones brand names. The Commercial Services part of the company manages restaurant and cafe facilities in supermarkets, department stores, shopping and leisure centres, as well as in museums, conference centres and exhibition halls.

In 1993, Compass gained exclusive rights to take the Burger King brand off the high street and market it in workplaces, hospitals, schools, colleges and railway stations throughout Europe in a deal with Grand Metropolitan which owns Burger King. This better known brand was used to replace the Casey Jones outlets at UK railway stations in 1994. The company also took over the airline catering subsidiary of the Scandinavian airline SAS, which gave it catering at 34 airports in eight countries, plus contract catering outlets in Europe. Compass aims to continue expansion into Europe and become one of the top two major European catering groups.

As part of achieving this aim, it established New Famous Foods to take its growing, exclusive portfolio of its own and franchised branded outlets to offer in any site where it provides catering facilities. In 1993, these also included brands such as Burger King, Le Croissant Shop, Café Select, Cooper's Taverns, Dixie's Donuts, Frank's Deli, Ginghams Coffee Shop, Harry Ramsden's Fish and Chips, Pizza Hut, The Lemon Tree and The Food Court, which was used to bring the various other outlets together – for example, in shopping centres or station forecourts. The various acquisitions in 1993 took its catering turnover above £600 million in 1994.

Source: Compass Group *Annual Reports* 1992 and 1993 and other company materials and *The Sunday Times*, 7 November 1993

Table 6.3 Sales in UK fast food and takeaway market by type of outlet, 1992

Type of outlet	Sales (£ million)
Fast foods	
Burger bars	960
Pizza bars	695
in restaurants	555
home delivery	140
Chicken bars	207
Hot potatoes	54
Others	28
Total	*1944*
Takeaways	
Ethnic takeaways	465
Fish and chips	870
Sandwich bars	1450
Total	*2785*

Source: Euromonitor Market Direction, London

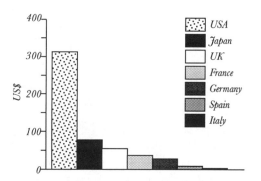

Source: Euromonitor Market Direction, London

Figure 6.5 Per capita spending in US$ on fast food in selected countries, 1992

erties are well researched and engineered to appeal to the majority of the market sector likely to buy from these outlets – often adolescents and young families. The product and the dining environment are highly predictable – clean and geared to families – as is the limited but fast service. Rapid throughput of customers is a key feature of these establishments.

The use of franchising maximizes the benefits of local management incentive, while providing the benefits of a large organization such as mass advertising and brand appeal. Wage bills are often kept to a minimum through the employment of part time youth labour. The London Food Commission in Britain found that such chains often paid considerably less than the average wage rates in the late 1980s. Sickness, holiday and severance pay were usually not available.

Yet these fast-food chains provide employment for the youth sector which has notoriously high unemployment levels. But do part-time youth workers displace adults from this sector of the work force? They also provide standard products that people can eat

wherever the chain is present, offering a comfort and familiarity factor to the consumer. There are variations between countries, however. Some are due to laws – for example, in Austria the fat content of a burger must be as low as 14 per cent, but it is much higher in most other countries. Others are due to local traditions and demand – for example, beer is served in McDonald's in Germany.

Hotels

In most civilizations, the hospitality industry emerged early, catering to the needs of travellers. These traditions have been much expanded during the twentieth century. Hotels vary from inexpensive farmhouse and charity accommodation, such as the YMCA, to five-star luxury hotels in developing countries where single meals cost more than the average monthly wage.

The big boost to hotels came with the advent of mass packaged holidays in the 1960s. Whole villages and towns on the coast of south-east Spain, for example, were transformed into urban canyons, walled by hotels for North European tourists. The food served in them became a North European version of Spanish dishes or simply North European fare.

Although many small hotels are still family run, the trend in hotels and hospitality has been similar to those in other parts of the food

---

Box 6.10
It's a Big Mac Now!

In 1954, Mac and Dick McDonald felt that they had enough on their plate with just one restaurant selling 'McDonald's Hamburgers' promising 'Speedee Service' in San Bernadino, California. Then a 52-year-old entrepreneur, Ray Kroc, persuaded them to appoint him as their first franchisee. He opened his first McDonald's in Des Plaines near Chicago the following year. By 1959 he had 100 and in 1961 bought the McDonald's concept from the McDonald brothers for US$2.7 million.

By 1989 sales exceeded US$17 billion and in 1990 the company broke through into China and opened the biggest McDonald's in the world in Moscow. It also served its 80 billionth hamburger that year. At the end of 1992, McDonald's had over 13,000 outlets in 63 countries, with plans to open 400–500 restaurants per year in 1992–96. Sales had reached US$21.9 billion, with just over 60 per cent of these in the USA. The company served 23 million people daily, world-wide, and made an operating profit of US$1.9 billion and net profit close to US$1 billion.

McDonald's intends to keep growing. Its strategy is based on adding restaurants, maximizing sales and profits at existing restaurants, and improving international profitability. It expects to achieve the last through economies of scale in individual markets and through benefits from the global infrastructure.

World-wide, McDonald's is the largest franchised food service organization, with 84 per cent of its 8900 restaurants in the USA owned and run by franchisees – one restaurant for every 28,000 people. The picture can be very different in other countries. In the UK, for example, of the 492 outlets at the end of 1992 only 53, or 11 per cent, were franchises.

The company's first UK restaurant opened in 1974 and the first franchised outlet in 1986, during which year it opened its 200th UK outlet. In 1992, turnover reached £578 million, producing a gross profit of £65 million and after-tax profits of £8.5 million. Nearly 63 per cent of its 31,373 UK employees were between 16 and 20 years old. Each restaurant employed an average of 61 hourly paid (average pay £3.30–3.58 per hour, depending on location), mostly part-time, staff and 4.32 salaried managers (average payroll and related costs per manager £13,839).

The company spent £33.7 million in total on marketing in the UK in 1992, mostly on TV and radio but including around £5 million on local store marketing and £2 million on national and community projects. Continued expansion by opening about 40 restaurants per year until 2000 is planned in the UK, but with franchises increasing so that they will account for 40 per cent of the total UK business.

The company provides training for all staff and has worked closely with its suppliers to obtain the products it requires. It had to set up its own subsidiaries to supply some products, but these later became independent. With Sun Valley, a subsidiary of Cargill, it helped with the introduction of new methods for producing its chicken products. It has introduced a new potato to British soil, the Russet Burbank, which gives a high yield, consistent texture and rectangular shape suited for its French fries. These are processed and distributed by McCain Foods, which deals directly with the growers.

Source: McDonald's UK (1993) *Fact Book, & Annual Review* London

---

system. Large national and international firms own or manage many hotels, using particular brand names, particularly at the upper end of the market of three-, four- and five-star hotels. Independent hotels, large and small, have found themselves, like small grocers, joining groups that provide some central marketing services and brand names.

While traditional craft-based approaches to catering remain at the top end of the market, and home cooking at the bottom, middle ranking hotels in the UK, for instance, increasingly buy in ready-prepared dishes from catering companies. But even in the top-flight hotels, traditional skills are mixing with high-technology to produce profitable businesses which are run more on industrial lines (see Box 6.11).

Today, chefs must know what they are spending on food and staff, as hotel catering has to pay its way in the business, with the restaurants marketed by the hotel. With around a quarter of a large five star hotel's costs going on staff, ways to cut staff costs, and staff, are being sought. One way is to buy in services that would have been provided by full-time staff in the past – from laundry services and night-cleaning to ice-carvings.

Airlines

International tourism blossomed when flying stopped being an adventure for a few and became a matter of putting bottoms on the seats of jumbo jets. Part of the flight package is catering, and providing food for such large numbers in such conditions posed special problems. In solving these problems, the airlines helped to develop methods that have come into much wider use in catering today.

In-flight catering is a complex affair, usually provided by specialized catering companies, often owned by the airlines themselves. In 1990, Qantas Flight Catering's 1250 strong work-force produced 100,000–120,000 meals per week for over 20 airlines from its Sydney base, working round the clock. Most meals were economy class and these were always pre-set on a tray – unlike some first- and business-class offerings which attempt a restaurant-like service. It was then that the company introduced a productivity-doubling computer-controlled system for loading the meal trays. They found that it ensured a consistent product which the airlines require; other companies use semi-automated systems for loading the food trays.

The food itself must appeal to a wide cross-section of people and be varied enough to provide interest during long, boring and often crowded journeys. Caterers also have to meet the needs of the minority of passengers with special dietary requirements (eg, medically prescribed diets or religious dietary prohibitions). Food service and menus are also designed to give the airline a distinctive competitive 'edge', especially in the first- and business-class markets.

Airline meals must be pre-prepared according to precise specifications, stored and reheated at some time later. Such catering demands high standards. The preparation of biologically uncontaminated food is a problem for all mass catering institutions. For example, chicken may be already infected with salmonella before purchase by the catering company. If it is not properly cooked, reheating may culture the bacteria nicely. Airlines, however, have added complications caused by busy flight schedules. Meals have to be provided according to very tight schedules and they have to be served in confined spaces to many people.

The response to this has been to develop a cook-chill system for food preparation. Remarkably, there are few food-poisoning incidents, but poor quality control has sometimes led to passengers suffering food poisoning. A single incident can have major consequences for those affected and for the airline in lost revenues and reputation (see Box 6.12). For safety, the pilot and co-pilot have different meals on board, and on the ground elaborate procedures have been developed to monitor food supply, production and delivery to maintain food safety.

Hospitals

Hospitals were among the first institutions to be involved in mass catering. In Britain's National Health Service (NHS) hospital catering is a massive undertaking – probably the largest food-catering service in Europe. In 1990–91, the 23,000 catering staff in the NHS produced about 180 million patient meals

Box 6.11
Haute Cuisine and High Tech – London's Park Lane Hilton

Hilton's five-star flagship British hotel had 448 rooms and employed nearly 560 people with an annual turnover of around £35 million in 1992. The executive chef, Vaughan Archer, headed over 70 chefs and oversaw a business worth about £10 million. The room service kitchen worked 24 hours a day, while other kitchens serviced two restaurants and a bar and could produce banquets for up to 1000 people with very varying tastes and expectations.

Even here, though, the days are gone when the chefs did absolutely everything, everyday. In 1993, the executive chef faced pressure on his payroll and used training and capital investment to supplement his chef's skills. A high-tech kitchen runs more like a production line with separate areas for meat, fish, raw vegetables and patisserie.

Batch producing and then blast freezing the pâtisseries, using the latest technology, makes more efficient use of the chefs' time. The Park Lane Hilton now supplies pâtisseries to nearly a dozen other Hiltons around London. The kitchen also uses sophisticated, programmable energy-saving ovens which can humidify the air to suit the dish and reduce cooking times by up to half.

Vaughan no longer has 20 chefs to prepare and cook vegetables for a 1000-person banquet. Now he specifices precisely what he wants, the variety, preparation method, etc; and contracts out production of ready-cut vegetables to speciality firms in London. He has even found a firm to do ice-carvings - used as centre-pieces in banquets – for a third of the cost of doing them in-house.

He uses cook-chill methods to produce the 5000 staff meals needed each week, but does not use this method for patrons. However, some *sous-vide* is used, for example, taking one day to produce different hotpots for each day of a promotion in one of the restaurants.

Vaughan sees restaurant food being led by what is happening in the retail market in the late 1980s and says that his food is often compared with that produced by the best of the cook-chill caterers for the multiple retailers.

The London Hilton is part of a world-wide chain owned by Ladbrokes, a company which started out in horse race betting and grew to include property and do-it-yourself retailing in its portfolio. Ladbrokes took over Hilton International in 1987 and operated in 47 countries with 159 hotels at the end of 1993. Hotels provided £1.7 billion of their almost £4.3 billion turnover, but £118 million of the £238 million operating profit from the four divisions. The group employed a weekly average of just over 28,000 people world-wide in its hotels in 1993 and had over 53,000 bed-rooms.

The Hilton brand name is used by Hilton International world-wide and by a different company, Hilton Hotel Corporation, in the USA.

Source: London Hilton and Ladbrokes *Annual Reports* 1987, 1988, 1992 and 1993

from some 1300 units and spent £435 million. The patient's average stay was between five and six days and they had two main meals and a snack-type meal per day.

Not only must hospital meals be safe, acceptable and produced to a budget, but many of them must also conform to stringent medical and nutritional requirements. For example, hypertensive patients usually require low salt meals, and diabetics need meals with specific amounts of complex carbohydrates. Clinical dieticians are responsible for working with caterers to ensure that these needs are met. The meals may also have to adhere to recommended dietary guidelines.

Another problem, unique to this sector, is that many of the eaters do not feel well. Special attempts may have to be made to

---

Box 6.12
Poison Aspic

Aspic glazes make canapés look good. They are also an ideal medium for the growth of food-poisoning bugs. They were the cause, in 1984, of the only food poisoning incident at British Airways (BA) between 1975 and 1993. A chef unwittingly contaminated an aspic glaze used for the canapes for first- and business-class meals. Some 760 people were affected by food poisoning, caused by *Salmonella enteritidis* phage type 4, 200 of them staff,. Sadly, two people died from complications allegedly as a result of their illness. In mass catering, one mistake can have widespread results.

It was a major disaster for BA, which had been, somewhat ironically, the first airline in the industry to employ a hygiene adviser in the mid 1950s. The chef, it turned out, was a carrier of the bugs who showed no symptoms. He had returned from a foreign holiday after having a tummy upset, but had not reported it and so had not been checked before returning to work. Not only had he contaminated the aspic glaze, but he had then used it over a whole morning, in breach of company regulations, in preparing canapés for all the flights.

After the incident, aspic came off the menu for good and tougher rules for staff returning from overseas trips were introduced, with medicals and stool samples taken after every overseas trip. The company also toughened up its rules for catering operations and suppliers and developed a total quality management approach. By 1993, BA was serving 63,000 meals per day world-wide with about 30,000 from its kitchens at London Heathrow.

Source: British Airways

---

make the food appetizing and to coax the patients to eat. Here, there is a gap between the caterers' role in producing appetizing and nutritious food and the medical staff's role in ensuring that it is eaten.

Apart from the ever-present problem of food safety (the prevention of biological contamination and of parasitic disease), hospital catering faces two opposing health problems – malnutrition and obesity. Malnutrition may occur with long-stay infirm patients who need a great deal of help with their eating. They may lack appetite or be physically unable to feed themselves. Although pleasant food may be provided, if the dietetic and nursing staff are not vigilant, the patients may eat too little. For many hospital patients, however, food may be abundant and exercise restricted, and weight gain may result. A related problem is the feeding of patients with inappropriate food. For example, patients recovering from a heart attack have been fed foods such as bacon and egg, which is rich in saturated fat. While this may not cause immediate harm, it is a

wasted opportunity for patient education.

Many hospitals do offer patients a choice between their usual (high fat) foods and tasty versions of healthier alternatives (eg, wholemeal bread versus white bread). Surprisingly, resistance to patient education programmes often comes from management and staff rather than from patients. Many health professionals do not believe that their job includes health promotion! Don Nutbeam, Professor of Public Health at Sydney University and formerly of Heartbeat Wales, found that many English hospital boards did not buy foods which were recommended by Government as being healthy alternatives (eg, they preferred to buy butter instead of polyunsaturated margarine and biscuits instead of fruit). The situation was remedied to some degree when the results of simple audits of the nutritional status of foods served in hospital were given to the local press.

Patients are not the only people to be fed in hospital. Staff and visitors also eat there, often at subsidized prices. Again, this is an opportunity for health promoters to demonstrate that healthy food can be tasty and inexpensive and to disseminate healthy recipes and food information. Similar problems can be found in school systems over the provision of unhealthy foods for meals and snacks.

Similar pressures face hospital caterers as face the rest of the sector, and they have adopted a whole range of methods – from cook-and-serve to cook-chill and cook-freeze – and the purchase of pre-prepared bases for dishes as a way of saving on labour and skills. In the early 1990s, food costs were about 49 per cent of the food budget in the British NHS and labour costs about the same.

The difficulty of ensuring food safety in such a large operation with the staff skills available does affect the technology used, however. *Sous-vide*, for example, where food is vacuum-packed into laminated plastic bags, cooked between 65°C and just below 100°C, cooled rapidly, stored under chill conditions (0–3 °C) and reheated before consumption, was not adopted in the NHS owing to the difficulty of ensuring that the safeguards required were in place throughout the whole system.

Contract catering

Since the early 1980s, catering in the NHS, which is organized locally, has begun to be contracted out to industrial caterers. Contract catering in the UK is expanding rapidly from its traditional industrial base to catering for the public. The total number of outlets grew from 8518 in 1990 to 13,354 in 1994, and the number of meals served grew from 486 to 940 million, according to the British Hospitality Association. Until the 1990s, most contract catering was in the work-place, but the percentage in business and industry fell from 80 per cent in 1990 to just over 50 per cent in 1994. About 60 per cent of meals served in this sector were provided by contract caterers. These meals often compete directly with high street retailers who sell sandwiches, itself a major growth area in the late 1980s and 1990s. Some 13 million sandwiches a week, with a choice of over 200 fillings, were sold in 1992 in the UK. These are increasingly made by large food industry operations supplying large chains.

A number of factors have affected this change, including Government policies to contract out catering from public services like the NHS and the buying power of contractors which can increase cost-benefits to clients. Another factor is the pressure on individual catering managers to meet increasingly stringent food safety rules and regulations (see also Chapter 9).

Even within the UK catering industry itself, however, most catering staff lacked basic training in healthy catering practice, which included using healthier ingredients, recipe composition, cooking practices and merchandizing, according to the chief dietician of a major contact caterer.

The Changing Balance of Power

The food system is dynamic, with the different actors having to adapt to changes in the world around them and each jostling to ensure their survival and development. Obviously, within each group of actors there are a variety of interests, and some changes will unite different actors. One example is packaging, where manufacturers and retailers came together to try to pre-empt the imposition of EC legislation which they felt would hurt them. The packaging industry, unlike the food industry, is still very fragmented. Although there are relatively few suppliers of raw material, many firms convert these into packaging, although their number is likely to fall. In the early 1990s, the top 100 firms produced about 50 per cent of the packaging, compared with food sectors where the top five or so firms produced over half the product in that sector. With packaging getting an increasingly bad press from environmentalists, and plans for a directive on waste to be issued by the EU, both industry and trade have co-operated to look after their interests.

The basic purpose of packaging is to contain, protect and preserve. The main driving forces in the packaging industry, however, are concerned with a product's image, cost, convenience and environmental impact. As consumers in Europe became concerned about the last issue, Dr Ernst Rapp, responsible for Food Law and Safety with CPC Europe Consumer Foods, was put to work lobbying to ensure that new EC directives on packaging waste would still allow the food industry and trade to use the packaging it wants, as he explained to The European Food Industry Conference in Brussels in 1992.

Some actors have spread their roles to other areas – for example, companies that began as traders have also become processors. However, all of them face the trend to concentration. Small farmers and workers must compete with large and powerful users of their products and services. Large manufacturers, especially in the UK, have found themselves supplying increasingly powerful retailers who are able to set terms and drop their products if they fail to meet the retailers' sales standards. Others have worked profitably with the buyers of their products to produce own-brand goods or, for example with McDonald's, to produce potatoes which met their catering requirements.

Among retailers, both competition and concentration are likely to grow in Europe as the existing multiple and discount retailers seek to develop the market share. Retailers themselves might find their role changing, however, with the use of the interactive technology now becoming available in the store and home. All this will influence retail formats and location, and will offer the opportunities for competition from new sources – directly from manufacturers or from communication industries. This may raise the question of who is the middleman. If interactive technology is added to the above trends, then radically different food retailing patterns could emerge, with consumers interacting with people right down the food chain. Indeed, the whole idea of a linear food chain is being discarded as interaction increases across the different parts of it.

Whatever happens there is a fascinating battle going on for who processes – in the factory, home or small business – the food that goes into people's stomachs world-wide. The new deregulated (liberalized) framework for trade agreed in the GATT Uruguay Round will facilitate the spread of these trends world-wide (see also Chapter 9). Already, rapid developments are underway in East and South Asia. It is a battle in which workers throughout the food system fear they will be squeezed out, including those working in supermarkets, as automated whole basket checkouts and perhaps tele-shopping are developed. A report on the restructuring and trends in Unilever produced for the International Union of Food Workers, which is believed to apply generally to the major food companies, says 'A large-scale reduction in job numbers, increased average turnover per worker and large scale automation and reorganisation of production and labour may be envisaged by the year 2000'.

Generally, companies are trying to relate with everybody in a more individualistic way – as consumers and also as workers – and are trying to act more internationally. Unions may be losing some of their limited power to influence trends in the food system, but another group may become crucial – consumers, and their power over the people who supply them. Despite their unorganized, individual nature, there are signs that pressures are growing on other actors in the food system from consumers who are unhappy with the contradictions of the present situation. Indeed, in the UK, it was consumer concerns in the late 1980s about a variety of food safety issues that led to a review of food legislation. It is to consumers that we turn next.

7

Consumers

Caveat emptor – Let the buyer beware

Although we all eat food, are we all consumers? Consumers are those who spend money on goods and services. With money, we can exercise what is called effective demand. We may all influence the decisions of consumers, however. And while each of us sees ourselves as an individual, the other actors in the system tend to group us by various means into target groups of consumers, or groups likely to influence consumers – for example, children.

When enough individuals act independently in a similar fashion or when they come together through voluntary groups, they can have a considerable impact on the other actors. Over the past couple of decades, many ordinary people, through a growing number of groups, have expressed concerns about various aspects of the food system, including: famine and hunger, the effects of diet on health, additives and pesticide residues in food, and intensive farming methods. These concerns reflect both an individual's self-interest – is what I eat safe? will it bring me to an early death? – and his or her broader social concerns – for example, the ending of hunger and famine in a world of abundance, improvements in animal welfare, and the use of production methods that are ethical and sustainable. Ordinary people are concerned about the safety, security, sustainability, sufficiency and nutritional value of their food. In this chapter, we discuss consumers as actors in the food system, how they are grouped by other actors and how consumer groups have begun to influence the system

The Idea and the Reality

The notion of 'consumer' is quite alien to many people. Traditionally, people grew or gathered the food they needed to eat and some bartered goods they made for food. This was almost always done within the family or local community. Often, individuals, and certainly their families, were multifunctional – they could grow food, catch it, process, cook and prepare it, sell it, buy it and eat it! Consumers do not do many of these things!

People in urbanized, industrialized societies are called consumers. This reflects the affluence and dominance of money in these societies and the importance of the continuing consumption and growth of goods and services in the market economies. But 'consumer' is a dehumanizing term that reduces a person to a single, somewhat passive, aspect of human life – the ability to use money to consume goods and services. Consumers are more dependent on services and resources outside themselves and their families than their forebears. For food, they go to a store because they do not produce a great variety of things. The main product many people bring into their homes is money. In this sense they are less skilled than their counterparts in history or in the poor countries of the world.

For consumers, the skills needed to select and prepare food to eat are changing. Control of what goes into foods and how they are treated has passed more to the other actors who operate between farm and mouth. The task of collecting foods to consume at home is becoming one of selecting prepared products for reheating and assembling at home. This has deskilled those traditionally responsible, usually women, for household food management and preparation.

Work and leisure

The separation of work and leisure is part of a long-standing development in industrialization exemplified by one of the masterpieces of modernism – the Model T car factory set up by Henry Ford. This broke down the skills of master craftsmen into its constituents and paid many 'specialists' to carry out just one or two parts of car manufacture; so one operative put the nut on the bolt and another turned it, and so on (see also Chapter 8, Box 8.11). For most people working under this type of system, there is a marked contrast between 'work', which is pointless apart from wages and undemanding, and 'leisure' when they can be really human and do interesting or intrinsically enjoyable things.

Just as Henry Ford broke car manufacturing into little bits, so in a similar way, capitalism as a whole has broken everyday life into simple tasks performed by specialists – we have all been deskilled to some extent. As a result, in our working hours we produce very specific products and services for a wage and in our 'leisure hours' we shop or 'consume' – buy, use and throw away. Today's market-based economic system is based on consumption as much as on production. If people could make their own goods and services or ceased to buy them, as happens in recessions, there would be little need for the

huge array of consumer goods that we see in shops.

Consumers buy all their needs and wants. 'Consumer goods' usually means all the products people buy like cars and washing machines. 'Marketing' is the way business tries to increase the sales of its products and services to consumers, which we discuss further in Chapter 8. Consumer, then, is a loaded term. Consumers are assumed to have money to spend and to be relatively incapable of meeting their basic needs by themselves.

Choosing a suitable selection of foodstuffs for a healthy diet, however, is much harder today, owing to the image marketing of products and the apparent variety available, with 15,000–20,000 lines on supermarket shelves. That variety – in both suppliers and products – is much less than it appears since we rely on a relatively limited number of basic foods and ingredients to make the products.

Food consumers

There has been continual extension in the reach of the market and an increase in the numbers of consumers world-wide. This has gone furthest in the industrialized countries. Food consumers – those that spend money on food – and food preparers are mainly women in many countries. About four out of five people who spend money in shops buying food for households are women. Women's traditional responsibility for the preparation of food from basic ingredients has been transferred to the world of supermarkets; it is usually women who take responsibility for meal planning and food purchases. They have to balance the household budget. A recent study at Melbourne University showed that in the 1980s women were more successful at balancing their household budgets than governments or big business.

Domestic work is ordinary, unspoken work which is done all the time and much of it is food related. The value of this work is immense and has rarely been considered by economists. Marilyn Waring has estimated that even at conservative minimal monetary values, housework represents by far the largest single segment of national Gross Domestic Product of most countries. She writes

In June 1970, Lisa Leghorn wrote to the public relations department of the Chase Manhattan Bank concerning a survey conducted with the families of Wall Street employees to determine 'what a wife was worth'. The Bank replied, 'The information contained below [see Table 7.1] was derived from an informal survey of Wall Street employees and their families which we believe demonstrates that maintaining a household often requires as many or more skills as required in jobs outside the home.'

Leghorn then demonstrated that by multiplying that figure by the number of weeks in a year and the number of housewives in the United States, one arrives at a figure of between $500 to $650 billion per year, which was (at that time) over half the declared Gross National Product of approximately $1 trillion, five to six times the military budget and twice the total government budget.

Waring, M (1988) *Counting for Nothing: What men value and what women are worth*, Allen & Unwin Port Nicholson Press, Wellington

Women's work?

No society can exist without this domestic work. The burden of this work, even in the most 'liberated' countries, still falls on women. Relatively few men do similar work and fewer do it for no monetary reward. Because it is so commonplace, there is a tendency to dismiss it as 'easy' and unskilled – even as 'mere women's work'! But it involves skills in planning, foresight, budgeting, observation of others' activities, balancing the needs and demands of other members of the family, working within financial and food preference limits set by others, knowledge of nutrition, cooking and, above all, the ability to perform and integrate several tasks at once. Compared with the average office job, this is very complex, demanding work.

Households themselves have changed considerably in the last 40 years in industrialized countries. First, there are more of them but each contains fewer people. This is a result of several demographic trends. Couples have had fewer children; women tend to wait longer before they have children; marriages have become less stable; people are living longer. Women, however, still tend to live longer than men. The result is that the 'traditional' family

Table 7.1 Chase Manhattan Bank Survey: What is a wife worth?

Job	Hours per week	Rate per hour ($)	Value per week ($)
Nursemaid	44.5	2.00	89.00
Housekeeper	17.5	3.25	56.88
Cook	13.1	3.25	42.58
Dishwasher	6.2	2.00	12.40
Laundress	5.9	2.50	14.75
Food buyer	3.3	3.50	11.55
Gardener	2.3	3.00	6.90
Chauffeur	2.0	3.25	6.50
Maintenance man	1.7	3.00	5.10
Seamstress	1.3	3.25	4.22
Dietician	1.2	4.50	5.40
Practical nurse	0.6	3.75	2.25
Total	99.6		257.53

Source: Leghorn, L and Warrior, B (1973) *Houseworker's Handbook* Women's Centre, Cambridge, MA: in Waring, M (1988) *Counting for Nothing: What men value and what women are worth*, Allen & Unwin Port Nicholson Press, Wellington

of two parents and children has become a minority household form.

Many families are headed by only one parent, usually the mother. This has resulted in the so-called 'feminization' of poverty. Single-parent families generally have less money to spend on food. In Australia, for example, with a total population of 17 million, around three quarters of a million children live in single parent households. At the other end of the age range, many households consist only of a single woman; women live longer than men and in old age tend to live by themselves or in their daughter's household.

Other actors in the food system have responded to these changes. For example, packages of some food products have become smaller to accommodate single-person households, although these tend to be more expensive and aimed at single working people. Frail and aged people often have great difficulty in opening modern food packages, many of which have been made deliberately 'child-proof'. Older people tend to have reduced taste sensitivity, so foods may be produced with enhanced flavouring, and because old people's eyesight is often impaired, labels may have to be printed in larger fonts. The poor, the old and those with limited mobility, however, find that access to modern supermarkets, which tend to be bigger and designed for those with cars, is denied to them. This forces them to rely on decreasing numbers of small shops, which are often more expensive, thus squeezing their food budgets..

An increasing workload

Related changes may have had even more profound effects, increasing many women's work-load inside and outside the home. Most women return to paid work outside the home within a few years of childbirth. However, they still feed, clothe and comfort their children and take the main responsibility for their health and well-being. They also tend to provide the same level of personal care for their husbands or male partners. Although many women now refuse to be solely responsible for domestic chores, for the most part men do not appear to take a major share in domestic work. Change on their part has been inhibited by negative attitudes to doing 'women's work', as well as inflexibility on the part of employers to the needs of young families (eg, over flexible shifts, part-time work and parental leave). Thus, many women are still expected – and expect themselves – to provide traditional main meals and nourishing food 'like their mothers did', while they take on full-time paid work outside the home. Recently, in affluent parts of society there have been signs of an emerging solution to this gender impasse. Working couples are beginning to employ professional domestic labour for household tasks, such as cooking, shopping and gardening. Domestic service is making a comeback!

Why do women work so hard? There are many possible reasons. The patriarchal ideology which makes domestic work appear 'natural' for women is still powerful. Caring for others has been part of girls' socialization for ages and many women feel that they are obliged to take the major responsibility for these tasks. A countervailing pressure has been the rise in women's consciousness and their increasing participation in paid work outside the home.

The fact that women go out to work *and* do the housework is having some remarkable effects on the food industry. It has created opportunities for the food industry to provide less time-consuming ways of preparing meals that are acceptable to the family, especially to the husband. Hence the rise of the microwave, the cook-chill meal and other technologies. These technological fixes are responses to the recent changes in gender relations and in particular to the low participation of men in domestic work. As one market forecaster has observed, if men do not take up some of the domestic work undertaken by women, then the food industry can help through the production of labour-saving and ego-protecting products. For example, microwaves can help to reduce the time taken to cook meals – they even help children and husbands to make their own meals from pre-prepared bought dishes.

Pre-prepared meals in a wide variety of forms – from pre-packed salads to exotic ready-cooked dishes – can allow affluent meal providers to show some flair and expertise, not in the cooking but in the selection of the meal. Just as wine buffs may question wine sellers about the exact details of a wine (in which area the grapes were grown, at which precise time they were harvested, and so on), so value-added meals enable providers to gain social recognition by displaying a detailed knowledge of the product (what vegetables are in the product, where they came from and their supposed nutritional or health value). According to Phil Ruthven, a market forecaster, these 'value added products' are valuable because of the information and 'image' that goes with them.

The major increase in eating outside the home is another effect of the participation of women in the work-force. Here, however, the decision as to what to consume may be less dominated by women. Fast-food restaurants, cafés, snack bars and sandwich bars provide simple meals at most times of the day. By the year 2000 in Australia, for example, one in three meals will be eaten outside the home. Long ago, Lenin promised Soviet women that communism would liberate them from household drudgery – food kitchens would provide meals for everyone. This promise was not kept. It is ironic that microwaves and McDonald's may be fulfilling Lenin's promise of convenience meals (see Box 7.1)!

The responses of other actors in the food system to social-demographic change have created many jobs in production, marketing and retailing, and made a great deal of money for the companies involved. It also demonstrates how they shore up the social status quo. Instead of men sharing domestic tasks with women, most women still remain responsible for them, 'aided' by new products and technologies. Industry therefore plays a conservative role in social change between men and women.

The effects of these demographic changes on food consumption are most apparent in those West European and North American societies which are based on nuclear families (in which only parents and dependent children live under the same roof). In other societies, such as those of Japan and East Asia, increasing longevity and greater participation by married women in paid work are also evident. However, the extended family still provides many services, and traditional forms of food preparation appear to be stronger in these societies, although similar innovations can also be observed. Rapid urbanization is occurring in Asia, which had only five cities with a population of more than four million in 1950. In the early 1990s, there were 18 such cities and by 2000 there will be 28, with a combined total population of 260 million – more than the US population in the mid-1990s. Household sizes are declining, with families no longer living 10 to 15 to a home with a big kitchen, but 3 to 4 people now sharing an apartment with a tiny kitchen.

Social trends and mind mapping

In the recent past, lifestyles have become more informal and more autonomous. This can be seen in the changes in traditional meals (eg, Sunday lunches) in which all the family ate together. Today, Sunday 'brunch' may include friends of the family, but not all of the family members. Children may eat at different times or eat different foods. Informality is also seen in the preference for 'light' foods, such as salads and crisp breads, over 'heavy foods', like red meat and bread.

In consumer societies the increase in autonomy has been in the form of family autonomy – each family behaves on its own terms separate from the rest of society. Members of the family may even feed themselves separately at different times. These trends reflect the individualism of rich countries, which is part of the market ideology, and their high levels of individual affluence. They are likely to spread as the highly developed Western market spreads.

Sets of deeply held beliefs underpin the consumption behaviours of people in affluent countries. These are the cultural and social val-

Box 7.1
The Trend Towards Convenience

The trend towards convenience appears to be a world-wide phenomenon. Professor David Buisson of the Department of Marketing at the University of Otago, New Zealand argues that:

Worldwide for many working women, the time available for both food shopping and for cooking has been substantially reduced suggesting time is now a critical dimension in the 'shopping' and 'cooking cycles' of major consumer groups.

Women require food products with substantially reduced cooking cycles – ie, those with an in-built convenience component.

We see predictions that by the year 2000, 90 per cent of food will be in some prepared forms and that it is quite conceivable that a proportion of the population will not know how to cook at all.

On the whole, convenience foods tend to be more expensive than other foods. However, for the working woman, the convenience aspect of shopping is becoming so critical that the function of price is much less significant.

Thus it would seem that recognition of the importance of 'time' in all respects is one of the keys to competitive advantage for food retailers in the next decade.

In Asia, the increased number of working women has resulted in a strong market demand for more processed and convenient food items. People used to shop twice a day, now the shopping is down. As Asian cooking is so labour-intensive, housewives buy things ready prepared.

The market for convenience food in Asia differs markedly from that found in the US, where microwavable or chilled/frozen food is normal. The extent and use of processed foods is very dependent on the economic group, although there is a shift across all income groups to forms of processed foods.

The first regionally accepted convenience food was the instant noodle and even in low income areas of Indonesia, Vietnam, etc, such products are found.

Strong market trends throughout the region that equate Western foods as being 'good and wholesome' help to overcome the general belief that goodness and fresh foods are related.

Overall, convenience foods are being demanded increasingly throughout the world, and this accounts for the amazing penetrations of microwavable items and the influence of convenience on new item successes is second only to that of health products. Using trend extrapolation, if the US microwave trend follows that of the TV, we can expect most US households to have two microwave ovens by the year 2000 and that they will be standard items in US cars by 2003. Similar household trends are being seen in other areas of the Western World.

The industry, I feel sure, will also develop multi-chambered microwaves to accommodate a meal that requires several microwave components at the same time.

At present, only about 2 per cent of the food dollar is represented by expenditures on microwavable foods, but it is estimated that in advanced industrial countries that figure will increase to 20 per cent by the year 2000.

The pursuit of convenience is causing a blurring of food service and retail, resulting in a monumental impact on retailers and manufacturers alike (Pillsbury Cor. Football field of food). How far all this will go remains to be seen, but the whole concept of the manufacturer's mentality toward sales and marketing may need to be changed. It is predicted that in the US up to 60 per cent of expenditure on eating outside the home could be in retail by 2000.

The entrepreneurial manufacturer and marketer will have to move to accommodate this blurring trend between food service and retail, and to produce products that are appropriate for them with the right promotional package.

Source: Buisson, David H (1992) 'Consumer Food Choices for the 2000s – The Impact of Social and Marketing Trends', paper presented at the *CSIRO Food Industry Conference*, Adelaide

ues, or life goals (sometimes referred to by marketers as 'visions of the good life'). Much effort goes into identifying these goals so that market researchers, who play a crucial role in how food is packaged and sold, can typify consumers (see Box 7.2 and Chapter 8). Their clients can then aim different sorts of products at these different groups, or 'market segments'. If they are able to group people into different market segments – for example, by income, social class, attitudes or psychological needs, then they can target particular products and market them to appeal to people in these groups.

Classifying people

We can all be classified in various ways. Some social psychologists suggest that many consumers are driven by their responses to anxiety – some repress or deny it, others want to reduce it and others are made helpless by it. Food products may be sold more easily if they promise to reduce people's anxiety. Thus, low fat foods may reduce anxiety if people think they will help to reduce the risk of early death.

Marker researcher Christine MacNulty, of Applied Futures Limited, divides consumers up into three types. One is an inner directed group. They derive their sense of direction from within themselves and are concerned not to exploit. They are hard to see, as it is their motivation that distinguishes them from other people, but they have been the long-term trend-setters for the past few decades. The outer directed group compare themselves with others and want to display what they achieve. She calls these people the power-house of society who create growth. The third group is the sustenance-driven group. These people like safety and belonging to a group with well-defined characteristics; they tend to put the break on innovations. All three groups exist in about equal numbers, but the number of the last group has been declining.

It is the inner directed group, acting on their long-term values, that have led the way with concerns about healthy eating, labelling, the environment, etc. According to Applied

Futures research, their key issues for the future are quality, meaning, taste and diversity. They will trade off speed for quality, says Christine MacNulty. Next comes convenience, health and well-being, and environment and safety. These three types are the three main segments of the VALS typology – Values and Life-Styles – originally developed by Arnold Mitchell of Stanford Research Institute International in the 1970s. They were further divided into a number of sub-groups (see Figure 7.1).

Many words ('psychographic' descriptions) have been coined to describe these different groups by different researchers. They often reflect contrasting themes, such as relative affluence or poverty – 'Battlers or Strugglers', 'Affluent', Comfortably Off' or resistance or acceptance of change – 'Traditionalists', 'Conservatives'. Another theme is social dominance or exhibition of materialistic wealth (eg, the 'Look at Me' category), and a fourth is a relative concern about fairness or equity. These market segments always have 'neat' names like those above. They are called 'psychographic' because they are usually derived from interviews with people about aspects of their personal life, their attitudes, interests and opinions. Their basis in social reality is usually clear. For example, people have certain physical resources, control over their lives or social prestige. Much of psychographics is about the attainment of the psychogenic goals mentioned in Chapter 4. What matters to those in consumer marketing is that people's perceptions of themselves and of the social world help to predict their buying behaviours, as Professor Buisson shows:

Psychographic trends worldwide, while more difficult to measure, complement changing demography (as a factor affecting consumers' likely buying habits). In the UK, for example, the Target Group Index has provided continuous information for the analysis of product usage to extract psychographic profiles of segments of the market.

Six attitude groups of women between 15 and 44 years of age, for example, have been established from this base as follows:

Box 7.2
Mind-mapping

Market researchers use many qualitative techniques to understand consumers and provide information to their clients. Some try to tap beliefs and feelings which are not always in the immediate consciousness of consumers, but which may affect product choice, like the 'image' of products. Projective tests may be used to identify underlying perceptions of products or the motivations of consumers. A person may be shown cartoons, drawings and other pictures and be asked to describe the action in them. The ways that these pictures are interpreted gives some indication of the person's 'hidden needs'.

The description and measurement of consumers' perceptions of products (or of people including themselves) is known as 'psychographics' (literally 'mind-mapping'). One useful set of psychographic approaches is called 'repertory grid techniques'. These were first used in marriage counselling; the roles of the different people in a couple's lives were thought of as theatre roles, hence the name 'repertory'.

Things or persons (in the jargon 'elements') are assumed to have different perceived properties ('constructs'). These can be elicited in various ways. For example, three foods (or their names) may be presented to a person. He or she has to say which is the odd one out and why (try it with cabbage, chips and banana – there are no wrong answers!). The subject may give an answer such as 'one is round, the others are not'. He or she will be asked to think of the opposite of round and may say something like 'long' or 'not round'. It does not matter, as long as the subject comes up with a 'bipolar pair' of opposites.

This first step in 'construct elicitation' is then followed by the presentation of another three 'elements' (say, chocolate, cabbage and apple) and the procedure is repeated again, only this time the answer to the previous 'triad' of elements may not be used. And so it goes on. Usually, people find it difficult to go much beyond eight to ten 'triads' This technique really plumbs the depths of one's thinking about any set of objects or people!

Next, subjects rate or rank the objects ('elements') on each of their 'constructs' – the bipolar pairs. So cabbage might rate five on a poor versus good nutrition scale, and chocolate two. The result is an element X construct 'grid' (in English, 'an object by perceived characteristics' grid or matrix).

Such grids are usually analysed by fairly sophisticated mathematical routines to yield 'cognitive maps' which, like geographic maps, plot 'places' in two or more dimensions. The main differences are that the 'places' are objects, such as foods and their perceived characteristics. The closer together points are on these maps, the more similar they are seen to be; the further apart, the more different they are perceived to be. Unlike geographic maps, cognitive maps often have more than two dimensions and can be constructed for individuals or for groups of people.

There is a variety of repertory grid techniques, which is only one set of psychographic methods used by market researchers. Frequently, more standardized techniques are used, such as sets of bipolar rating scales, paired comparison techniques (to identify the similarities and differences between products), or theoretically derived questionnaires and inventories, such as social attitude and values inventories.

The main advantages of psychographics are that the psychological characteristics of consumers and the perceived characteristics of products are made explicit and so can be used by market researchers and their clients. They may be able to link products with desirable social characteristics, identify confusion between products or even recognize that no existing product satisfies certain psychogenic needs. This might suggest the desirability of inventing a product (or product image) to cater to these 'needs'.

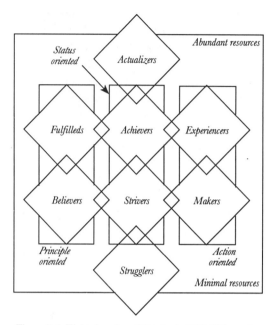

Figure 7.1 Eight American lifestyles in VALS™ typology

Notes:
Briefly, the key characteristics attributed to the various groups of consumers lumped under each heading are:

• Actualisers: high self esteem; can indulge any or all self-orientations; image matters to them as an expression of their taste, independence and character; their consumer choices aim for the finer things in life; highest incomes.
• Fulfilleds: mature, responsible, well-educated professionals; leisure activities centre on their homes but they are well informed about the world and open to new ideas and social change; high incomes.
• Believers: conservative and predictable consumers, favour American products and established brands; lives centred on family, church, community, the nation; modest incomes.
• Achievers: successful, work-oriented, get satisfaction from their jobs and families; politically conservative, respect authority and status quo; favour established products and services that demonstrate their success to their peers.

• Strivers: similar values to achievers but fewer economic, social and psychological resources; style very important to them as they strive to emulate people they admire and wish to be like.
• Experiencers: youngest group, median age of 25; lots of energy, put into physical and social activities; avid consumers, spend a lot on clothing, fast foods and music, particularly on new products and services.
• Makers: practical people who value self-sufficiency; focused on the familiar – family, work and physical recreation – with little interest in the broader world; appreciate practical and functional products.
• Strugglers: too few resources to be included in any consumer self-orientation; oldest segment with median age of 61; tend to be brand loyal consumers; lowest incomes.

Source: Riche, Martha Farnsworth (1989) 'Psychographics for the 1990s' *American Demographics*, July, pp24–26ff

1. *Self-aware – concerned about appearance, fashion and exercise.*
2. *Fashion-directed – concerned about fashion and appearance, not about exercise and sport.*
3. *Green goddesses – concerned about sport and fitness, less about appearance.*
4. *Unconcerned – neutral attitudes to health and appearance.*
5. *Conscience-stricken – no-time for self-realization, busy with family responsibilities.*
6. *Dowdies – indifferent to fashion, slight interest in exercise, dress for comfort.*

Psychographic studies are rendering obsolete simplistic demographic classifications.

How these psychographic groups change will be of major concern. International lifestyle considerations will become a major consideration in international food manufacture where previously, profitability was largely based on cost economies of standardisation.

Such strategies may to a great extent be unrealistic in the future.

Buisson, David H (1992) 'Consumer Food Choices for the 2000s – The Impact of Social and Marketing Trends', paper presented at the CSIRO Food Industry Conference, Adelaide

Consumers' Concerns

Unfortunately for those interested in consumers' behaviour, the concerns of food consumers are difficult to categorize because they relate to the complexity and changes in people's everyday lives. If half-a-dozen shoppers sit down together to discuss any aspect of food, there is every likelihood that within minutes a long list of feelings, views, experiences and comment will be evident. Shoppers, especially those caring for families, have to juggle many issues every day. Along with price, value for money, quality and availability of foods, they may consider many other aspects, including questions such as: Is it good for his heart? Will my child gain or lose weight? (health and nutrition); Can I make a nice meal out of it? (preparation); Will it make me fat? (appearance); Could it harm us?; Is it fresh? (safety); Is it good or bad for the environment? (sustainability); Will we like it? Will I like it? Will they like it? (taste); Do the producers of this coffee receive a fair price? (equity).

Many people in the food industry say that public opinion is fickle – as soon as one issue has been dealt with (eg, dolphins destroyed by tuna fishing), then another takes over the public limelight (eg, Alar in apples), and then another and another. The mass media have a hand in all this, as they tend to pick up and drop issues. But to get the huge response that they do, it seems likely that some themes underlie public concerns. We believe that these themes are concerns about us as individuals and as members of larger communities – what we might call concerns about 'me and mine', 'you and yours', and the world in which we live.

Shoppers' concerns

In a survey in Sydney, Adelaide and Melbourne in 1991, 900 shoppers were asked to rank their degree of concern about 28 issues. In general, the shoppers were concerned about many food and health issues such as additives, harmful bacteria in food, clean handling of food in shops and the honesty of food labels (see Figure 7.2). Narrow nutrition issues such as the links between cancer, heart disease and food did not appear to be very high among the public's priorities (they ranked 8 and 13 respectively), despite the great amount of publicity they receive. Other issues to do with personal safety (eg, additives and harmful bacteria), honesty and concern for others (drift-net fishing and poverty in Australia) were ranked more highly.

The shoppers, who were mainly women, saw the issues mainly in terms of threats to personal well-being (eg, safety of drinking water, harmful bacteria, clean food handling) or in terms of concern for others and the environment (eg, drift-net fishing, starvation overseas, food packaging). They seemed to distinguish between health issues over which consumers' might exercise some direct control (eg, nutrition issues, such as children's TV advertising, heart disease and cancer links, eating too many fatty foods, but also chemical additives and the safety of take-aways) and issues that appear to be beyond an individual shopper's direct control, such as drift-net fishing, imports of foreign food, cruelty to animals and the irradiation of foods. They also distinguished between industry issues (eg, retail ownership, imports, packaging, fruit waxing and the safety of imports) and ethical and personal well-being issues.

These and other findings from the USA, Scandinavia, New Zealand and Germany suggest that the complex manifestations of consumer concern may be based upon more basic beliefs about food and health. Although consumers may express concern about one issue (eg, irradiation) at one time and then appear to drop it in favour of another (eg, the safety of fast foods) at another time, these may be only parts of a more pervasive, deeper concern about personal safety and well-being. These findings, together with the concern expressed in the mass media over personal

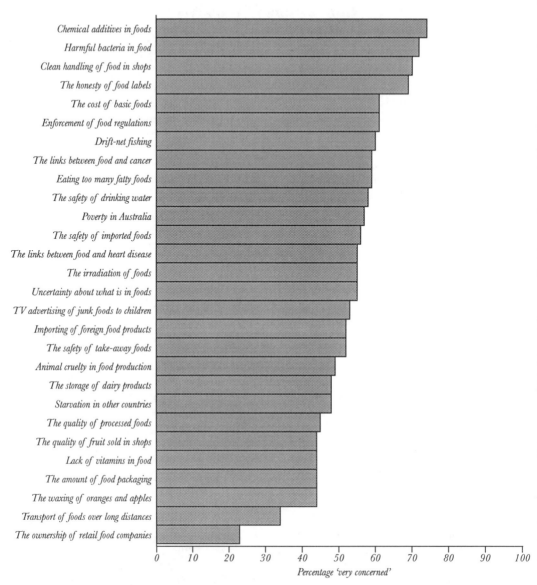

Source: Worsley A and Scott, V (1991) 'Consumers concerns about food and health', Food Policy Research Unit, Division of Human Nutrition, CSIRO, Adelaide, SA, unpublished manuscript.

Figure 7.2 Rank order of 900 Australian shoppers 'Very Concerned' about food and health issues in 1991

health and safety and the huge donations and effort given by millions of citizens to humanitarian and ecological appeals (eg, Band Aid), suggest that consumers do care about 'me, you and the world'. The question which remains is how these concerns may be translated into positive actions throughout the food system.

Experts and Information

People need information to take informed decisions. Indeed, the theory of the competitive market-place is based on people being able to make fully informed decisions. But what is the role of experts in informing us and what sort of status should expert knowledge have? Clearly, a wide range of specialist knowledge and experience is used by the different actors in the system. By definition, almost, experts tend to know a lot about a narrow subject and, as we have seen, consumers' concerns encompass a wide range of divergent subjects.

Experts are often used by those actors with the most economic clout to justify or defend their actions and needs. Experts, especially economic experts, do not always agree and the accepted wisdom changes. In nutrition, for example, the agricultural policies of whole countries have been affected by nutritional doctrines which subsequently have been rejected – for example, the idea of widespread protein malnutrition and the consequent emphasis on animal products in development policies.

Expert views are often expressed as authoritative and used to defeat broader policy debate about the principles behind the issues in question. An informed public can make up its mind about these issues. But the need for an informed public has educational implications and requires a willingness by experts to explain more clearly both the basis of their views and the limits of their competence.

Whose views count?

There is a continual tension between meeting the needs and demands of various actors in the food system, those of scientific experts and those felt by consumers. Pat Shannon, a leading New Zealand sociologist and author of *Social policy: current issues in NZ society*, believes that the felt needs of consumers tend to be fairly vague, but that they convert into demand for goods and services over time. Governments, however, tend to take much more notice of experts' views, and these tend to dominate government policy making. Consumers rarely participate in policy-making, although they are occasionally 'consulted' through surveys and suchlike. The experts tend to be male, white, middle class and to have quite different daily lives and responsibilities from the majority of food shoppers – women with family responsibilities. Recent Australian research suggests that experts employed by bureaucracies also tend to hold fairly conservative views of society and humanity in general.

All institutions, of course, develop bureaucracies, although the term bureaucrat has come to be associated with public service. However, in both public and private sectors, bureaucracies have their own agendas which can either inhibit or promote change. From a bureaucratic point of view, public debate, greater transparency in policy-making and the involvement of the population in determining food policies can be uncomfortable. The bureaucrats have to engage in open debate, look beyond their specific, usually narrow, specializations or interests, and justify their actions.

Unlike many governments, some sectors of the food processing and retailing industries have much closer contact with consumers' feelings through extensive market research, and they take them very seriously. The retail market in many countries is so competitive that if one company does not cater to consumer concerns, another will.

For individual consumers, the difficulty is in balancing the competing demands of a variety of views from experts and lay people. Experts tend to promote only one viewpoint (eg, the scientific or their industry's view), whereas consumers have a broad range of interests: financial, social, ethical, ecological and scientific. The evidence from market research suggests that many people are wary of information from many sources, but especially from food manufacturers (marketers) and even governments. A British study in the mid-1980s found that people were more likely

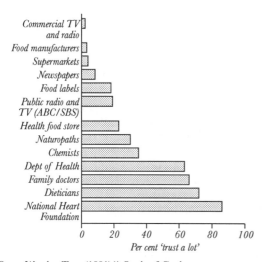

Source: Worsley, Tony (1991) 'A Study of Canberra shoppers' trust in sources of nutritional information' (unpublished)

Figure 7.3 Canberra shoppers' trust of sources of nutrition information

to trust retailers than manufacturers or the Government. Recent Australian studies also produced similar findings although, unlike in Britain, Government agencies appear to have been highly regarded (see Figure 7.3). Obviously, the amount of trust expressed depends on many factors such as the topic involved, the precise nature of the source, the history of consumer relations within a country and the ways in which surveys are conducted and research done.

Sources of information

In many countries today, traditional folk ways are unable to provide people with much information about food. The natural environment is so dangerous that evolution has built in to us a number of processes which protect us from harmful compounds. These range from our innate rejection of bitter tastes (often signs of toxic substances) to learning mechanisms which teach us to avoid foods that have made us ill in the past (so-called taste avoidance learning). Cultures also transmit information about foods and food practices from generation to generation by word of mouth or

through religious and other social rules.

Today, however, these social and cultural transmission mechanisms have broken down. This is partly because the contents of foods have been altered during manufacture and also because the transmission of information has altered radically with the rise of mass literacy, mass education and mass media. It is a combination of these that influences consumers' views and behaviour, tempered by the economic circumstances and the range of issues that each consumer faces. Advertising in the print and electronic media, which we disucss in more detail in Chapter 8, is widely used to influence the consumer directly and indirectly through appeals to those who affect their decision-making, such as children. Packaging and labels also play a considerable role.

A major source of information for food consumers is the label on a food product. About two thirds of shoppers claim to read the information on new or unfamiliar products to check their contents. People are interested in a range of food constituents, but especially in added chemicals (additives, pesticide residues, colouring and flavouring agents, and preservatives), fats and sugars. Most labels are checked inside the shop before purchase. In an Australian survey shoppers indicated their satisfaction with the current Australian food product label, which is similar to those found in many countries. They were most interested in the degree of freshness (use by date), additive content, ingredients list and storage instructions.

Unfortunately, this general interest in labels does not mean that people understand them. In another survey, only 80 per cent of shoppers, after reading the ingredients list taken from a white bread wrapper, could identify it as bread, only 40 per cent could name the second ingredient by weight, almost half did not know what the 'numbers in brackets' (food additives, 'E numbers') were and almost all of the shoppers did not know about one or more items on the list. But if consumers are a little vague about the ingredients, they are often more puzzled by the nutrition information panel which appears on many food product labels (see Box 7.3).

Box 7.3
What's in a label?

Label information is usually designed by scientific experts. A prototype nutrition label produced by the WHO/FAO Codex Alimentarius Commission, which advises on international food standards, has been followed by Food Standards Committees around the world. Energy, carbohydrate, protein and fats are listed according to their amounts per serving and per 100 grams. This format is now quite old. The Coronary Prevention Group in London found that most consumers have little idea of what a 100 gram serving of a food is, or indeed of what is an 'average' serving. Most people cannot understand numeric descriptions of amounts of nutrients and most are unable to compare amounts between two products. In one British survey, almost half of the respondents could not handle the percentage sign in statements such as '30% of this product is fat' but they knew what '30 grams out of every 100 grams is fat' meant. As exercises in communicating nutritional information to shoppers, most labels fail.

The Coronary Prevention Group claims that simple bar graphs of amounts of ingredients (and nutrients) and 'verbal bands' (eg, high, low fat) would enable more people to understand label information. Consumer psychology studies suggest that labels should be in standard formats, have large print, use bright colours and shapes, and order information according to its perceived importance to most consumers. Risk information – for example, about 'negative' nutrients such as fat – needs to be listed in the most prominent position.

In the early 1990s in the USA, the Food and Drugs Administration clamped down on the use of health or nutritional claims such as 'high fibre diets may prevent bowel cancer' and 'low fat'. Now, labels may include only one of a small number of health claims under very strict conditions, with severe penalties for breaching the rules.

Nutrition scientists also have problems with terminology – for example, 'carbohydrate' covers many compounds, such as sugars and starches, with quite different health-related properties. Moreover, important areas of interest to consumers, such as ecological and ethical issues, are not included on current labels. When shoppers are asked about what they would like on food labels, we get a much broader and different range of items, even within the narrow area of nutrition. In a 1991 survey, Australian shoppers were presented with a list of health information items and were asked to indicate how useful each would be on a food label. Information about additives topped the list, followed by health claims, fat and calories or kilojoules, cholesterol and the irradiation status of the product. It is quite a different list from that produced by Codex Alimentarius. Shoppers considered that information about energy content, dietary fibre, complex carbohydrate and price per 100grams was less useful than the experts thought. They believed that a wider range of information would be useful on a label. This included 'orthodox' nutritional concerns like fat and salt content, as well as issues to do with food safety.

This comparison raises a central problem: who should have more influence over food labels, those knowledgeable in science or those who are experienced in everyday life? Nutritionally speaking, information about dietary cholesterol is rarely as important as information about dietary fibre, but many people think that it is, so should it be on the label? Similarly, should information about pesticides and preservatives be on the label because most consumers believe that it is important information? Negotiations are needed between the two groups and nutritionists could clearly do more to explain their viewpoints to the public. Perhaps that is where advertising should focus its attention.

Buying food – how do we choose?

For food manufacturers, what matters is not whether you continue to eat food X but whether you keep buying their versions of food X; for food retailers, what matters is that you buy food X from them and not from their competitors. Many consumers often have far less choice than it may appear over whether, where or from whom they will buy, owing to lack of time, money or access to different shops.

So where and how do shoppers shop? As we saw in Chapter 6, people shop increasingly in supermarkets for food. Market researchers have some idea about the answers to this question but there is relatively little publicly available information because this is very valuable knowledge, saleable to the other actors in the food system who are keen to influence consumer behaviour. A survey of Sydney shoppers in 1991 found that during one week, shoppers may make up to half-a-dozen visits to the shops for food, in addition to eating outside the home. There appeared to be a strong preference for a major shopping excursion either weekly or two-weekly, often towards the end of the week. This main shopping took about two hours, and half of the trips were made with family members, especially children. There were other shopping trips closer to home – to greengrocers, to bakeries, to local small food stores. These trips occurred throughout the day and evening and it took 15–20 minutes by car to get to the main shopping area. The shopping pattern depended very much on the number of people in the household and the occupation of the main shopper. About 80 per cent of main shopping trips were done by women.

Less is known about what actually happens in preparation for shopping and during the process of shopping. About 60 per cent of the Sydneysiders prepared a shopping list for the main shopping, although they did not always stick to it. What they think about in shops is not really known, although there are several theories, discussed further below, about the ways in which they may weigh up product features such as price, quality and health benefits.

One marketer described *going* shopping as the most popular recreational activity in the country, but he also noted that *doing* the shopping was the most unpopular activity, partly because supermarkets can be unattractive places which everyone wants to get out of as fast as possible! Shopping can also be a heavy job, especially if the shopper has small children.

It makes good business sense to know more about the detailed tasks people perform in shops. For example, parents with young children can have a bad time (as can other shoppers) if confectionery (advertised frequently on TV) is put in the reach of children, or if displays are attractive and vulnerable (eg, pyramids of cans). Such displays, however, are marketing techniques to sell the goods; so how far does some consumers' discomfort affect buying behaviour and sales? People with disabilities, with poor eyesight or in a wheelchair, can find shopping extremely difficult and frustrating. The elderly and frail often find reaching high or bending low for products extremely unpleasant. Enlightened retail managers try to put themselves in the shoes of their customers – if shoppers' problems are not addressed by them, other stores may do so and take away their business. One training scheme for young managers in England puts blinkers around their eyes, stuffs plugs into their ears, puts weights on their arms and then asks them to try to go around a supermarket – many frail people have similar disabilities.

Supermarket layout, the width of the aisles, and the readability and accessibility of displays can all be encouraging or off-putting to shoppers. The geographic location of the shops also needs consideration. There is a trend towards very large, out-of-town supermarkets or hypermarkets, which often sell food at competitive prices, built with ample parking space. This can take the supermarkets out of the reach of people without cars and can adversely affect town centres. If public transport is poor or non-existent, then people without cars may miss out on cheaper shopping. Even with public transport, they have to be confident of its safety if they are to use it, and carrying heavy shopping home on the bus is not much fun.

Why do we buy certain things?

The word 'consumer' conjures up a picture of passivity. Many consumers buy without much thought. Indeed, much advertising money is spent to encourage consumers to buy on 'impulse' – that is, without thinking. It might seem that there is not much to buying food – it is only a matter of buying what you need. But there is a big difference between what you *need* to feed yourself and your family and what you *want* as food.

As we saw in Chapter 4, relatively simple foods will satisfy our biological and taste requirements, but our wants are, for practical purposes, infinite. This is because humans have psychological needs – for example, for power, status, love, recognition, sociability, and so on. Foods are marketed by means of implicit promises based on those needs: if you drink beverage X you will have the respect of your peers, if you buy breakfast cereal Y you will be a good mother, and so on. If you do not think much – and advertisements try to entertain so that you will not think too much – then you may be drawn into buying more expensive food products which provide similar ingredients to unpromoted foods.

A range of factors influences buying. The importance of these varies for different types of consumer (see Box 7.2). Some people think a lot about the consequences of eating food – for example, whether it will help to give them a heart attack or give their children allergies, others rarely do so and are more interested in enjoying it. In most cultures, the thinking is often done by women and most men take a passive role in the acquisition and preparation of food. Armed with detailed knowledge about people – and companies pay heavily for this – companies can set out to tempt buyers to try their product.

So, a great deal of effort has gone into finding out how consumers buy food, but surprisingly little of this commercially sensitive information has emerged. Nevertheless, we do know that several factors are important in food purchasing.

Decisions, decisions

Shopping is a form of decision-making; however, the vast majority of food shoppers do not have the time to evaluate thoughtfully all the properties of between 10 and 50 or more food products. Many shoppers use discounting forms of decision-making. First, we do not buy certain things. For example, we know that we need bread but not vegetables, so we bypass the vegetable section and head for the bread shelves, probably selecting other needed items which appear on the way. When we select a product, we may focus on one or more internal or external attributes of the product, discounting others. Some of us are interested in the brand and ignore the price, or we may believe that most brands of a product are similar and choose the cheapest.

Some shoppers are impulse buyers, with little awareness of what and how much they will buy before they enter the supermarket. However, supermarkets are usually laid out to prompt the memory, and the management try to influence us to buy through displays and price changes. Thus the shopper may see 'specials' for baked beans, for example, and buy several cans in order to stock up and take the benefits of the lowered price. Manufacturers also try to prompt us to buy through special offers on packs and advertising.

Supermarket shopping, however, is not the only form of food shopping, even in industrialized societies. Indeed the 'high street of small specialized food shops' (eg, bakeries, meat, fish and wine shops) is making a comeback in North America, parts of Europe and Japan. In the small shop, personal service is more evident and the social nature of purchasing is much more important.

Product attributes

Internal product attributes are the inherent characteristics of the product which become fully apparent when it is consumed. Apples may look good, but you only know what they taste like by eating them! In most supermarkets, we cannot sample these internal attributes until the items have been bought, in contrast to most village markets. Instead, we have to use external attributes for decision-making – for example, the quality of the packaging, the image conveyed by the advertising (a 'superior' product), the ingredients list and health messages (low fat, sugar reduced). These attributes give some guide to the key properties of the product, such as quality, convenience, price or value for money, and, for food, satisfaction.

Most of us can make some distinction between high and low quality products, but the attributes that make up quality appear to vary across groups of consumers and certainly between manufacturers and consumers. To a manufacturer, a high quality cheese may be one made from the finest ingredients with a long shelf life. To the customer, a high quality cheese may have a particular taste and contain less fat than other brands. Each of us uses a variety of attributes to judge the quality of a product. A central attribute is reliability or consistency of quality – brand X baked beans are *always* the same. This is a highly valued feature of modern food production. Once you have experienced a product, you know what you will get next time.

Comparing prices of similar products at first sight looks simple, but it is not. First, the price may be unclear to shoppers. In stores with electronic scanning, prices may not be marked on individual items. The unit price may not be given (eg, $/£ per 100 gram weight) so comparisons may be too difficult to work out for products of differing weights. In fact, many shoppers cannot recall the prices of common items immediately after purchasing them! Secondly, price is a relative concept. Products may be overpriced relative to what the shopper considers them to be worth. Thirdly, price is often perceived to be related to quality. If a product is seen as having a low price, then it may be deemed to be of low quality ('cheap'), although there is little evidence that this is always the case in supermarkets with own-brand products.

Convenience is widely promoted as a product attribute. It conjures up the idea that the preparation of the meal in which the food is served is quick and easy. Many processed foods have freed much time for women – and men. The trade-off for convenience, however, is a loss of personal food preparation skills, lowered status of food preparation and possibly lowered self esteem among women who have traditional sex-role orientations (those who see themselves primarily as wives and mothers). For some, adding an egg to a cake mix is almost as satisfying as making something from scratch.

The expectation of likely satisfaction is very important. Will the food taste nice? Will it appeal to other household members? It is estimated that five out of six new product launches fail because they fail to meet customer taste preferences.

These product attributes are manipulated to varying degrees by supermarket chains and other food outlets. For example, in the poorer parts of cities, stores are likely to be stocked with lower priced items, which are often of lower perceived quality than those in more affluent areas. Food marketers often differentiate market segments according to their degree of affluence. Stores in affluent areas will have more value-added products (high-cost, high-image products) with low-sales volume; those in poor areas make their profits largely through high-volume sales of low-profit products like canned foods. Sometimes, the same product may be priced differently in different branches of the retail chain. Sometimes, whole retail chains may market themselves as low-cost food suppliers, sacrificing the range of goods for lower prices and a limited number of lines.

Shopping is not a simple activity. It is a complex set of behaviours carried out in various ways by people from differing social and motivational backgrounds. Shopping is an active decision-making process which involves many product attributes. It is not surprising, then, that many consumer organizations try to help consumers with their shopping decisions.

Consumer Organizations and Power

There are many consumer organizations which to some degree increase the power of individual consumers. Most rich countries have at least one national, independent consumer organization which deals with a range of consumer issues. They often have extensive research and publication arms so that goods and services can be monitored and their members (and through the mass media, the public) are kept informed (see Box 7.4). Many belong to the International Organization of Consumers Unions (IOCU, see Box 7.5), which has its headquarters in London. This calls for a number of rights for consumers – the right to basic needs, safety, to be informed, to choose, to be heard, to redress, to consumer education and to a healthy environment. It also acknowledges five consumer responsibilities: critical awareness, action, social concern, environmental awareness and solidarity (see Annex 5).

Some organizations deal with food as one aspect of their work, while others are devoted exclusively to particular food issues. Several of them have developed consumers' charters for food issues, such as labelling, food additives and children's TV advertising (see Annex 5). Some groups, such as the Baby Milk Action Network and the Hyperactive Children's Association may be more or less single issue groups which pursue relatively narrow goals, such as the removal of artificial colouring from foods or changes in the marketing of baby milks. Most focus on education and political lobbying both directly and through the media, but some seek to influence company policy through boycotts and other means (see Box 7.8). Still other organizations, such as the Save the Children Fund, the Salvation Army and conservation foundations, are concerned about particular aspects of the food supply as part of their general mission.

Box 7.4
The Consumers' Association

Britain's Consumers' Association (CA), formed in 1957, modelled itself on the Consumers' Union of the USA which was established in the 1930s. In 1992–93, the CA's 562 employees produced a turnover of over £45.5 million and membership stood at 810,000. Members were those people who subscribe to one or more of its various magazines, *Which?* being the flagship, which had a total of almost 1.2 million subscriptions.

The CA began life by testing independently and publishing comparative information about consumer goods; the first results appeared in *Which?* in October 1957. Since the early 1970s, it has also campaigned actively for consumer rights and seeks to represent the consumer to Government and industry. Articles on food and health are among the best read in *Which?* and the CA has been lobbying for meaningful, clear information on food labels. Although known as the Consumers' Association to the public, it actually consists of a limited company of that name which does the campaigning, and The Association for Consumer Research, a charity that does the research.

It co-operates with other national consumer organizations in its testing and campaigning work through the Bureau of European Consumer Unions (BEUC) and the International Organization of Consumers Unions (IOCU). As well as lobbying, it makes extensive use of the media in campaigning and spreading the results of its work to inform the general public

Source: Thirty years of Which? – Consumers' Association 1957–87 and The Association of Consumer Research *Annual Report 1992–93.*

Box 7.5
Consumers Around the World

The International Organization of Consumer Unions (IOCU) had 180 member organizations, large and small, in 72 countries at the end of 1993. It is dedicated to the protection and promotion of consumer rights and interests through research, information and education, and to the representation of consumer concerns on international policy making bodies. Founded in 1961, the IOCU is independent, non-profit making and non-party political.

It has a head office in London (UK), regional offices in Penang (Malaysia), Santiago (Chile) and Harare (Zimbabwe) and also runs programmes for the Developed Economies and for Economies in Transition. The IOCU co-ordinates the skills and experience of consumer groups throughout the world, particularly giving support to those from the developing world and new market economies. It has developed various areas of expertise in supporting the work of its members and building the world consumer movement. These include model legislation for consumer protection, development of consumer magazines, support of product testing, media and communications, consumer education, and national and international advocacy guidelines.

The IOCU runs information networks, seminars, workshops and has a triennial world congress. It initiates research and action on many international issues such as trade, product standards, food safety, health services and sustainable consumption. As an international non-governmental organization (NGO), it has official representative status on many international bodies including the UN Economic and Social Council (ECOSOC), Codex Alimentarius Commission, FAO, WHO and the OECD.

Source: IOCU, London, 1994. IOCU has since been renamed 'Consumers International'

These groups are predominately supported by middle class consumers. Perhaps as a result they can summon a lot of resources and exert some pressure for change in government policy and by food companies. The reform of food safety in Britain in 1990, for example, can be attributed in part to the activities of these groups and widespread media coverage of food issues for several years (see Chapter 8 and Annex 3). It appears to be a very diffuse power, however, and may require much effort to secure even minor changes in the food system.

Changing styles

Tim Lang, Professor of Food Policy at the Thames Valley University and former director of first the London Food Commission (now The Food Commission) in the 1980s and then of Parents for Safe Food, identifies three different phases and styles in consumer movements in the post-war period. First, he says 'the focus was on value-for-money, information and labelling. Is this food or fridge or washing machine better value for money than that? Can the purchaser get legal redress or their money back if things go wrong?' This phase was very successful, but was based on wealth and consumers as smoothers out of the operation of the market mechanism.

The second wave was perhaps launched by Ralph Nader in the USA and his investigative work into unsafe cars that were knowingly produced by a multinational company. Consumer organizations in this phase assumed that large corporations could not be trusted to look after consumers' interests. This would be done if such groups, often called public interest groups or non-governmental organization (NGOs), championed the individual. 'The logic is simple but realistic' says Lang. 'If we are citizens, and they (the companies) want us to consume (for why else do they

Box 7.6
Consumers Boycott a Food Giant to Help Babies

In the year after the Church of England General Synod joined a boycott of Nescafé instant coffee, its UK sales fell by 3 per cent, reported Baby Milk Action in 1993. At the same time Nestlé, producer of Nescafé, increased the money spent on advertising by 27 per cent to £19 million for all its Nescafé brands. It increased advertising of its main Nescafé brand to £14 million in 1993, a rise of 75 per cent. Over the longer term, from mid 1991 to 1994, Nescafé sales increased by 4%.

The Church Synod suspended its boycott of Nestlé in July 1994 but many groups are still boycotting this famous product of one of the world's largest food firms as a protest against the company's marketing policies, particularly in developing countries, for another of its products – baby milks.

Baby Milk Action is spearheading the campaign in Britain to get Nestlé to change its marketing policies which they claim contravene the WHO/UNICEF Code of Marketing of Breastmilk Substitutes agreed in 1981 after a previous, long-running consumer campaign. UNICEF, in its report *State of the World's Children 1991* (Oxford University Press), stated: 'Reversing the decline in breastfeeding in the developing world could save the lives of an estimated 1.5 million infants every year. Bottle-fed babies, who are often given powdered milk over-diluted with unsafe water in unsterile bottles, are several times more likely to die in infancy. Breastmilk is the complete nourishment, safe, hygienic, inexpensive, and helps fight common infections'.

The code recognizes that the usual marketing practices are unsuitable for baby milks and the WHO and UNICEF have called for an end to all promotion of them, not just the ending of free samples in hospitals to mothers. Baby Milk Action is one of 14 national co-ordinating offices for the boycott, and is part of the International Baby Food Action Network, which includes over 140 citizens' groups in 70 countries. They do not accept Nestlé's claims that it observes the code, and will continue the boycott until independent monitoring shows that Nestlé has ended all its bottle-feeding promotion.

In 1991, the WHO and UNICEF launched a Baby Friendly Hospital Initiative aimed at reversing the trend away from breast-feeding. UNICEF called on all companies marketing baby milks to stop giving free or low-cost supplies to hospitals, but the companies refused unless they were required to do so by governments, perhaps fearing their competitors would capture the market if one of them stopped first.

Source: Baby Milk Action, UK and UNICEF

---

try to mould our consumer consiousness?), the task of consumerism is to encourage people to see themselves as active citizens, not just passive purchasers'.

In the 1980s came the marriage of environmentalism and citizenship to produce a third wave, sometimes called the new consumer movement, the kind of organization Lang has been in. It has taken two main forms – the ethical and the ecological. 'In general', he says, 'the ecological consumer asks about a product's quality. What is its impact, not just on the purchaser's pocket, but on health, the environment, distant parts of the world? The ecological consumer recognizes the limits of individual solutions to global problems.' This third wave raises the question of *need*, of why the product or process is needed, who says so and on what grounds. The ethical wing of this third phase asks the consumer to buy appropriately and invest ethically, and has spawned both ethical investment research services and funds. Monetary cost is not the only or main criterion for buying. Indeed, this third wave of groups questions not merely the products, but the production processes and the methods used by the other actors in the system, and raises questions about their sustainability and safety.

As well as a growth in self-selecting membership consumer groups, there has been a growth in attempts at local community building. The Healthy Cities movement – a WHO supported movement aimed at making city environments more health supporting – adopted the methods of community developers in attempts to build strong networks of people in specific localities. Since people who live in a locality often share the same problems (like the absence of a supermarket), they are in a position to act in a concerted manner to overcome problems which are shared by everyone in the locality (eg, attracting a local supermarket or using local food suppliers to form food co-operatives). This can only work to the extent that the food supply is locally based, but it is an alternative which is being adopted in many forms around the world.

More than mouths, less than citizens?

The thrust of today's consumer movements is that people are more than mouths to be sold to but, as yet, less than citizens able to make well-informed decisions about what they eat and how to affect the system that produces it. To transform consumers into active citizens requires addressing several issues, including lack of knowledge, loss of basic skills, and, for all too many people, poverty.

The complexity of the consumer agenda is a complicating factor. People are far more complex than simple consumers of products. We have a changing complexity of wants and needs. In contrast, many professionals in the food system have rather 'one-eyed' viewpoints and find themselves selling products by implying that these needs can be met simply by consuming them. This can make it difficult for consumers to negotiate with experts, especially if the latter are in positions of power. It also makes it relatively easy for lobby groups to push their own narrow points of view. Simple

issues are easier for governments to act on than the complexities and vagueness of everyday life.

Lack of knowledge

Many consumers know little about the food system or the food they eat. This stems from a lack of education about food in the home and school, coupled with the amount and type of information beamed at consumers. This ignorance is sometimes accompanied by unrealistic perspectives in several areas. For example, knowledge of the relative dangers posed by the natural toxins that occur in fresh clean plants, or by artificial compounds such as colouring agents, flavour-enhancers or pesticide residues, or by a high fat diet is so minimal that most people cannot distinguish serious risks from the trivial. They are vulnerable to exaggerations and emotional claims from a variety of media personalities and equally emotional denials and knee-jerk reactions from representatives of the food industry.

A furore may be created by revelations that a baby food contains one billionth of a gram of alar, while a steady regimen of death and debilitation from cancer and heart disease or death and injury on the roads provokes little outcry. The very idea of a toxic chemical often provokes intense emotion. Research into risk assessment suggests that lack of control over a risk factor (such as in the manufacture of canned baby food) is important, but perhaps a greater influence is that few people have a working knowledge of the relative risks of food components.

The uncertainty about relative risks is compounded by the lack of sufficient studies or conclusive evidence of the possible dangers of, say, pesticide residues in food or additives plus the very different 'expert' views expressed. So, for many consumers, the safest option is to avoid them. The suspicion that it is in manufacturers' interest to deny any danger, does not help. The awareness of conflicting interests and the complexity of the subject – how do different people's metabolisms react to foods with many different additives over the long term, for example, when toxicological testing tends to focus only on the effects of one substance – may also affect scientific objectivity. It has also made food stories increasingly attractive to the media.

Perhaps more significant is the consistent and long-standing unwillingness of the other actors to be completely open and above-board with consumers and to present them with all the facts about both products and processes and let consumers make up their own mind. Whether it concerns use of water as an additive, mechanically recovered meat in meat products, milk produced using the synthetic hormone BST to increase yields, irradiated food, or a host of other issues, the pressure from the other actors is almost invariably for minimal, rather than maximal, information for consumers. And even where information is given, as in the case of nutrition labelling, it is done in ways which are often least comprehensible to most people (see also Chapter 8).

Education and the loss of cooking skills

As parents cease to turn basic ingredients into meals and as cooking skills drop out of schools' curricula, a new generation is emerging without the basic knowledge and cooking skills that are needed to be able to use fresh, unprocessed foodstuffs and turn them into meals. Thus, the consumers become dependent on processed and pre-prepared foods and often no longer have the alternative of cooking for themselves. This can be especially important for poor people who lack the money to buy sufficient pre-prepared foods. People who cannot cook are denied one of the essential elements claimed for the consumer culture – choice. They no longer have the choice of deciding whether to make the meals from basic ingredients or from ready-processed, generally more expensive, foods.

The loss of cooking skills and food knowledge is a concern found in varying degrees across Europe. In France, there is an annual Day of Taste, when chefs go into schools to teach young children the joy of being able to distinguish between different tastes. In Britain, over 170 organizations, including MAFF and the Department of Health, launched a '*Get Cooking!*' project in 1993 aimed at increasing children's cooking skills. A survey of 7-15-year-olds commissioned for the project found that barely half could boil an egg, but 75 per cent of the children did want to learn more about cooking (see Figure 7.4).

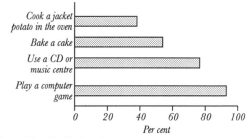

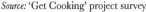

Source: 'Get Cooking' project survey

Figure 7.4 Percentage of British children able to perform certain tasks in 1993

The organizations promoting the scheme endorsed six reasons for learning to cook:

1. Home cooking tastes better.
2. Cooking your own meals is easy.
3. Cooking for yourself saves money.
4. Being able to cook gives you more choice.
5. Cooking is creative and satisfying.
6. Cooking for yourself makes healthier eating easier.

While greater scientific knowledge is potentially useful for consumers, its usefulness remains to be demonstrated. For example, people may reject scientific descriptions for a variety of emotional and ideological reasons (such as its alleged ties with industry). Similarly, even the best information disseminated by vested interests (such as food manufacturers) is likely to be rejected by many. There is a role for independent intermediaries such as the education and health professions, consumer organizations, and the media to find and disseminate scientific knowledge in ways which are acceptable to most people.

The Coronary Prevention Group's work on the use of numeric food labels suggests that educated people have greater self-confidence but do not perform numerical tasks any better than less educated people. This underlines the specificity of much of the knowledge and skills that are required for successful functioning in the food and health area. Literacy and numeracy skills are essential in modern consumer culture, as are sets of specific knowledge about food safety and other matters, as well as cooking skills that give people more choice.

Poverty

Poverty is a major constraint on people's food choices and habits. In many countries, access to a healthy, plentiful food supply is not guaranteed. Even in rich societies like the USA and the EU, for people below the poverty line – for example, some pensioners, the homeless, unemployed, many single parents and their families – buying food can take well over half of their income. The proportions of household budgets spent on food and drink by poor and more affluent sections of the Australian population in 1988–89 varied from 56 per cent for

the bottom 10 per cent to 12 per cent for the top 10 per cent (see Table 7.2). Similar patterns are found in other affluent countries. For people on low incomes, access to healthy food is full of problems.

People on low incomes have to meet continuing fixed demands on their budgets – for example, rent (or mortgage), transport and telephones are essentials in modern life. Often poor people live on the fringes of urban areas. This means that they have to travel long distances to find work and to shop for food. They may have to use infrequent and possibly dangerous public transport to do these tasks. The food budget is relatively flexible so that it can be reduced to pay for a household's fixed costs. The nutritional consequences of poor diets are long-term and take less precedence than maintaining a roof over one's head.

Budgeting for food on low incomes is stressful. To run a tight food budget continually for years on end, with no room for mistakes, is extremely wearing. Evidence from Britain and Australia suggests that poor people (mainly women) are very efficient at choosing nutritious foods for their households and trying simultaneously to make meals pleasant and attractive. However, although it might be

Table 7.2 Household spending on food in Australia, 1988–89 (in Australian dollars)

| | Average weekly household gross income and expenditure | | | | |
	Lowest decile	Fifth decile	Sixth decile	Highest decile	All households
Income	92.29	484.30	595.41	1680.50	636.05
Expenditure					
Food and non-alcoholic					
beverages	46.10	90.07	97.44	159.26	95.83
Alcoholic beverages	5.77	15.36	14.42	34.37	16.90
Total food and beverages	51.87	105.43	111.86	193.63	112.73
Percentage of income					
spent on food and					
beverages	56%	22 %	19%	12%	18%

Note: The decile groups are 10 per cent groupings of the estimated population when households are ranked in ascending order according to each household's total gross weekly income. The lowest represents the bottom 10 per cent, the fifth decile 51-60 per cent, and so on.
Source: Heywood, Peter and Adams, Margaret Lund (1991) 'The Australian food and nutrition system: a basis for policy formulation and analysis', *Australian Journal of Public Health*, vol 15 no 4, pp258–270

nutritionally feasible to exist on a limited range of foods, the social enjoyment provided by food demands that we all 'splash out' occasionally. Parents of young children often feel compelled to feed their children take-away foods and party foods because that is how other (more affluent) children eat – and their children want to be just like other children! Such 'treats' are part of human warmth, but they can ruin tight budgets and lead to feelings of guilt and worthlessness among financially deprived parents (who more often than not are single or divorced mothers – the 'feminization of poverty').

Changing Roles

A variety of influences affect people's food habits. For most people, buying and consuming foods are deeply engrained habits which they follow routinely week after week and month after month. Two questions arise:

1. Why do people try new or different products?
2. Why, having tried a product, do they continue to buy it regularly?

Obviously, there are changes in the regular pattern, but these tend to occur incrementally, unless there is a discontinuity – for example, getting married or leaving home. People stay with a product, according to consumer behaviour experts, so long as the product meets their needs and wants. Dissatisfaction with the product is often raised through advertising, and concern about the needs for products and processes used in the food system is raised by reports about the possible costs and dangers involved in these.

Although consumers are the least organized of the actors involved in the food system, there is the potential for greater pressure from consumers concerned about the direction and nature of the food system. However, how far individuals as consumers can influence substantially the policies and practice of the other major actors in the system, is far from clear. They may stand a greater chance of affecting change in their role as active citizens, through democratic processes and democratically instituted legal and regulatory frameworks which guide the system to meet certain basic objectives. It is to these and other control mechanisms used by the various actors to influence their activities, consumer habits and food policy, and to the issues in food policy for the future, that we turn in the final part of the book.

Part 3

Food Control

The various actors in the food system use different tools to try to control their particular activites so that they bring them as many benefits as possible. Some actors see nature as a gift to be worked with, while others see it as something to be transformed for their benefit. Some see culture as the basis for their approach, while others see culture as something to be moulded to suit their needs. In Part 3, we first look at three major tools – science and technology, information, and management techniques – used by different actors to increase their influence on and control of the outcome of their activities. Secondly, we examine the legal framework established by the society in which these actors operate, whether as individuals or legal entities. This framework helps to determine how the benefits produced by those in that society are distributed. Of course, it relates to the political ideologies of those in power who make government policies.

The laws, rules and enforcement mechanisms that make up the framework within which the food system actors operate may benefit some and harm others. For example, rules in Britain which have closed small abattoirs in the name of hygiene have made it more difficult for small-holders with just a few animals to have their animals killed. The various actors may try to influence the making of rules and policies to suit them or they may even try to avoid complying with regulations.

Those with most political influence have the power to affect policies most and so affect what benefits are generated and how they are distributed. In the 1980s, in the industrialized world, there was a policy switch from public to private interests, which is reflected in the food system. Finally, we draw together our discussion of the food system by considering food policies and the issues we should be addressing as we approach the twenty-first century.

8

Tools for Control: Science, Technology, Information and Management

He that pays the piper calls the tune

N ew scientific knowledge and technological developments may lead to
great changes throughout the food system. As we saw in Chapter 3, new
refrigeration technology in the nineteenth century, for example, permitted new trade
patterns for beef and bananas, while the introduction of domestic refrigerators and freezers
in the twentieth century affected the amount
and types of products that most people buy
and increased the time for which they can be
stored. Without the domestic technological
revolution, there would be little point in much
supermarket shopping.

Today, information technology, computers and satellites are creating more powerful
public and private information networks and
giving them a global reach. These are combined in management strategies and
techniques that change the ways that companies operate, their requirements for labour and
production methods.

Although science and technology perme-
ate the whole system, it is worth considering
them as a factor themselves. Science and technology are not the same thing. Historically,
technological developments have not necessarily depended on a correct scientific
understanding of why a particular technology

works. Trial-and-error invention produced
many new technologies before the science
behind them was understood. Increasingly,
improved scientific understanding underpins
modern technological development, such as
that in biotechnology.

The different actors in the food system
finance a wide range of research and use
increasingly sophisticated technologies. The
ability to affect the direction of research and
use the results is of vital interest. Those who
can introduce innovations first stand to gain
the most benefit. Technology, however, is not
simply a matter of tools, but also concerns the
organization, management and other knowledge that is embedded in those tools. Using all
three can greatly affect an actor's performance
in the food system.

The central questions are: what research
and development (R&D) is done and for
whose benefit? This is related to other questions, such as who funds what science, who
owns the results, where do priorities lie, and
do the purveyors of new technologies have any
responsibility for their effects? The forces that
shape the R&D agenda affect what is discovered, focus research in particular directions,
and provide products that suit particular ends.
Today, most scientists are essentially hired

hands who work on what the grant givers or employers are willing to pay for. If the areas of research are increasingly those that improve profitability, then the important questions are, whose profitability and how is it defined? It is becoming much harder for scientists to pursue their own 'pure' scientific goals.

Science and Technology

For whom?

Until quite recently, our scientific knowledge was treated as a public good. The results of scientific research were shared openly for scrutiny by other scientists and formed part of the common heritage of humankind. Other researchers could check and use the results of work produced by scientists. Such scientific knowledge increasingly underpins the development of new technologies, although using trial-and-error experimentation (empiricism) to develop new techniques and processes without understanding the science of what occurs is still possible.

Many researchers in the food system deal with the agricultural and horticultural sciences. They are employed by government agencies, such as the US Department of Agriculture, Ministries of Agriculture and Departments of Primary Industry, to find more efficient and profitable ways of producing primary products, (eg, disease-resistant wheat and green revolution rice). Other scientists are employed by agricultural supply and food companies in product development, preservation and the development of new techniques.

Much research has been financed from public funds – through publicly owned research institutions attached to government ministries or through universities and other academic institutions. This has been especially true in agriculture where individual farmers could not afford costly research programmes. In Britain, for example, a complex network of research stations and experimental farms was part of the Ministry of Agriculture, Fisheries and Food (MAFF). This Ministry also employs staff in a separate extension service, the Agricultural Development and Advisory Service (ADAS), to spread the research findings and new techniques to farmers. Similar services were developed in the USA and other developed countries, and have been taken up widely in developing countries, too (see Box 8.1 and Table 8.1).

In 1990–91, about 9 per cent, or £246 million, of the total civil R&D budget of £2.79 billion was spent on agricultural and fisheries research in the UK through MAFF and the Agricultural and Food Research Council (AFRC). During the 1980s, however, research priorities were reoriented. Budget and staff cuts were made after a long period from 1947 to 1982–83, when the real resources devoted to agricultural research by the public sector in the UK increased about eightfold. The Government decided to fund less applied, or 'near market', research and increasingly to sell advice to individual farmers who would have to rely more on the results of research carried out by companies (see Box 8.2). By 1988, real research expenditure in the public sector had fallen by about 18 per cent from the 1981–83 peak and the AFRC had lost a quarter of its scientific staff. The number of AFRC institutes had been cut from 29 in the mid-1970s to 7 in 1990 and some, such as the Plant Breeding Institute and the National Seed Development Organization, were privatized. The AFRC was incorporated into a new Biotechnology and Biological Sciences Research Council in 1994.

One of the crucial differences between science and technology was that while technological inventions were often closely guarded until they could be patented, scientific understanding was not. That is now changing.

As large food businesses developed, they began to fund their own research. They can keep their results private until they are sure that they can capture the benefits from them. This is of particular concern in the development of biotechnology, where there are attempts to use patents to control the use of discoveries about the fundamental building-

Box 8.1
Public and Private Agricultural Research

In the more developed countries, an increasing amount of research is financed by the private sector. In the mid-1980s, for example, private expenditure in just four countries was almost US$1.9 billion for agriculture. A further US$1.3 billion was spent on privately-financed food research. Recent figures suggest that about 60 per cent of agricultural research is private in the US and UK.

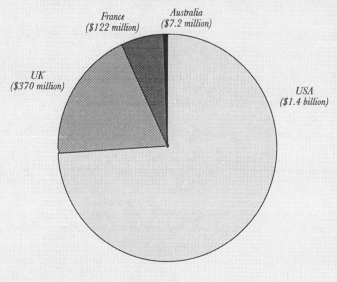

Private agricultural research expenditure

France ($122 million)
Australia ($7.2 million)
UK ($370 million)
USA ($1.4 billion)

Although figures are not easy to come by, privately financed research almost certainly accounts for less than 10 per cent of total research in the less-developed countries. Most of that is concentrated in a few large countries like Brazil, Mexico, Argentina and India. The research is done mainly by local and multinational firms working in the farm input supply industries: seeds, fertilizers and other chemicals and machinery. As the number of purchased inputs in farming in developing countries grows and the value added in agriculture moves off-farm to the marketing and processing sectors, incentives for private research are expected to increase.

Far more is spent in the developed countries on a per head of population basis and per scientist employed than in developing countries, even when the figures are adjusted to reflect local purchasing power of local currencies (see Table 8.1).

Three sets of factors affect the type and amount of private R&D taking place:

1. market factors, such as expected growth in demand for the products, demand for inputs and factor prices (ie, for land, labour and capital) facing farmers and agribusiness;
2. the ability of firms to capture benefits from new technology;
3. the technological opportunities for producing new products.

The balance between public and private research as well as the direction they take are influenced by various policy instruments, such as education and training for people able to do such research, and the patent and plant variety protection laws. Depending upon how they are framed, these laws can encourage private R&D by enabling firms to control the products produced by their research and capture the benefits from it.

Source: ISNAR (1992) *Summary of Agricultural Research Policy: International Quantitative Perspectives*, ISNAR, The Hague; and the book of the same title published by Cambridge University Press, 1991

Table 8.1 Annual agricultural research personnel and publicly financed expenditures, regional totals, 1981–85

	Research personnel (full-time equivalents)	Research expenditure (Millions 1980 PPP US$)[a]
Sub-Saharan Africa (43)	4,941	372.3
China	32,224	933.7
Asia and Pacific, exc China (28)	22,576	1,159.6
Latin America and Caribbean (38)	9,000	708.8
West Asia and North Africa (20)	8,995	455.4
Total, less-developed countries (130)	*77,737*	*3,629.8*
Total, more-developed countries (22)	*56,376*	*4,812.9*

Note: The numbers in brackets are the numbers of countries in the group; the figures may not add up exactly owing to rounding and do not include private sector research.
a PPP is purchasing power parity dollars and represents a synthetic exchange rate that attempts to reflect what a currency will buy locally in terms of US$ equivalents. It is not a simple conversion of local currency to US$ at the prevailing rate of exchange, which would generally give much lower values for less developed country expenditures.
Source: *Summary of Agricultural Research Policy: International Quantitative Perspectives*, ISNAR, The Hague, 1992; and the book of the same title published by Cambridge University Press, 1991

blocks of life and in some cases about living creatures themselves, such as a mouse that was genetically engineered in the USA to develop cancer or genetically engineered soybeans.

New techniques such as ultrafiltration and gene-splitting will produce plants, animals and foods which so far have only existed in the realms of fiction. Such products include novel foods, such as firm, super-large tomatoes or 'crispy' tomatoes with four chambers instead of three; designer foods such as calcium enriched low fat milks (produced by ultrafiltration techniques), substitute foods like non-calorie sweeteners and fat substitutes, and functional foods which supposedly affect some bodily functions. Other research is carried out on the cooking technologies, packaging and preservation methods, as well as into consumer buying habits and perceptions.

Why research?

The reasons for research vary. Much publicly funded research is aimed at increasing basic human knowledge, developing techniques, especially for farmers, and increasing the wel-

fare of society. For companies, science and technology are used to create competitive advantages in a quest to seize market niches.

The Danish food industry has a massive research investment to create a series of competitive advantages over its international competitors. It is combining technological research with consumer science research to make 'value added' products. In the past, pure technological research has led to spectacular market failures, such as the low fat margarine which contained three times more water than fat. It failed because it fell between two market sectors: one group wanted to use the new margarine like butter, but it could not be used to fry foods, so they were disappointed; the other group wanted a really low fat product but found that it was too 'buttery'. The result was that hardly anyone bought it. Similarly, the crispier four-chamber tomato failed in the market-place because shoppers could not distinguish it from the supposedly less crispy three-chamber tomato.

Given the range of problems available for study and unmet human needs, the fruits of much research into product development can tempt us to think that it is largely a waste of

Box 8.2
Wellie Boots and Farming Research Priorities

Professor Gordon Dickson of Newcastle University's Department of Agriculture was dismayed by the narrow approach in funding agriculture research in Britain in the 1990s, with the focus on blue-sky molecular biology. As MAFF withdrew from 'near-market' research and the AFRC stayed up in what he called the 'research stratosphere', the funding for applied research in one of the most applied of subjects was unavailable.

Research interests have changed. A decade ago researchers were only really concerned with physical production. If there was high output, then farmers got a sound economic performance. 'Now production methods relate to quality, animal welfare and environmental impact,' he said.

Although commodity levies have been introduced to fund research, for example on cereals, where they have brought in £4.5 million in the Home Grown Cereals Authority, they make it harder to find funds for whole farm research or research which may not bring benefits to particular industry sectors or suppliers but which may help farmers and the environment.

Dr Peter Rawlinson, also in the same department at Newcastle, wanted to work on systems making greater use of legumes in grassland which could help to cut the use of nitrogen fertilizers and thus the run-off into the rivers but it was 'hard to get funding for that'. Instead, he has been working on the controversial hormone, BST, which boosts milk production nearly 20 per cent. The use of BST will help to increase the production of milk, of which there is already too much, but he worked on BST as that was where the funding was.

'At the Wellington boot end of agriculture,' said cattle specialist Dr Malcolm Willis, 'we don't yet know how to pick a good beef bull.' He laments the loss of animal breeders in research as funding has switched to molecular biology, when in fact farmers are constantly breeding animals and need a more scientific basis on which to breed.

It is a similar story with silage, say Dr Rawlinson, while Professor Dickson says that there is no clear definition of quality for meat, so breeders cannot ask geneticists to produce this or that characteristic.

What farmers breed for and whether they keep the animals in a humane way has become more important in research over the last decade, says Dr Mike Ellis who is trying to marry the profligacy of a Chinese breed of pig with the meat quality of an American breed to produce the mix of flavour, juiciness and tenderness desired by consumers.

Dr Eric Evans, senior lecturer in crop production at Newcastle, believes that technical progress would help farmers to respond to the tremendous variability in their fields and apply fertilizer and pesticides in variable amounts over the land according to its yield potential. Computer-controlled sprayers, for example, would turn on and off as they passed over the field, introducing optimum levels of inputs. 'We need research to show how to increase the uptake of N [nitrogen] from 60 to 100 per cent efficiency', said Professor Dickson.

Other researchers, such as Dr Philip Cain, wanted to work on low input farming, but as it would not benefit the industries supplying agriculture, he was finding it hard to finance. He contrasted the UK position with that in the USA where the US Department of Agriculture had what it called a small programme with 90 projects.

Source: Revised version of article from *The Times Higher Education Supplement*, 24 April 1991

effort. Does it really matter that one tomato is slightly crispier than another in a world with the range of problems ours has?

The answer probably depends on your perspective. For the Dane, in a country with few natural resources other than farming, it is important to maintain a share of world trade. The Danish share of the world food export market has fallen in recent years, hence the major research initiative. This matters for the Danes, both because of the income generated by exports and the jobs it provides.

From another perspective, it may seem a waste of talent and scientific resources to spend so much on more products for the saturated markets of industrialized nations, when people can only eat so much. This sense of foolishness is heightened when the ecological problems associated with industrialization (including industrialized farming) are counted. Optimists among ecological scientists give the planet about another 200 years before the human population crashes. Would the scientific resources spent on 'added value' products be better expended developing less ecologically destructive ways of producing good food for the world's human population? That is the kind of question that needs to be asked in developing food policies and research goals.

High-tech, low labour production

The aims of those who are developing technology – whether empirical or science based – can have a great impact on the type of technology developed, work organization and the effect of technology on the different factors of production and society (see Box 8.3). Much technological development aims to increase the productivity of labour and bring greater returns for capital. This usually means that it helps the better resourced. Much less effort has been devoted to R&D into, for example, making low-input farming techniques more effective or sustainable environmentally.

Firms doing R&D want to make a profit out of it – to cover their costs in developing techniques and products and to gain some competitive edge for their business. Part of the drive to understand processes and improve technologies, stems from a desire to reduce risk. Farmers may benefit from many agricultural innovations and manufacturers may gain from substitute product and process technology, which can remove the need to rely on cyclical, nature-dependent supplies. This is one of the attractions of the new biotechnologies.

By finding new processes, firms can cut their costs and increase profitability compared with competing firms, until they, too, adopt similar practices; by producing new products, they can increase sales until others compete with them. The increase in privately financed R&D affects the openness with which results are discussed and the areas in which work is done. If it cannot patent the products, or gain a manufacturing advantage from a particular line of research, then a firm is unlikely to pursue it.

This has had particularly important consequences for anything the market does not value, which means anything for which there is no effective demand. It has hit poor developing countries where very little research has been done on subsistence crops of interest to small farmers, such as sorghum and millet, for example, compared with crops of commercial importance, such as tobacco and wheat. These countries have neither the trained manpower nor the wealth to develop substantial research operations and have been served to some extent by the charitably and publicly funded international research centres (see Annexe 1).

Overall, there is a shift in the balance of R&D taking place, with less at the farm level and more for the food industry and input suppliers. Moreover, there is a growing tendency for the enclosure of previously public knowledge as the need to patent or otherwise capture the benefit from research becomes essential for the private firms doing or financing research.

The dynamics of science and technology in the food system have come under scrutiny recently through a development with major implications – biotechnology. Like the steam engine and electricity, as Paul Kennedy says in his book *Preparing for the Twenty-First Century*:

Box 8.3
Hybrid Corn, a Golden Goose!

What are said to be fundamental discoveries about the nature of life often mask simple commercial relations that provide a powerful impetus for the direction and subject of research. The best documented example we have of purely commercial interest driving what is said to be a fundamental discovery about nature is in agriculture. It is commonly said that the invention of hybrid corn has resulted in immense increases in the productivity of agriculture and the consequent feeding of hundreds of millions of people at low cost and great efficiency. Whereas in the 1920s an average acre sown with maize in the corn-belt of North America might have yielded 35 bushels per acre, today it can yield 125 bushels. This is widely regarded as one of the greatest triumphs of basic genetics as applied to human welfare. But the truth is more interesting....

The invention of hybrid corn was, in fact, a deliberate use of the principles of genetics to create a copy protected product.

Hybrid seeds, such as maize (corn), have two advantages. First, they yield more than either of the two true-breeding varieties that were crossed to produce the seed and more than the open pollinated maize from which the true breeding lines were produced. It takes plant breeders years to self-pollinate many lines of maize until they have many genetically uniform lines. These are then crossed. The many crosses produced are monitored to see which give improved yield and other characteristics such as disease resistance that will make them attractive to farmers.

The second advantage is that farmers have to go back to the suppliers each year for more hybrid seeds; this is of great commercial importance to the seed company. Unlike ordinary crops which reproduce themselves, so that if farmers find a good variety they can keep seed from one harvest to plant the next season, the hybrid seeds will not breed true. If seeds from the hybrid maize crop were planted, for example, farmers would lose at least 30 bushels per acre in the next crop. Thus, the hybrid-seed producer has found a way of copy protection for his product, which means, says R C Lewontin, Professor of Zoology at Harvard University, that the company can charge a price for the hybrid seed that is equivalent to the amount the farmer would have lost – the market value of 30 bushels per acre, say – had he not bought more hybrid seed from the seed company.

This second advantage has ensured that wherever it is possible to breed plant or animal hybrids, these tend to be bred commercially at the expense of older varieties that reproduce themselves on the farm. The Pioneer Hybrid Seed Company, for example, was set up in 1926 to develop hybrid maize and its founder, Henry Wallace, whose father had been US Secretary of Agriculture in 1920, was appointed Secretary of Agriculture in 1932, after which pressure for the introduction of hybrid corn became 'irresistible', according to Professor Lewontin. But if hybrids are the best way to get higher yields, does it matter? Yes, he argues, because they are not.

The nature of the genes responsible for influencing corn yield is such that the alternative method of simple direct selection of high-yielding plants in each generation and the propagation of seed from those selected plants would work. By the method of selection, plant breeders could, in fact, produce varieties of corn that yield quite as much as modern hybrids. The problem is that no commercial plant breeder will undertake such investigation and development because there is no money in it.

One of the most interesting features of this story is the role of agricultural experiment stations like the state agricultural experiment stations in the United States or the Canadian Department of Agriculture. These institutions might be expected to develop alternative methods since they are not connected with profit and are working at the public expense. Yet the US Department of Agriculture and the Canadian Department of Agriculture are among the strongest proponents of the hybrid method. A purely commercial biotechnology seems likely to introduce a new historical era and greatly change the way people live. It offers new products and improved ways of creating new ones. It opens new markets, reduces the costs of many manufactures and services, and might alter the pattern of international trade. It could change the way national economies are structured, how investment capital is allocated, and the spectrum of scientific knowledge. It will create many new jobs and eliminate many traditional ones.

Biotechnology

In a system so dependent on biological processes, modern biotechnologies have truly revolutionary implications. They confer a generic *technical capacity to undertake selective genetic intervention in existing forms of life and to create novel life forms. The genetic code can now be manipulated and nature refashioned according to the logic of the market place. As a result, biotechnologies have suddenly new alternatives and paths of development for* all *major actors in the food system: farmers and input suppliers, primary processors, final food manufacturers and consumers. It is this crosscutting, polyvalent capacity of biotechnologies which makes them such a potent force in restructuring the food system...*

The uses to which these innovations are put and the direction taken by the food system will depend on the balance of forces controlling policy and research agendas.

Goodman, David and Redclift, Michael (1991)
Refashioning Nature – Food, Ecology and Culture
Routledge, London, pp167–88

Definitions vary, but essentially biotechnology harnesses biological processes and techniques. We have been using biotechnology for centuries in brewing, wine- and bread-making, but new techniques now allow far greater manipulation of plants and animals than ever before, in particular, genetic engineering.

Biotechnology is one of the fastest growing areas of technology. It allows scientists to manipulate animals and plants by taking genes for a particular characteristic, for example drought tolerance or resistance to a certain pest or pesticide, and transferring them to another plant, not necessarily in the same species (see Box 8.4).

World-wide, hundreds of millions of dollars are being invested annually by industries and research councils in biotechnology research to modify plants and animals. Research is underway to make herbicide-, insect- and virus-resistant strains of staple crops like maize and wheat, to affect the keeping qualities of fruit and vegetables, and to increase the productivity of animals, as with bovine somatotropin (BST) for increasing milk yield in cows. Already genetically-engineered organisms, such as yeasts, are being used in food ingredients.

There are hundreds, possibly thousands, of research projects underway which will modify, perhaps profoundly, the plants and animals around us. Most of the world's major chemical companies are involved. In 1988, the agricultural biotechnology R&D expenses of the top 25 companies were estimated at US$330 million. Monsanto topped the list with US$55 million and Shell followed with US$25 million for plant biotechnology alone. Many new companies have been set up in the 1980s and 1990s

Box 8.4
New Genes for Cotton

In Australia, genetic engineers have introduced herbicide resistance and a built-in pesticide into one widely-grown cotton variety, siokra, using techniques that could be used on other crops.

Researchers at the Plant Industries Division of the Commonwealth Scientific and Industrial Research Organization (CSIRO) have developed a cotton with tolerance to the herbicide 2-4D, which is sprayed on wheat. This means that spray drift will not harm the cotton, and 2-4D could be used to deal with the weeds around cotton which otherwise is weeded expensively.

2-4D is biodegradable; it can be broken down in the soil in a few weeks by many naturally occuring bacteria and fungi. The researchers found one that breaks it down into less toxic compounds. They took the gene that deactivates the herbicide and modified it so that it would work in plant cells.

They tried it first on quick-growing tobacco plants to check whether the resistance could be transferred, and found that it could. When the tobacco with the new gene was produced and sprayed, it was 30 times more tolerant of 2-4D than normal tobacco. Next, they transferred the gene to cotton.

Another bacteria was used to place the desired gene in the plant. This one causes crown gall disease by transferring part of its genetic material (DNA) into the genetic material of the plant's cells. The team replaced this disease-causing part of the bacteria's DNA with the gene that breaks down 2-4D, so that when the modified disease-causing bacteria is mixed with chopped up seedling stems or leaves, it transfers the 2-4D breakdown gene to the plant cells.

These modified cotton cells are germinated in nutrient media and then planted in soil and grown to maturity; this takes 12–18 months. The next step is field testing.

The team has also introduced a gene from another bacterium (BT) to produce a pesticide inside the cotton plant. This will reduce the number of pesticide sprayings during the growing season; currently, cotton is sprayed 12–14 times and the pests are becoming resistant.

BT produces a toxin which poisons the cotton budworm caterpillar. The toxin is harmless to humans. Plants with the BT-toxin gene produce a poison which the insect eats when it nibbles the plant and this kills it.

Monsanto have already synthesized a gene to make BT-toxin in the plant and CSIRO have a licence to use this artificial gene. Monsanto will get a percentage of the seed price assuming field tests of the cotton prove its worth.

Source: Revised from *Gemini News Service* feature, London, 23 March 1991

to work in biotechnology, and older companies have expanded into the field or bought into it. In 1993, ICI, a major UK-based transnational, split itself in two, retaining its name for the chemicals business and shifting its bioscience activities, which cover agrochemicals, pharmaceuticals, seeds and specialities, to a new company, Zeneca. This new company employs 33,000 people world-wide and had sales of nearly £4 billion in 1992. For the companies, biotechnologies provide a great opportunity to create new products which will develop their markets and which they say will help to feed a growing world.

Critics of present trends in biotechnology research, however, fear that farmers could suffer and the vulnerability of food supplies increase (see Box 8.5). Firms involved in biotechnology promote a new form of enclosure, no longer of common land but of the common heritage of humankind in the genes of plants and animals which are needed in much genetic engineering work. They seek patents and plant breeders' rights over newly

Box 8.5
NGOs Question the Direction of Biotechnology Research

Farmers could suffer from present trends in biotechnology which might also increase the vulnerability of food supplies, according to about 50 non-governmental organizations (NGOs) from 15 European countries. They met to discuss genetic resources and biotechnology in Barcelona in June 1991. The 70 participants represented farmers, consumers, researchers, environmental groups and Third World development project agencies.

The danger comes from the way that the technology is being controlled and the direction research is taking, believes Henk Hobbelink of Genetic Resources Action International (GRAIN), which organized the meeting. Most biotechnology research, he believes, will benefit large multinational companies in the industrialized countries. These have already bought up many small seed companies following the introduction of Plant Breeders' Rights and are now pressing for patent legislation to be extended to living organisms so that they can control the technology.

Farmers could find themselves locked into using pesticide tolerant plants which tie them to particular chemical treatments, and having to pay to resow seed they have grown themselves if the plant contains a gene put in by a company, says Hobbelink. Some developing countries could find their products no longer needed as companies develop substitutes that can be factory produced – for example, substitutes for vanilla and cocoa butter.

Need should be a crucial criterion in determining research strategy, believe the NGOs, with need broadly defined in relation to the environment and society as a whole. The push for patent protection – originally designed for industrial products – for living organisms is privatizing further a common global asset that was previously freely shared, they argue. The plants used as sources of genes for new varieties have often been developed by small farmers in developing countries, who get no reward, or they grow wild in developing countries and are part of the natural heritage.

The NGOs want to see biotechnology research geared to reducing input requirements and enhancing genetic diversity. Too much public research money is benefiting private companies through various programmes like the EU's Eureka project, which aims to increase the EU's technological R&D base, while other types of research that would help more of today's poor farmers are neglected, says Henk Hobbelink. 'Simple mass selection to improve local varieties is one example of under-supported research,' he says. 'Work on enhancing multiple cropping and rotation techniques, rationalization of the use of wild plants in local diets and the upgrading of traditional crop protection practices, are just a few others.'

The NGOs saw the development of new biotechnology-based clones and varieties as posing a further threat to the world's genetic resources. They could increase the vulnerability of food supplies – for example, if resistance to pests or disease based on a single gene is introduced to a wide range of crops. The NGOs want farmers and growers more involved in protecting the world's gene bank.

In developing countries, thousands of local varieties are being swept away by the flood of uniform plant varieties and animal breeds emerging from modern, industrial agriculture. Relatively high-tech national gene banks have been set up – basically large cold stores for seeds – with help from the Rome-based International Plant Genetic Resources Institute, formerly the International Board for Plant Genetic Resources. Although gene banks are necessary as a last resort, they are not enough, according to Rudiger Stegemann of African Seeds of Survival. Such gene banks put all their eggs in one basket and the careful regrowing of seed to maintain its viability can be difficult even in rich countries. Only 28 per cent of the seeds tested in the US's central gene bank at Fort Collins were healthy; the rest of the collection had too few seeds to be tested (45 per cent), was dead or dying (8 per cent) or had been not tested for five years (19 per cent), says Henk Hobbelink.

Stegmann wants to see seed from the many varieties preserved, produced and improved in farmers' fields as well as kept in gene banks. This keeps the variety alive and allows it to be developed further by farmers.

Source: Revised version of article from the *Financial Times*, 13 August 1991

developed organisms to be able to exploit whatever products they might develop.

Biotechnology and its implications stretch far beyond the farm and, for some crops and products such as vanilla, could replace them entirely. Already, maize starch can be turned into high-fructose corn syrup which has replaced sugar in nearly 50 per cent of the US market by the mid-1980s. A major concern over the future direction of R&D is that the economics which underpins these developments is too narrowly based and ignores external costs which may be borne outside the firm or organization involved or by future generations.

An Information Maze

The main difference between [private and public media] is that private media are used mainly to transmit data and documents, often under conditions of extreme security, while public (mass) media are used mainly to broadcast 'entertainment', always under conditions of the greatest visibility to the consuming public. Both forms have been revolutionized in the last few decades by computerization which has led to the development and dissemination of new information and communication technologies, such as international networking, cable and satellite TV, video, and the total packaging of cultural products. The same technical means thus reinforce the internal unity of the global capitalist system while increasing its power to split producers from consumers, as well as to isolate consumers from one another.

Sklair Leslie, (1991) 'TNCs and Development: a Global Perspective', paper presented at the Development Studies Association conference, Swansea

The ability to monitor, use and control information is a key to success for the different actors in today's food system. Scientific and technological developments have also transformed information systems – both public and private – offering a degree of complexity, immediacy and control undreamt of only a few decades ago. Today, retailers depend heavily on turning the raw data from sales scanning at laser tills into automated stock-control procedures, and grain traders depend on information about growing conditions and price levels around the world. This information is often very specialized. It is produced and must be expertly processed to turn it into useful knowledge.

Consumers and farmers tend to rely on publicly available information while other, larger actors use more private sources. Some they produce and control themselves, others they buy in from outside. This information may be in the form of R&D results, market research or expert advice. Individual shoppers find themselves facing many thousands of items in a supermarket, many with implicit or explicit nutritional claims. They face tinned, frozen, chilled and fresh, ready-prepared products, requiring different keeping and cooking methods, sometimes involving new and poorly understood cooking technologies such as microwaves. They also have a wide range of concerns to consider – from individual concerns over the affordability of what they wish to buy to the desirability of a product from a health, ecological, acceptability or animal welfare point of view.

An individual's ability to select and use information to guide his or her decision making about food depends partly on education, partly on the availability, truthfulness and intelligibility of information about food, and partly on specific needs. The capacity of the main actors in the food system to gather, interpret and use information – now linked to powerful computer-based gathering and analysis mechanisms – is much greater than an individual's. This is again something of a David and Goliath situation, whereby the Goliaths of the system can study individuals and what makes them tick and use this to influence them, while the individuals find themselves in an increasingly complex information maze with insufficient tools to make a sound judgement.

We have divided those dealing with information about food into three groups. These groups play a considerable role in exercising food control, but the boundaries are not hard and fast :

1. the peddlers – in education, the print and mass media;
2. the pundits – in market research, specialists used by different actors as experts;
3. the pushers – in marketing, advertising and public relations, and lobbyists.

The peddlers

The peddlers trade knowledge – to inform, educate or entertain, or sometimes all three.

Education

Education is both formal – ie, schooling – and informal ie, socialization in the home and community. Each of us first develops skills, attitudes and knowledge about food in our homes as children. In Chapter 7, we discussed the gradual loss of the skills needed to grow or cook basic foods to provide a balanced diet in modern industrialized societies. If basic knowledge about food and the skills needed to deal with it are not learned at home, they may be passed on in schools – provided there is space for them in the curriculum and the facilities to teach them.

Food education has usually been the job of home economics teachers, but practical cooking skills are falling off the curriculum. In the UK, for example, food is being dealt with, if at all, in the technology and design part of the national curriculum. Even when it was in the curriculum, it suffered from being seen as a low status and women's subject.

Education should provide children not only with the skills to cook, but also with the skills that are needed to select foods in the increasingly complex food system. Adults have

to be able to judge the various claims made by different actors about food and to assess the issues and risks involved in future developments. Such education would help, too, in assessing the 'expert' views paraded by different actors and too often expressed as completely authoritative. These may be used to deflect broader policy debate about the principles behind the issues in question – for example, what risk is it acceptable to face? An informed public can make up its mind about these issues, but an informed public requires sound education and a continuing flow of accurate, sufficient and wide-ranging information about food. Some consumer groups in the UK are worried about the independence of the materials available in school, where an increasing number are produced by particular companies and lobbies for school use. Others are concerned about the independence of research workers in higher education.

Those working within the food system also need more specialized education and training. This, too, is changing (see Box 8.6). But people are not only informed and educated through schools and colleges. The media also play a major role. In the 1980s in Britain, the media have found a public interested and concerned about food issues.

The media

The various communications media play a major role in information dissemination in modern consumer societies. They are part of informal education and both reflect and influence public attitudes. TV, in particular, plays a major role in informing and influencing children today. As the number of media channels increases – TV, radio, newspapers and magazines – it may become harder to reach a whole population with any particular message, as the population is divided into segments which only consume a few of these media channels.

The media need good stories to keep up their sales and audience ratings. Recipe publication has long been a regular interest to many newspaper and magazine readers, but more

Box 8.6
New Menus in Food Higher Education

A new menu for food-related higher education is developing in Britain in the 1990s aimed to attract more students. Course structures are becoming more à la carte or 'modular'. Many are putting Europe on the menu by offering students a chance to spend part of their course abroad.

More graduates for an increasingly complex food industry are essential, according to Chris Haskins, Managing Director of Hull-based Northern Foods. Industrialists and academics are worried about both the quantity and quality of those attracted into food. 'Less than 10 per cent of any graduates are even interested in a career in food,' says Nick Craig from the Institute of Grocery Distribution, which, along with Northern Foods and Marks & Spencer, is talking to careers officers about opportunities in the food industry.

'Each year the number of graduates produced is way below the needs of industry,' says Dr Peter Warren, head of the school of Food and Fisheries at Humberside College of Higher Education (now the University of Humberside). To attract more students, the college is offering broader introductions to the subject and more choice, including a link between agriculture and food technology, with final-year food resources students spending 40 per cent of their time at an agricultural college. At the college, to match changing times, they now teach conservation farming courses and will farm 25 acres organically.

Public interest in food in the 1980s may have made it easier to fill places in the School of Agricultural and Food Sciences at Nottingham University, believes Professor Peter Buttery. Despite public interest, some of the students training to be dieticians, home economists and catering managers at Leeds Polytechnic [now Leeds Metropolitan University] found the basic ignorance of their fellow students about food and health disturbing. All their courses include current recommendations for producing a better balanced diet.

Industry funding for research predominates in the food sector, which worries Richard Lacey, Professor of Clinical Microbiology at Leeds University. He sees the techniques used in food production, storage and distribution changing very fast, with too little research into them. 'Apart from this department, which has no significant funding from the food industry, there is no other independent department of food micro-biology,' he says. Professor Lacey believes that central funding for research on the basic requirements of living, such as food, is essential to avoid dependence on vested interests which are now expected to pay for research but may be adversely affected by the results.

It is a view supported by former consumer lobbyist Dr Tim Lang, Professor of Food Policy at Thames Valley University. He is worried by the small number of independently minded academics, since so many work under contract or in liaison with food companies. That would not be a problem, he believes, if there were an equal number of independent consumer-oriented groups working on other issues. Professor Lacey wants something similar to the US Food and Drug Administration established in Britain which can promote education and fund research.

One key problem with food is the need to pull together disparate interests who need to talk a common language, according to Dr Patrick Hayes, a food microbiologist at Leeds University. Often, he says, food processing industrial designers have no idea of food hygiene. His point was illustrated at a conference on poultry organized by the Food Policy Research Unit (FPRU) at Bradford University. Conference participants heard how an engineering change to increase the rate at which chickens passed along the killing and defeathering lines caused increased contamination with salmonella.

Dr Verner Wheelock, past director of FPRU, is interested in managing the whole food chain. He emphasizes the need to be interdisciplinary and to take an international perspective. For him, the central food issue in the UK is the continuing high incidence of heart disease, which is the major risk associated with poor nutrition.

Source: Revised version of an article from *The Times Higer Education Supplement*, 23 February 1990

recently a much wider range of food stories has been seized on. These focus on the activities of consumer and health groups, academics, governments and industries. Journalists report their investigations of practices like intensive egg production or the health consequences of high levels of saturated fat consumption. The media's interest has often been heightened by the seeming reluctance of government agencies to disseminate information of public interest (see Annex 3).

A high-minded right to know is mixed with a commercial need to sell by interesting readers or viewers. Nowadays, newspapers and television companies report on famines, animal production, food additives, diet and its health consequences, and the variety of enjoyable foods and beverages. This coverage appears to have had a profound effect on food buyers' opinions and choices (see Box 8.7).

The specialized and mass media perform somewhat different roles.

Specialised media come in three main forms, may be sources for stories in the mass media, and tend to serve relatively narrow market niches:

➤ Scientific and technical journals report scientific investigations and the proceedings of scientific and professional associations. They help to develop the sciences, which underpin much of the global food system, but rarely provide information about the factors which influence the direction of research. They tend to be subscription based and carry little or no advertising.

➤ Trade newspapers and magazines inform members of different sectors about events in their industry – for example, new product launches, staff movements, conferences and exhibitions. They are very informative about their specific areas of interest and even the advertising can give a good picture of the latest developments in an area. They contain both editorial and advertising and may be distributed free to qualifying readers, sell on subscription or at newstands.

➤ Consumer magazines cater to an increasingly subdivided segmented consumer market, which targets audiences very carefully, such as 15–17-year-old women, young mothers with children, career women, and so on. Many women's interest, food and general consumer magazines write about some aspects of food and often contain recipes, which can help to influence the types of foods cooked and the ways of cooking them. Most depend on advertising revenues to survive.

The mass media are dominant in today's society and their entertainment programmes help to promote a global consumer culture. They exist in many forms and are difficult to censor, at least on a global scale. The use of computers and satellites could make mass media more powerful and freer than they are today. However, the major global media empires are controlled by relatively few people and companies, with networks of TV and print interests spanning the globe.

Although print mass media such as newspapers and magazines tend to be nationally based, this is less so with television, with the USA providing a major share of programming for countries all over the world. Both print and electronic media provide advertising space for the food industry, from the basic newspaper advertisements giving price specials at local supermarkets to long-running sophisticated brand advertising. Precisely how the media affect people's views has been a matter of considerable debate (see Box 8.8).

As news-gatherers and informers of the public, sections of the media investigate the food system on behalf of consumers – ranging from chicken farms through famines to biological laboratories. Such reporting feeds off public opinion and feeds it. The news carried is important for the actors in the food system. One headline, for example, linking cancer to product X can spell its end on the supermarket shelves and affect sales of products associated with it or similar to it. Many scientists and many trade associations wish that the media could be more 'responsible' in their reporting.

Box 8.7
The Media and Consumers' Opinions

In 1986, William Morrison's supermarkets, a Yorkshire-based (UK) retail chain, decided to run an education campaign about 'healthy eating'. Leaflets, shelf talkers and other media were used to convey health messages about foods, mainly with an emphasis on eating less fat, salt and sugar, and more dietary fibre.

Group discussions were held with samples of customers before the campaign started and three months later. During this period, the national media carried many stories about nutrition, food and health. The researchers were struck by the finding that the customers' comments directly reflected what was currently being publicized in the mass media. For example, before the campaign started, they referred to fruit and vegetables as 'healthy' foods, but three months later, in line with media coverage, they were questioning the 'healthiness' of these foods. Similarly, additives, pesticides, animal production methods and irradiation were discussed in much more detail during the second series of group discussions; people referred to the television programmes they had seen recently which raised these topics.

Although the shoppers had welcomed the supermarket's initiative, the investigators concluded that 'Perhaps the most obvious finding demonstrated the power of the media in influencing areas of dietary concern'.

Source: Slattery, J and Wright G (1986) 'The consumer's reaction to a healthy eating initiative', Food Policy Research Unit, University of Bradford, UK

However, the media thrive on controversy and bad news. This helps to keep much of the food system on its toes, however unpleasant that posture may be. Like retailers, the media sell what their readers and viewers want.

The media have the happy knack of 'sniffing out' maladministration and attempts to keep information from the public (see Annex 3 for examples). Publications relying on advertising revenue may also be affected in what they cover by the effects it could have on their advertisers. Concern to maintain their independence is one reason why the Consumers' Association, publishers of *Which?*, accept no advertising and do not allow their findings to be used in advertisments.

The pundits

The pundits are those people whose work gives them special knowledge that other actors in the system draw on and which is used to inform public debate and private practice in the food system. Some are scientists whose specialist expertise is used by governments on advisory committees and by the media in reporting and explaining food issues. The pundits, or experts, put their expertise before others in recommending what should be done. One concern, as we discussed in Chapter 7, is over the independence of these people. At the very least, their potential biases and conflicts of interest should be made clear. One special group of pundits are the market researchers, who are usually employed by the major actors in the system to give them guidance about both the type and acceptability of products they wish to develop and market. Very often their views are expensive and are not publicly available. They may help the key actors from making costly mistakes in their product development.

Market researchers

Market research is a highly developed industry that employs many thousands of people and uses numerous methods. It has developed sophisticated theoretical models of human

Box 8.8
How Does the Mass Media Work?

Three major models of mass communication have been used to try to answer the above question:

The direct effects model
This widespread view, much loved by politicians and some health promoters, is sometimes called the 'hypodermic syringe' model. It assumes that messages can be injected into society and most people then behave accordingly. It probably came about because of the apparent success of the early advertisements (eg, Pears soap in the UK) and of British propaganda during the First World War. This propaganda exhorted men to join the military and women to take up wartime work. It is only with hindsight that we can see that people had little choice but to do as they were told by an all-powerful government.

The step models: opinion leaders and followers
In the 1930s, classic work by American agricultural extension workers confirmed the inadequacy of the direct effects model. Many people simply did not pay attention to mass media. Other people, however, appeared to be very well informed. Among farmers, they tended to be the first to adopt new innovations such as the use of tractors and fertilizers, and became known as opinion leaders because it seemed that other people followed their lead.

The mass media were thought to work through a two-stage process. First, they influenced the opinion leaders and these leaders then influenced other people, 'the followers'. This model seemed to explain many of the effects of the mass media and advertisers tried to portray opinion leaders or group leaders using their products in the hope that people would copy them.

Three things are wrong with this model, however: first, the assumption that everyone knows an opinion leader; secondly, the assumption that opinion leaders or innovators would want to show others new ways to do things (they may be in competition with them); thirdly, above all, the assumption that the majority of people are passive followers with few opinions of their own.

The active human information processing model
Research in the 1940s, 1950s and 1960s showed clearly that people have many opinions about most subjects. They were not blank slates but interpreted events and the mass media in distinct ways, largely according to pre-existing attitudes. The question becomes not 'what do the media do to the people?' but 'what do the people do with the media?' In reality, people reject, accept and distort information according to their own views and attitudes.

This does not mean that the mass media (and advertising) are powerless, but that the media are unlikely to influence strong opinions. Instead, they work best where opinions are either unformed or where statements cannot be verified. Thus, as most people do not have strong opinions about various brands of tomato sauce, media exposure to one brand may well influence people to associate that brand with tomato sauce.

This probably explains why advertisers often say that they only follow trends in public opinion – they do not make them. If a message runs counter to a strong attitude which is based on personal experience, then it is likely to be rejected. The media are just one source of information which can be overwhelmed by the opinions of our families, friends and colleagues. This is one reason for carrying out market research, so that messages can be targeted to reinforce certain attitudes and views and not others.

The media, however, do have some role in defining what is normal and what is important. They help to set the climate of public opinion. A good example of this 'agenda setting' function was the coverage by the British Press of salmonella contamination of chicken (see Annex 3). Salmonella had existed for years before the story was broken by the Press, but few people knew about it. As a result of continued publicity (fanned by clumsy Government denials), many people in Britain became so alarmed that the sales of eggs dropped drastically. Salmonella and chicken was now a public issue. The media coverage did not influence people directly, but it did bring into play strong pre-existing attitudes towards personal safety.

behaviour derived from psychology, economics, sociology, statistics and other disciplines. Its chief aim is to identify and describe the likely or actual market for products.

Researchers spend much of their time finding out the probable demand for products that can be given a particular image and promise of satisfaction – that is, appeal to a particular market segment. A key approach is 'formative evaluation'. They may want to explore potential customers' ideas and feelings about a product or a product category, such as teenagers' views of soft drinks. So they try to sketch the main wants, frustrations and life goals of teenagers, and their views of the product category. They may use unobtrusive observation to see what is important to a particular target group; a researcher may spend time talking with teenagers, or may simply listen to them and observe them in their haunts – for example, in fast-food establishments.

Another well-used approach is to conduct focused group discussions ('focus groups') with members of the target group or people who know them well. Sophisticated content analyses of these discussions can identify subtle underlying themes raised during the discussion. They can yield valuable information about the nature of consumers' concerns, motivations and perceptions. Often, though, the resulting information has a sense of 'obviousness' about it. For example, work on bread (consumption of which is declining in many countries) has shown that customers are not so worried about the nutritional content of bread as about the kinds of fat spreads used on it. They feel that all you can do with bread is to put butter or margarine on it. This is obvious enough, but since it is a key point in focus group discussions, perhaps bread manufacturers should suggest alternative, healthier spreads for bread.

Market researchers use many other qualitative techniques such as mind-mapping (see Box 7.2) and social trend analysis, the results of which were discussed in Chapter 7. Some market research agencies run their own social research projects unrelated to any company's products. Taylor Nelson in the UK, for example, have conducted surveys of social trends for several years. Two other widely used market research methods are sensory testing and surveys.

Testing food products using sensory evaluation panels enables manufacturers to test whether their new products are going to appeal to the tastes of consumers. If a product does not taste (or look) nice (ie, have appropriate 'bliss points'), then consumers are unlikely to buy it more than once. Most new products 'fail' because of poor sensory characteristics (in the jargon: 'poor organoleptics'). The testing is often done by specialized laboratory-based 'perceptual psychologists'.

Testers usually use 'taste panels' who have been selected as being representative of the population of consumers to whom food companies wish to sell their products. Often, taste panel members receive some prior training in sensory evaluation – for example, which words to use to describe particular sensations. Usually, product samples are presented to the taste panels 'double blind', with neither the subject nor the experimenter knowing which sample is which product at the time of testing.

Qualitative studies of consumers are often followed by extensive quantitative surveys to establish the range or prevalence of consumers' behaviours, views, concerns, motivations or opinions. These may relate either to a product or product category, or to the characteristics of consumers themselves. Some surveys concentrate on consumers' use of a class of products (eg, cream cheese) to find out about competing brands and the possibility of generating new 'niches' for the client's brand. Others are more concerned with consumers' views of presently available and 'ideal' brands so that the client's product can be presented in the 'ideal' light.

Food preference surveys may be carried out by large scale caterers such as the armed forces, hospitals and airlines to investigate people's actual food choices ('How often do you eat A...Z?'), or their liking for various foods ('How much do you like A...Z?'), or they may present potential choices ('Would you rather have potatoes mashed or roasted?'). One of the main aims in carrying out such surveys is to reduce waste, and also to maintain or increase customer satisfaction.

The pushers

Market research provides important information that our final group of people use to 'push' or promote the interest of those whom they represent. By far the largest actors in this area are the marketers and advertisers, but others, such as trade associations, professional lobbyists and voluntary consumer and other public interest groups, also seek to influence public opinion and behaviour in support of the products, policies or causes they represent. The media are a major vehicle used here, especially by the advertisers, to affect consumers' behaviour, but product packaging, labelling and the like are also used. More private routes to secure a particular actor's interests may also be used, especially by lobbyists and trade associations who seek direct contact with those involved in the decision making that affects their clients' interests.

Marketing and advertising

Marketing, public relations and advertising go hand in hand to sell products and maintain the value of branded products. Advertising can be seen as a branch of marketing, which concerns both the strategy and tactics involved in the development and maintenance of a com-

pany's markets, from its public relations to its promotions and free offers, from its logo to its advertising activities. Marketers and advertisers use very expensive and powerful weapons to push their messages out to the public.

Advertising communicates images and possibly some facts about products, raises awareness of product categories (especially brands) and of previously dormant 'wants' – for example, 'I really want to look like Madonna' – and tries to associate that idea with the idea that 'to do so I should eat brand X like she does'. However, marketers are increasingly using softer forms of advertising, through sponsorship of the arts, sports and TV programmes, developing public relations, and providing educational materials distributed through schools and doctors' surgeries. They also do a lot of research to try to understand our individual feelings and then target their materials in ways that we are likely to accept. A breakfast cereal may be enjoyable to eat and it may be nutritious, but you might not buy it. If, however, it is also recommended by the National Heart Foundation or by an eminent heart transplant surgeon, then you might buy it – especially if you feel vulnerable in some way to heart disease.

Marketers, then, have good reason to be interested in public opinion and in people's motivations, needs, wants, desires, relationships, aspirations and life goals. If they can claim that a product will help people to realize some of these factors – whether they are consciously stated or not – they will stand a better chance of making sales. And with this knowledge, they can design their advertising better to attempt to create leading brands. One example comes from KP Foods, part of United Biscuits (see also p113):

KP Foods' development of a premium crisp brand is one of the snack market's great success stories, and represents Britain's most successful snack launch ever. It resulted from KP's need to break out of the low margin, standard crisp sector by creating a product with strong brand imagery capable of generating high returns.

The task was far from simple. Initial research had highlighted the attractiveness of the thick cut sector, which, during the 1980s, had claimed a 40 per cent market share in the USA. After a three year programme of consumer research and innovative technical development, KP developed a thick cut crisp product, modified for the UK consumer by combining a crunchy initial bite and a light texture.

The branding was paramount. The product's characteristics needed to be communicated to the consumer in a bold statement that left no doubts that there were no imitations. The solution was 'The Real McCoy's', based around 'hard-working middle America' imagery. This branding was endorsed by premium quality packaging, an authentic American flavour range and stylish advertising which successfully portrayed the McCoy's imagery – 'The Big Chip from the Big Country'.

After test marketing, the brand was rolled out nationally. The rest is history: sales of McCoy's were worth over £40m in 1991 and now account for 27 per cent of the premium crisp sector.

Quoted with permission from *United Biscuits Annual Report 1991*

Marketing activities can be viewed positively or negatively. On the positive side they enable products to be made which will sell because people are disposed to buy them; the manufacturer tailors the product to suit the consumer. On the other hand, marketing, especially in the form of advertising, can often be seen as a misleading form of manipulation of people's desires. Just where the truth of these opposing views is, varies from product category to product category and from company to company.

Food is eminently suited to marketing activities – it is involved in most aspects of human existence and so can be transformed to meet many human wants and desires. Food marketers are driven by the fact that people's biological needs are satisfied by relatively small amounts of a basic range of foods. So, main-taining and maximizing companies' profits depends to a great extent on the creation of 'niche markets'. This, in turn, means identifying the psychogenic needs of particular groups and targeting products at them.

Advertising has many forms. There are newspaper advertisements for cheap groceries, TV spots for chocolates, breakfast foods, lollies and soft drinks, sponsored TV programmes, advertorials (text dressed up like the independent part of the editorial content of a publication) and magazine inserts. Some is indirect, more subliminal, which was evident a few years ago when Coca-Cola bought one of the Hollywood studios and managed to have its famous product positively displayed in the scenes of several movies. And there is advertising which is not advertising, seen in many sports grounds around the world where hoardings just happen to be picked up by TV cameras on many occasions during the events.

A great deal of money is spent on food advertising – in Australia, around A$400 million each year. Spending on advertising in economically booming South-East Asia is growing rapidly – it trebled from around US$2 billion to US$6 billon between 1984 and 1990. In the UK, some £523 million was spent on advertising food and soft drinks on TV in 1992 – more money than on any other product category (see Figure 8.1). Most food advertising on television was for processed products – just £4.5 million was spent in the UK on fresh fruit and vegetables in 1991. In the USA, food firms are the largest group of advertisers, spending about US$12 billion in 1990 on direct advertising. Eating and drinking establishments and food stores spent a further US$1.7 billion. Mass-media spending by food processors was US$6.3 billion and consumers redeemed over US$2 billion worth of manufacturers' coupons. However, the United States Department of Agriculture (USDA) reported the following in 1991.

All advertising is not targeted at consumers, however. With heightened competition for scarce retail shelf space, food processors increased their trade promotions aimed directly at food retailers. Going into the

1990 s, maunufacturers' promotions aimed at retailers and wholesalers was estimated to be about twice what was spent on promotions to consumers.

To recover some of the costs associated with stocking new items, for example, some retailers charge manufacturers a slotting fee, a one time charge for shelf space. Fees per new item per store range from $15,000 to $40,000. Manufacturers might pay $2–$3 million in slotting fees to introduce a new product in stores across the country. Because of the proliferation of new products and the incidence of product failure, some retailers also charge 'failure fees'. These fees help retailers recoup costs of removing failed new products from their distribution system.

Gallo, A E (July 1991) 'The Food Marketing System' *Food Review*, USDA, p39

Consumer advertising tends to distort real life rather than reflect it. It has been called 'the distorted mirror' and its effects may be profound. Its distorting effects are clearly seen in the use of female models to sell everything from clothes to cars. The models shown are slimmer than at least 95 per cent of all women. So if slimness is fashionable, advertising accentuates the trend and portrays extreme slimness. This has been linked by several researchers to the increased incidence of dieting disorders among adolescent and young women. Whether it is a causal link or not is a matter of debate.

It may seem that the role of advertising is to tell consumers about the advantages of products. However, the basic functions of foods are pretty obvious, so more than the product has to be advertised and sold. Food advertising often uses images of happy, healthy, social, successful models rather than the foods themselves.

There has been much research into the effects of advertising and many authors have commented on the different effects it has or they believe it has. Because advertising works to persuade people to act in certain ways and to introduce new products, many see it as a process that helps to change attitudes, behaviours and values. Advertising also occurs within a cultural context and helps in the selective reinforcement of those styles, roles and values which are readily commercialized,

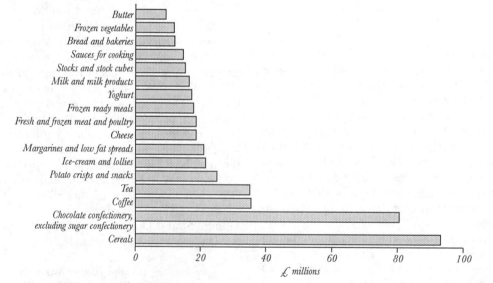

Source: Dibb, Susan E (1993) *Children: Advertisers' Dream, Nutrition Nightmare? The Case For More Responsibility in Food Advertising,* National Food Alliance, London, figures from Register-MEAL, in Table 5

Figure 8.1 UK spending on advertising of food product categories in 1992

easily linked to products, dramatically visualized and reliably responded to.

Advertising is also characteristically pervasive and persuasive. It is used in many media, penetrates everyday life, is relentless over time and professionally executed. Some of the presumed unintended effects of this are:

➤ profound, in that they affect social, political, cultural, moral and spiritual values, not just personal habits;
➤ environmental, in that they are hard to detect and measure, impossible to avoid and affect all (despite the myth of personal immunity);
➤ intrusive and dominating, in that they help to set agendas and goals, specify alternatives, specify criteria for choice and prompt passivity.

Because advertising promotes goods (objects), it is seen as materialistic, promoting the belief that consumption is the route to happiness, a meaningful life, and the solution to most personal problems. Although advertising is usually easily understandable, it may also reinforce social stereotypes, trivialize language and promote perpetual dissatisfaction (see Table 8.2).

Table 8.2 The types of appeals advertising makes and what these promote

When advertising appeals to:	It also promotes:
A. Mass markets	A. Conformity
B. Status	B. Social competitiveness
C. Fears	C. Anxieties, insecurities
D. Newness	D. Disrespect for experience, tradition and history
E. Youth	E. Reduced family authority, disrespect for age
F. Sexuality	F. Sexual preoccupations and dissatisfaction, pornography

Source: Pollay, R W (1986) 'The distorted mirror: reflections on the unintended consquences of advertising', *Journal of Marketing*, April, pp22–23

Children's telefood

A key concern for many parents is the content of children's advertising, although direct advertising is just one of many ways that children are targeted (see Box 8.9). While it can be argued that advertising any product to young children who are legally incapable of making a contract is unethical, food advertising has posed additional problems. It has been shown repeatedly that television advertising aimed at children is predominately about food products which contravene national dietary guidelines. Around 80 per cent of advertisements are for confectionery, snacks or fast foods, which have extremely high amounts of fats, particularly saturated fats, salt or sugar.

Heather Morton, a researcher from the University of South Australia, is an avid viewer of children's television. She has conducted several studies into the food content of children's programmes and advertisements in Australia where the self-regulating commercial channels carry many advertisements for foods in children's viewing hours (see Figure 8.2) There is a considerable discrepancy between what is advertised and the national dietary guidelines. For example, breakfast cereals and rice products represent the broad bread and cereal group of foods which is supposed to underlie Australian dietary habits, and fruit and vegetables – those other dietary

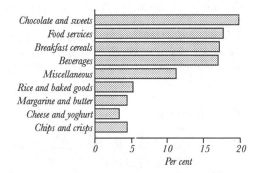

Source: based on data from Morton, H (1990) 'Television food advertising: a challenge for the new public health in Australia', *Community Health Studies*, vol 14, pp153–61

Figure 8.2 Percentages of food advertisements shown in Australian children's viewing hours (4–8pm) by food category

mainstays – along with meats and their alternatives are scarcely seen. In contrast, foods that are high in fat and sugar, such as chips, crisps, cheese, margarine and chocolate dominate the TV screens during these hours. The advertisments promote an inverted pyramid, promoting most the foods that children should eat least of.

Does this mix of advertising, which is found in many countries, have any bad effects on children's diets? Some parents and nutritionists think so, but most food advertisers do not. They point out that most children distinguish between the fantasy world of advertising and nutritional reality. So why do manufacturers continue to spend large amounts on advertisements if they are so inconsequential? Epidemiologists can only point to the rising obesity found among many groups of children, but is this linked to the balance of television food advertising, inactivity or both?

Whatever is the case, it does seem a pity that vegetables, fruits, cereals and meats cannot get a share of the fantasy action. At present, marketing of many horticultural products is severely hampered by the fragmented nature of horticultural production – producers cannot afford TV advertising. Should governments, which are responsible for the community's health, give a helping hand to these products, or curtail advertising of the others? In 1994, health campaigners in the UK were calling for a new code of advertising, possibly including a television watershed to protect younger children, as in Canada, so that commercials did not undermine the Government's nutrition targets. Some changes to food advertising during children's television times were introduced in 1995.

Why advertise?

The main explanation for these major advertising expenditures concerns brands and maintaining a brand image and share of the market. Many processed foods have very similar nutritional and sensory properties. To sell, they need unique images to meet the psychogenic needs of the target group.

Competition exists between the different brands of these very similar products to gain as large a market share as possible. Image advertising helps to make the consumer aware of their existence and of their possible social or psychological connotations. Indeed, for some sectors and products, advertising is an essential element in their success, as economist John Sutton showed in his study of the food sector in *Sunk Costs and Market Structure* (see also Box 8.10).

Advertising, then, varies not only with respect to the medium which is chosen to carry it (newspapers versus television), but also according to the 'information' which it portrays. The two-page list of specials from Bloggs Supermarkets in a local newspaper does provide real, hard-nosed information – about price reductions. The TV spot for Bloggs Cola, however, almost never mentions price or nutritional details. Instead, it makes social and emotional promises, such as,' if you drink Bloggs Cola, you could be big and popular like these people' or even,' people like the people you want to be with drink Bloggs Cola'. Consumers vary in their responses to 'image' and 'benefit' advertisements. Those who tend to conform closely to the demands implicit in social situations prefer image advertisements. Others who maintain close correspondence between their personal roles and their behaviour, respond more to advertisements that promote the 'real' benefits of products. Neither type of advertisement is likely to portray reality, but this is not the aim.

In reality, women still do most shopping. They are increasingly in the work-force and have more money but less time, since they still do the majority of domestic chores, including shopping. Shopping and choice are also becoming more confusing with more and more information to process. Signposts to guide shoppers through the clutter are useful. One industry response to this is to develop a family of products from a well-known branded product. The hope is that people will buy a product because it has the same brand name as something they already buy and trust. Building up trust in a brand and increasing

---

Box 8.9
Targeting Children

In Britain, market researchers have identified two types of children who influence what their parents buy: trolly loaders who are not discovered popping things in the shopping trolley until it is too late at the checkout, and naggers or pesterers who exasperate their parents into buying products. These tend to be savoury snacks, sweets, soft drinks and cereals, the same foods that are commonly advertised.

Marketers use six major ways of targeting children:

1. Computer games – some manufacturers make sure that their products are featured in computer games by sponsoring them, which results in long-term exposure to products as children play the games. Penguin biscuits in the UK, for example, sponsored a game called James Pond 2: Robocod, in which a fish like character collects the biscuits. Researchers for the company promoting this kind of deal have shown that 18 per cent of all 5-14-year-olds have played this game, and each game is played for an average of 25 hours.

2. Character licensing – children are brand conscious by the age of four and can be influenced by cartoon characters. Companies license the use of such characters on their products to increase sales to children. Some US$100 billion is estimated to be spent on character licensing products in all areas, not just food. In the early 1990s, Nestlé signed a £70 million deal with Disney to use Mickey Mouse, Goofy and friends on all its food products across Europe for 11 years.

3. Television advertising – much of the TV advertising aimed at children is for food products, mostly those high in fats or sugars or both. One UK study found that if children watched one hour of commercial TV in an evening and all of Saturday morning, they would see 92 food and drink advertisements in a week – nearly 10 per hour.

4. Promotions and packaging – manufacturers and retailers use a mixture of clubs, sponsorship of events, promotions involving free gifts, and packaging with lots of detail and wrapping aimed at them.

5. Sponsorship – the Scout movement has been used by one margarine company in the UK to build future brand business by sponsoring cook badges.

6. In the classroom – producing materials for use in schools. Some materials carry overt plugs and others are misleading. One project pack funded by the British Egg Information Service listed eggs first under protein foods and did not include meat and poultry, left fruit out of the section on vitamins and minerals, which was also topped by eggs, and did not cover cereals or grains in the section on fibre.

Source: Consumers Association (1993) 'Food Frenzy: marketing food to children', *Which? Way to Health*, Consumers Association, London, February, pp14-16

---

the service element in a product – for example, providing free phone numbers for information – are important marketing issues.

With the development of international markets, and an EU wide market following the Single European Act, the marketing of products across Europe requires new advertising skills. Judie Lannon, the former research and development director for the international advertising agency J. Walter Thompson, found that there are five broad sources of advertising ideas:

1. Those rooted in national lifestyles, such as French boules, English pubs, or Spanish bullfighting, that are idiosyncratic and local.

Box 8.10
The Real Thing?

In many important ways, the world's markets are also becoming more alike. Every corner of the free world is increasingly subjected to intense and similar communications: commercial, cultural, social and hard news. Thus, people around the world are today connected to each other by brand name consumer products as much as by anything else. Tokyo, London, New York and Los Angeles resemble each other today far more than they did 25 years ago, in large part because their residents' tastes in consumer products have converged.

Roberto C Goizueta, Chairman, Board of Directors and Chief Executive Officer, The Coca-Cola Company in *Size, Scope Power: The Global Coca-Cola System Delivers*, The Coca-Cola Company, 1991 Annual Report

If you knew that this box was about Coke from the headline, it is perhaps proof of Coca-Cola's claim to be the most famous world brand, and the most understood words world-wide except 'okay'. It has a global philosophy and practice to match. With a market value of over US$58 billion at the end of 1993 and a 44 per cent market share of the world-wide soft drink industry, the company sees itself as an international corporation with a headquarters in the USA as opposed to a US company with a large international business. It has built a global business system which had US$66 billion in retail sales in 1991 in over 185 countries.

The company sees opportunities to increase sales as the world's markets become more alike, and as disposable income around the world increases, with more and more young people outside Europe and the US, and with markets becoming easier to reach, especially Eastern Europe and the former USSR. It has even paid for a new Chinese TV soap opera, provided that the only soft drink in it is Coca-Cola – soft advertising to the world's largest single market.

To reach the world, it believes that a company must have, build or buy a powerful trademark. It also needs a global business system to reach consumers, which must tailor products and messages to local markets. It also needs something less tangible – a central theme or idea that binds them together. For Coke, that has been built using advertising and marketing to encourage consumers to associate Coca-Cola with the best feelings and memories.

The Coke business system is built on the manufacture, marketing and distribution of soft drink concentrates and syrups. Its main brands are Coke, Diet Coke, Fanta and Sprite, with Tab Clear, Powerade (official sport drink of the 1994 Olympic Games and World Cup) and Nestea (joint venture with Nestlé) being developed. The company is essentially a wholesaler, selling syrups to bottlers, canners and post-mix drink fountain users, although it does have stakes in many bottlers. It also processes and markets citrus and other fruit juices, notably through the Minute Maid trade mark.

In 1991, 650,000 employees earning US$7 billion in wages scattered all around the world, bottling partners, vendors and customers formed the physical part of the system. It used 115,000 company or bottler-owned vehicles with a capacity of 22 billion cases of soft drinks to reach its 8 million customers. These can range from major multiple retailers to individual street vendors. They then sell to the consuming public.

By 1992, Coke held 61 per cent of the world-wide cola soft drink market, 36 per cent of lemon-lime drinks and 32 per cent of orange soft drinks. Mexicans drank most in 1993 – with an annual average per person consumption of 306 servings (8oz), beating even the Americans at 303 servings. The Australians drank 250 servings, the Germans 180, the British 95, Thais 49, Russians 2, Chinese 2, and Indians none.

The company spends around US$4 billion annually in marketing and runs 20 million pieces of advertising as well as supplying 6 million pieces of sales equipment such as post mix fountains and 2 million vending machines. It aims to put a cold Coke and its other soft drinks 'within an arm's reach of desire' of the world's over 5 billion potential consumers.

All this produced a net operating revenue for the company of US$14 billion in 1993. Gross margins were 63 per cent and net margins 22 per cent. The cost of goods sold was nearly US$5.2 billion, leaving a gross profit of US$8.8 billion, and the cost of selling them US$5.7 billion. The last was made up of administrative and general expenses of US$1.3 billion and sales expenses of US$4.4 billion, which included US$1.1 billion for advertising. Coke net profits passed the US$2 billion mark for the first time in 1993 at almost US$2.2 billion.

Already the world's largest operation of its kind, the company foresees enormous further potential for expansion to the 3.2 billion people it hardly reaches, as well as further expansion in the markets it is already in. By the end of 1992, its products reached 195 countries. It moved into India in late 1993, acquiring the local brands that accounted for 60 per cent of Indian soft drink sales and planned to use the network of 60 bottling plants to introduce rapidly its international brands such as Coke.

Source: The Coca Cola Company, 1991, 1992 and 1993 Annual Reports

2. Those that arise from news, fashion, etc that are increasingly global in scope.
3. Those rooted in the emotions and experience of the human condition – for example, children, animals, love, greed, etc, which are perhaps the most important.
4. Those rooted in myths and archetypes that strike a universal chord and seem to transcend the local and cultural, such as the myth of the Marlboro man.
5. Those that demonstrate what the product does, the oldest and most transferable category.

Essentially, all of these strategies reflect human values in some way. She believes that a strategy for marketing products across nations has to acknowledge national differences. Marketers can also learn lessons by looking back over time; advertising for a particular brand has to be adapted to changing times. Long-running products were given a personality which changed over time but for which the image remained clear.

The aims and effectiveness of advertising

Businesses use advertising to maintain and increase sales and profits. Manufacturers or retailers may commission the advertising, but food consumers pay for it through food prices. Money is spent on advertising mainly to increase brand shares, although there are some deliberate spin-offs from this. For example, some campaigns have been aimed at improving positive attitudes towards one food ingredient and others may be directed in part at employees within a large industry in an attempt to build a corporate identity.

Advertising is effective. In 1982, the Australian sugar industry launched a long-term campaign to promote the naturalness of sugar ('Sugar a Natural Part of Life'). After three years, the number of those questioned with a positive attitude to sugar had almost doubled from 24 per cent to 46 per cent. However, there are major qualifications to the effectiveness of advertising, depending on how effectiveness is defined. Advertising campaigns certainly do change brand shares, although details are hard to obtain. The amount of change may not be impressive in percentage terms – say, 2 or 3 per cent – but in some countries, this may amount to increases in turnover of several millions of dollars per year. Even so, many new food products fail to make sufficient profits for manufacturers or retailers, despite extensive advertising, and are dropped.

Recent laboratory studies of television viewers also suggest that many people simply switch off their attention when advertisements appear. Increasingly, they do this by videotaping programmes and then speeding through advertisements when they play the tape. This is a major source of concern for advertisers and network television companies. Newspapers are now advertising their space as a better place to sell than TV, and TV commercials are being designed to be still intelligible when fast forwarded on video playback.

There are more subtle effects of advertising. These relate to the roles of the mass media in society. One view is that the mass media create the 'climate of public opinion'. They put certain issues on the public agenda (like food irradiation, ecological and ethical concerns). They can also influence the public's notions of what is normal behaviour. The promotion of cigarette smoking among American and European women by advertising in the 1920s is one example. Prior to that period, women did not smoke in public, but thereafter it became 'normal'. The promotion of cigarette smoking and alternatives to breast milk in poor countries are other examples in which unhealthy or unnecessary practices have been encouraged.

Children with little experience of the adult world, who may be unable to distinguish reality from fantasy, may be particularly prone to the deceptive promises of advertising. A night's prime time commercial TV viewing might easily lead a child to conclude that is entirely 'normal' to consume chocolate bars most hours of the day or that nutritious foods come only from fast-food chains!

Advertising, however, is just a set of (powerful) communications. Changes in sales require much more. They require infrastructures. Manufacturers can advertise produce all they like, but they will not sell unless the products are on the retail outlets' shelves. This requires co-operation from retailers, distributors and others. Part of the success of wine, ice-cream, soft drinks and tobacco companies is that they have developed intricate networks by which they distribute their wares. Some ice-cream manufacturers in the UK, for example, provide small outlets with freezers to stock their products and do not let their competitors use them. So consumers are never very far from their products – they do not have to go to much effort to get them.

Distribution networks are very important. Companies have been known, for example, to place soft-drink vending machines in schools and on railway platforms, after they have arranged to have water fountains (the competing product) removed. To quench their thirst, children and travellers then have no choice but to drink their product or go without! Advertising is only part of marketing and selling.

Lobby groups

While advertising is a visible means of influencing behaviour, lobbying is usually a much more private affair, except when those doing it seek to use the media to gain publicity for their cause. It has a similar function to advertising in that it seeks to secure benefits for the lobbyists, but its target is usually different – the policy makers, politicians and government members and officials that set the framework in which the actors operate and which we deal with in the next chapter.

Some individual actors hire firms to lobby for them in the corridors of power in Washington, London, Brussels and other capitals, providing lunches and briefings for important officials, and so on. Most have bodies, such as trade associations, that try to look after their interests as a group, even though they may be competing within the group.

Trade associations often consist of many divergent private company interests which come together at times when external threats (eg, the Government) might affect them all. Thus, food manufacturers often have competing interests with each other, but they may unite to further their common interests (eg, for tariffs on imports, against tariffs on imported machinery, or to forestall legislation on packaging or labelling that may place responsibilities on them). These trade associations often have a small full-time secretariat or executive. Member companies agree to pay relatively small sums of money to further their common aims. In this respect, trade associations are rather like statutory marketing boards or marketing co-operatives. They have a steady supply of money with which they advance fairly broad aims.

Management

The technologies and understanding of people's behaviour that have developed in the past hundred years have affected the way that production is organized and processes and people are managed. Increasingly, scientific understanding and technological development are incorporated in the management of enterprises in the food system in ways that are aimed at increasing control of the business. This is now changing the whole work organization.

Changing work organization

Work organization has shifted from craft-based, small-scale production through a large-scale, mass-production phase, which is still dominant, to a newer, lean production phase, which is likely to dominate in the future (see Box 8.11).

These trends towards just-in-time manufacturing are emerging in the food industry, with giants like Unilever and Coca-Cola adopting similar practices to those pioneered in the car industry. These new trends concern the food workers' unions, who point to a new £60 million high-tech, low labour, soft drinks factory in Wakefield as an example. It can produce 240,000 cans per hour and 40,000 PET-type plastic bottles per hour in a highly automated computer-controlled factory operating 24 hours per day for 364 days a year. It uses a just-in-time system involving minimal stocks of raw materials and finished products. Here, the company has negotiated a very flexible working, single union, no-strike deal with the 150 employees and closed four other factories with a loss of about 600 jobs. From the company's point of view, this minimizes risk, but it is using a technology that only produces non-returnable packaged drinks – ie, in cans or PET bottles – which may be socially undesirable given the environmental costs.

The unions also see lean production as linked to new approaches to worker management – now called human resource management – which seeks to have workers identify themselves as an integral part of the company and identify their interests with those of the company. They fear that such an approach at best will marginalize and at worst destroy the unions. The question arises as to how far a company will identify its interest with its workers when the next capital investment decision comes along, when global considerations may mean that relocation elsewhere is more profitable for the business. This approach could be damaging for both the workers and the company if the processes adopted by the company are such that the effects on the environment – for example, wasteful use of resources, disposal or recycling costs, and so on – are left for others in society to deal with.

These trends are also evident in retailing, with retailers influencing suppliers, working with them to meet quality standards, using just-in-time stocking and computerized information systems which allow them to control these processes. Use of computer technology may allow a much leaner operation by eliminating manual stock control.

Finance

New technology, coupled with the removal of exchange controls in many countries, has helped to produce a profound financial revolution to create a computerized, global capital market operating 24 hours a day. The banks and accounting houses that operate this market play a major role in the food system as they provide the finance that keeps the system working. If a farmer wants to adopt new technology, or a multinational wishes to take over another company, banks will usually be called on to provide the necessary finance.

At an international level, banking systems have had profound effects on the food systems of poor countries. The policies of international banks in the 1970s led in part to the massive debt burdens of many countries in Africa and Latin America. Their encouragement of sin-

Box 8.11
From Mass Production to Lean Production

Mass production, sometimes called Fordism, was developed in the car industry and named after Henry Ford who pioneered the assembly line. He wanted to break the strangle-hold on production that was held by the craftsmen mechanics. He and F W Taylor, the father of time and motion study, analysed the components of their skills (via 'work study') and gave unskilled labourers each a task to perform. Decisions about the speed of the assembly line and the sequence of these simple operations were entrusted to staff managers. Henry Ford paid his unskilled labour twice the going rate in return for their performance of extremely monotonous tasks. Quality control tended to come at the end when faulty vehicles were rectified.

The Fordist model of production was so successful in its day that many industries copied it, as did many government organizations. Specialists and specialization was the key, so organizations developed many tiers of management and specializations. Labour became so specialized that demarcation disputes often held up production, and top managers had little influence over daily production schedules. The effect of such organizational complexity is to distance decision makers from the results of their actions – they tend to have power without responsibility.

Fordist approaches also affected government structures. The Senior Executive Service of the Australian Public Service, for example, has now reached the pinnacle of Fordism in its philosophy of 'content-free' management – managers only have to know how to manage people and require little specialist knowledge. Specialists or consultants are brought into the organization when required. However, the managers may not always know when the experts are required! This is quite the opposite from Japanese managers who generally have trade or professional training and experience before they enter management and who, like their workers, tend to be multi-skilled.

Fordist methods served Anglo-American industry well during the early part of this century and underpinned the approach taken throughout the food industry. However, its days may be numbered by the new competition coming mainly from Japan, Italy and Germany. These countries skipped much of the mass-production phase of industrialization and took many of their practices straight from medieval craft organizations and from family work groups which did not distinguish between work and leisure or between 'functions'. They have become formidable competition for the older Fordist ways, mainly because their organizations are more efficient.

In the 1990s, the car industry, this time Japanese, is again leading the way to the next phase of manufacturing with lean production. This has developed from the Toyota manufacturing system, which involves flexible team working, total quality management, and very low employment levels with capital-intensive, high-tech plants.

This system uses just-in-time manufacturing, with just a few hours' stock held at the plant. The specifications required are given to suppliers, and those suppliers are chosen who can produce the best quality, will agree long-term deals, and will adopt similar manufacturing procedures to ensure quality. Toyota will help them to reorganize to meet their standards. This approach cascades down the supply chain and each time any price savings have to be passed on up the line to Toyota.

gle commodity agriculture (eg, coffee or cattle plantations) weakened the ability of poor countries to feed themselves by making them over-reliant on world commodity markets. They required income from these commodities to import basic foodstuffs. Many of these internationally supported schemes have also brought about considerable environmental damage – for example, cattle farming in southern Africa, and the destruction of the rainforests in Brazil.

In the 1980s and 1990s, international

bodies such as the World Bank and the International Monetary Fund have exerted pressure on the governments of poor countries to adopt austere economic programmes and to promote exports, often of similar commodity crops, while at the same time doing little to reduce their debt burden. This has resulted in major reductions in education, welfare and health services in developing countries such as Tanzania and Zambia.

Crucial to financial institutions are their sources of information. How do they decide what is a good investment? Do they look at all the costs involved when they make financial decisions? Clearly, in many cases damage to the environment is not considered as a cost; similarly, damage to community infrastructures has not been included in many investment decisions. This is partly because investors are simply not expected to consider anything other

than short term prospects. It is also related to the prevailing ideology that places more value on the workplace and technology than on unpaid work and non-economic values. Feminists such as Marilyn Waring see this as one aspect of patriarchy.

Times are changing, however. Even conservative journals such as *The Economist* have pointed out the dangers of pollution and its long-term effects on the ecosystems upon which we depend. Environmentalists are calling for what were previously called external costs to be taken into account in business practices, trade, production methods, and so on. As awareness of the fragility and importance of social and ecological systems grows, we might expect more emphasis to be placed on long-term social and ecological factors in financial decision making.

Summary

At the same time as production and retailing technology are becoming more complex, and management and marketing tools are becoming more powerful, so too is the technology developed for food storage and preparation in the home. Some analysts predict that by the year 2000, 60 per cent of retail food products in some countries will be prepared meals. Fridges, freezers and microwaves complete the cold chain which can extend the reach of the food factory into the home by providing the technology to store and prepare its products easily, and with minimal skills, at home. In this process, many household food-preparers lose the basic skills needed to transform a relatively few, inexpensive ingredients into meals. Instead, they face a bewildering array of items in a supermarket, mostly processed, from which they must select and from which they will produce meals at home. The information they use to make their decisions becomes increasingly important.

The result is a food system in which the products are increasingly complex and require careful handing. Consumers and societies are left with the responsibility to deal with the environmental costs of the products and processes and with the burden of interpreting the wide range of information necessary to make suitable decisions about how to feed themselves well. Yet the framework in which these actions take place is socially determined through the legal framework and political institutions which society develops to deal with these issues. It is to these that we turn in the next chapter.

9

Food Law and Food Policy

Food laws were among the earliest of enactments known to man. Governments over many centuries have endeavoured to provide for the safety and wholesomeness of man's food by legal provisions and appropriate punitive action when necessary. Over the years also, crude forms of fraud, such as adding worthless substances to food or extracting valuable constituents from it, have been followed by sophisticated methods of adulteration more difficult to detect. The birth of modern chemistry in the early nineteenth century made possible the production of materials possessing properties similar to normal foods which, when fraudulently used, did not readily attract the attention of the unsuspecting purchaser. Later, better analytical methods were used in food control work to detect these adulterants. When scientists demonstrated that some were dangerous to health, an aroused public demanded laws which would both protect their health and prevent fraud.

FAO/WHO (1976) *Guidelines for developing an effective national food control system*, FAO/WHO, Rome

Laws, rules, regulations and the related enforcement mechanisms constrain the behaviour of the actors in the food system. Such laws are normally designed, in democratic societies, to protect individuals in society from the unfettered exercise of power by one set of actors and to protect law-abiding traders. These laws are human creations, however, and they result from the interplay of all the actors in society, not just those in the food system. Those with the power to make laws design them to implement particular policies and to serve particular interests. Those who are adversely affected by them can obey them, try to get round them, try to change them or ignore them.

When new policies are agreed, the legal framework in which to implement them follows. There has been a patchwork of laws everywhere and some rationalization is being attempted against a background that favours deregulation. This is only likely to work in the public interest if the policy within which these decisions are being made, and the costs and benefits to different actors, are made clear. In this chapter we look at the laws, rules and regulations and the bodies involved in enforcing them. We examine policy making and food policy in particular, and then discuss the elements needed for effective implementation of policies and the role that information plays in this.

The Legal Framework

For hard-nosed investors, according to a leading investment analyst, the primary determinant of the structure of the food industry, or any other industry, is the legal framework within which it operates. This varies from country to country, of course. The USA and Europe, for example, have historically had very different legal frameworks with different planning laws. In the USA there is an absence of strict planning legislation which makes it easier for retailers to open new stores and try new formats. In Europe, however, there are much greater planning controls and legislation to protect small grocery retailers in some countries. Local planning authorities can affect greatly the type of food retailing and have an important duty to ensure that people have access to food supplies, especially those who do not have the advantages of good health, cars and adequate incomes. Some regulate opening hours in the interests of shop workers, of social conventions, or even people on low incomes and/or those with young children who are unable to shop during working hours. In Britain, MAFF sees it as its task to ensure food supplies are available, and local authorities are responsible for food law enforcement.

Whatever the legal framework, however, food businesses still attract new investors, especially in times of recession, because food is an essential commodity. Throughout the 1980s and into the 1990s, some companies whose traditional businesses have not included food, such as tobacco companies, have bought into food businesses and built them up as part of their core business. They want to secure their investment and have as favourable a legal framework as possible where they operate. Even competing businesss may come together, as we saw in Chapters 6 and 8, through trade associations or other bodies, to influence laws and regulations – for example, the packaging directive of the EU.

A wide range of laws affects the food system, from contract law to legislation on pollution control to health and safety at work laws. Specific food control legislation developed, as we discussed briefly in Chapter 2, because enough people and companies put their private good – increased profit or business survival – above the public good. Such firms or individuals adulterated food, gave short weights, caused food poisoning, and so on, deliberately, or by neglect or incompetence. Laws, regulations, controls and enforcement mechanisms were needed to ensure that the public got wholesome food of the kind they expected and to protect honest manufacturers from unfair competition.

The legal framework can also help to equalize the David and Goliath relationship between the individual and large institutions. As the world becomes more complex, and institutions become larger and their reach global, there are growing pressures to harmonize internationally more laws, rules and regulations. The question is what kinds of laws and policies are needed to protect individuals and the public good. Are effective laws, regulatory and control mechanisms more necessary, or less so? Today, very tough rules could benefit some big companies, who can afford to meet them; small new companies that cannot make the capital investment required to meet such rules could be prevented from competing. Conversely, deregulation could open up the food system to the kinds of sharp practices that brought forth the demand for controls in the first place.

Different countries also have different food standards. Many of these national laws and standards, which often developed over hundreds of years, became a bone of contention in the past decade or so as the food system became more international. In the EU, the creation of a single market has produced a drive to remove tariff and non-tariff barriers that inhibit trade between the member states. Conflicts developed over national food laws which defined what could be called beer, or ice-cream, or a sausage in different parts of Europe, which led, in the end, to the European Court of Justice agreeing that a legal product in one country could be sold in another (see Box 9.1).

From regional to global laws

With the development of major regional and global markets, pressures are growing to harmonize laws, rules and regulations, and to permit mutual recognition internationally of products produced in one country. Partly, this comes from the growth of regional country groupings with single markets, such as the EU and the North American Free Trade Agreement (NAFTA, which includes Canada,

Box 9.1
Eurofood – 1992 and The Single European Act

At the end of 1992, the Single European Act (SEA) came into force. It aimed to remove physical, fiscal and transport barriers to trade between member states of the EU. It is particularly significant in facilitating the movement of food across national EU boundaries.

There are two ways to sweep away remaining protective barriers between EU member states. Either all countries adopt the same standards, called harmonization, or each country accepts the standard of any other country, called mutual recognition. Over the previous 15 years, decisions of the European Court of Justice went along the road of mutual recognition, and the SEA consolidated them.

The first famous test case in 1979 involved a French liqueur called Cassis de Dijon. When a firm wanted to sell it in Germany, a licence was refused because the alcohol content was too low for a 'liqueur' as defined by German law. The European Court established the precedent that 'any product lawfully produced and marketed in one member state must, in principle, be admitted on the market of any other member state, subject only to narrow exceptions, objectively justified ... public health is such an interest.'.

In a similar case, Germany refused to allow the import of foreign beers that contained certain additives, as it conflicted with their own pure beer law – the ancient Rheinheitsgebot. They lost in the European Court. The Court did not accept the German argument that the additives were potentially harmful.

Finally, in July 1988, the European Court delivered a long-awaited decision on pasta. The Italians maintain that pasta can only be made from hard (durum) wheats. A German company had been blocked by the Italians from importing pasta made from a mixture of soft and hard wheats. The Italian defence was that this mix would mislead the consumer. Their argument was tantamount to saying that anything less than durum was adulteration.

The Court sided with the German company, thereby reasserting the principle of mutual recognition in trade. According to this, a product sold legally in one part of the EU cannot be excluded from another part. Inadequate labelling or risk to public health are the only permitted defences. The thrust of the Court's decisions is that consumers will be protected by informative labelling and thus new products can be developed as compositional standards are phased out or kept only by member states for their own national producers.

The rush by business to remove national anomalies has fuelled concern by trade unions and public health and consumer groups that standards designed to protect public health and food quality could be undermined and reduced to the lowest common denominator. The SEA (Article 100A, Paragraph 3) claims to set out to maintain standards: 'The Commission, in its proposals laid down in paragraph 1 concerning health, safety, environmental protection and consumer protection, will take as a base a high level of protection'.

Standards are being set in a few areas relating to safety and labelling. Thus, for additives, for diabetic and baby foods, and for wrappers the laws will be harmonized. Hence, there will be some standards for certain contaminants, although no overall standards for food itself. So there will be no minimum amount of lean meat required in meat pies, nor limits on the amounts of excess fat or water in many products. The question of just how meat is defined also applies to other basic 'natural' ingredients, such as fish, fruit and vegetables.

Mexico and the USA), and which are made up of nations with their own laws and regulations. The Commission of the EU, for example, develops directives on a wide range of food issues, from additives to the weight of pre-packaged products, to eliminate differences in technical standards between countries and so facilitate the free movement of goods. These directives have to be implemented by each member state before they become law. Such a process can make national action on food-related issues impossible. Denmark, for example, had legislation to put deposits on bottles and encourage the reuse of containers. This was fought by industry which prefers recycling so that it does not have to deal with the containers again. EU wide rules, however, made it impossible for Denmark to enforce such legislation.

Globally, the international laws governing food and agricultural trade face the biggest shake-up ever following the agreement at the end of the protracted Uruguay Round of talks in GATT. GATT has attempted to regulate world trade since 1948. These trade rules have tended generally to favour rich nations over poor nations. The new agreement, which took seven years to reach, deregulates much of world trade, including agriculture, banking and services, and will affect national laws world-wide. One result will be to open up the markets of developing countries to large, mainly OECD-based transnational firms. In agriculture, the agreement aims to reduce the use of government subsidies in the trade in food commodities such as wheat, other cereals and beef. This might benefit some agricultural exporting countries through higher prices, but would hit food importing countries.

Food and agriculture were previously excluded from GATT rules but have been a central issue between the world's two major agricultural trading blocks, the USA and EU. In addition to measures to regulate food commodity subsidies, there are also harmonization measures which would affect national laws on food safety, and animal and plant health regulations (sanitary and phytosanitary measures in GATT-speak!) as part of the process of removing non-tariff barriers to trade. Countries party to the GATT agreement – nearly 110 in 1993 – are encouraged to base these measures on international standards, guidelines and recommendations and recognize those of other countries. A country could

not prevent products entering it, even if it had higher standards, without scientific justification or evidence of clear risks. A relatively little known body, the Codex Alimentarius Commission will become, in effect, the world body setting food standards (see Box 9.2). Critics want the bodies making rules to do so much more publicly, to reduce the input from industry and academia and to increase that from consumers and trading standards officials who deal with fraud.

Proponents of the GATT agreement to deregulate trade by removing non-tariff barriers to trade, such as different standards, and making them clearly visible tariffs applicable to all trading partners, believe that it will increase economic growth world-wide and benefit consumers through greater choice and lower prices. Critics feel that the GATT agreement makes freedom to trade as the traders wish the prime criterion for settling disputes between countries, which could override national laws and actions to protect the environment, workers or disadvantaged social groups. Some countries, for example, might want to ban imports of tuna caught using nets which trap dolphins, but will be prevented from doing so as the regulations call this an illegal impediment to free trade (see Box 9.3). They fear that the newly created World Trade Organization (WTO, see Annexe 1) established by the agreement, could thus put environmental, ethical and other questions concerning food out of court by their rulings.

From laws to regulations

A whole host of governmental organizations regulate various features of the food system and usually codifies laws, whether international, national or local, into sets of rules and regulations. They regulate complex issues such as trade, safety, consumer information, production techniques, quarantine measures, pricing and much more. Most people are unaware of their activities, but they affect much of everyday life such as the contents of breakfast foods, the label on the packet, the

Box 9.2
Codex Spans the Globe

To harmonize sanitary and phytosanitary measures on as wide a basis as possible, members shall base their sanitary or phytosanitary measures on international standards, guidelines or recommendations, where they exist, except as otherwise provided for in this agreement...

GATT (1993) 'Agreement on the application of sanitary and phytosanitary measures', *Final Act embodying the results of the Uruguay Round of Multilateral Trade negotiations*, GATT, Geneva, MTN/FA II-A1A-4, para 9

The Codex Alimentarius Commission – the name means 'food code' – is run jointly by the World Health and Food and Agriculture Organizations of the UN. It was founded in 1963 to facilitate world trade in foods, to establish international quality and safety standards and fair trading practices. By the mid-1990s, the Commission consisted of government-appointed delegates from over 130 nations, although the main trading nations, principally the USA, Japan and the EU, tend to dominate. The Commission works through a range of committees. Some deal with standards for specific commodities, such as sugars, processed fruits and vegetables, edible ices and cereals, pulses and legumes. Others deal with more general subjects, such as food labelling, food additives, pesticide residues and food hygiene.

Codex recommendations range from the safety of production techniques to the format of food product labels. It has produced over 200 standards and 40 codes of practice and guidelines for good manufacturing practice. These are advisory and it is up to governments to turn them into law. Its recommendations tend to be followed because of the influence of its members, although many countries, especially in the OECD, have higher standards than those proposed by the Commission.

Some consumer groups criticize Codex as being too much the creature of industry. They say it fails to represent the broad range of interests concerned about food standards. The National Food Alliance in the UK, for example, analysed attendance at meetings over two years – 16 committees and one full Commission meeting. They found that more corporations were represented than countries with 140 companies sending representatives to Codex, compared with 105 countries. Out of 2578 participants in Codex meetings, 660 represented industry interests and 26 represented public interest groups such as consumers. Industry representatives formed 49 per cent of the participants on US delegations and 61 per cent on Swiss delegations.

The critics want Codex reformed radically before it takes on a greater share of responsibility for regulating food trade and defining standards following the Uruguay Round GATT agreement. Their recommendations include:

➤ a UN review of the Commission's operations to encourage greater participation by developing countries and public interest groups;
➤ the use of broader assessment criteria in setting standards;
➤ the inclusion of government officials only in delegations, with industry and public interest groups as observers;
➤ broader consultations at national level on new standards.

Sources: FAO (1987) *Introducing Codex Alimentarius* FAO, Rome, and Avery, Natalie, Drake, Martine and Lang, Tim (1993) *Cracking the Codex: An Analysis of Who Sets World Food Standards* National Food Alliance, London.

Box 9.3
Dolphin Friendly Tuna – Not on GATT's Menu

Rules and regulations and how they are interpreted affect food practices. As the rules become fixed and interpreted internationally, national and more local control may become more difficult. Some fear that the new World Trade Organization, set up as a result of the GATT Uruguay Round agreement, may overrule ecological, social and other criteria in the name of free trade. They base their concerns in part on past experience in GATT. The issue of dolphin-friendly tuna catches illustrates these concerns.

Since 1972, the USA has had a Marine Mammal Protection Act (MMPA) designed to protect dolphins from the worst excesses of tuna fishing in the East Pacific where dolphins and tuna often swim together. Despite the MMPA, over seven million dolphins have been killed by fishing, so in 1988 amendments to the MMPA set conditions on the importation into the USA of tuna caught by the encirclement method. Under these amendments, countries exporting tuna to the USA, such as Mexico, were allowed to kill 1.25 times the US dolphin kill, but the rate was exceeded by Mexico, Venezuela and Vanuatu and the USA embargoed imports of their yellow-fin tuna. Mexico brought its case against the USA to GATT in February 1991 and it was decided in its favour in August 1991. The ruling by the three man panel, which met behind closed doors, gave new interpretations of GATT Articles 20 (b), on conservation of natural resources, and 20 (g), on the protection of human, plant and animal health, which have been seen by a wide range of US environmental, conservation and animal protection organizations as a direct threat to their efforts to promote simultaneously sustainable trade and conservation.

The panel interpreted GATT rules 'very restrictively' and this has implications for the environment far beyond the specific dispute. It decided that GATT prevents countries from taking trade measures to protect the environment or natural resources beyond their national boundaries. The implications of this for international environmental treaties or conventions are disastrous. It 'unequivocally' rendered illegal trade measures applied by individual nations to conserve wildlife and ecosystems beyond their national boundaries, including the high seas and all other global commons. This judgement removed the only compliance mechanism – economic sanctions – available to protect these habitats and natural resources....

Import restrictions based on production methods were also barred, thus as far as GATT is concerned, environmental impacts associated with the production methods used are irrelevant. Thus GATT could be used to challenge trade measures which discriminate between, for example sustainably or unsustainably produced tropical timber, or similar manufactured goods from pollution-controlled and highly polluting industries.

Source: Lang, Tim and Hines, Colin (1993) *The New Protectionism – Protecting the future against free trade*, Earthscan, London, pp65–67

size of the packet and the origin of the cereal that goes into the packet.

Many governments have Consumer Affairs departments to protect the interests of consumers. They regulate fair trade, inform consumers of their rights, and often provide mediation in disputes between industry and consumers. However, because they are government departments, they tend to adopt 'neutral' roles between consumers and industry. Their activities are rigidly limited by legislation. So there remains a need for consumer organizations which represent consumers' interests only.

In the USA, the major government food regulatory body is the US Food and Drugs Administration (FDA, see Box 9.4). This large bureaucracy regulates the types of foods allowed on to the US market, the labelling of foods, the examination of advertisers' claims about foods, and much more. It was originally set up in the late 1930s to protect fair trading by ensuring (among other things) that products contained ingredients they might reasonably be expected to contain (eg, that jam should contain a certain proportion of fruit or that tomato sauce should contain tomatoes). It has expanded its scope and directions considerably over the years.

One of these new avenues has been nutrition labelling. From the mid-1970s, the FDA and Codex standards called for printed statements giving the amounts of energy, protein, fats and carbohydrates whenever a nutritional claim was made (plus details of relevant vitamin and mineral content). After years of debate, this sort of nutritional information has been made mandatory in the USA. But, as we discussed in Chapter 7, consumer studies have shown that such types of labelling are not particularly effective ways of communicating to the vast majority of consumers. Its continuation on food labels in this form owes more to scientists' influence on government committees than to consumer needs.

In the UK, most of the food responsibilities that in the USA are within the semi-autonomous FDA's remit lie within the Ministry of Agriculture, Fisheries and Food (MAFF), which also promotes the interests of food producers and retailers, and its counterpart Departments in Scotland and Northern Ireland (see Box 9.5). Ministers in MAFF receive advice and recommendations from a

---

Box 9.4
Milestones in US Food and Drug Regulation

1784   Massachusetts enacted the first general food law in the USA.

1848   Import Drugs Act, the first Federal statute to ensure the quality of drugs, was passed when quinine used by American troops in Mexico to treat malaria was found to be adulterated.

1850   California passed a pure food and drink law.

1879   Chief Chemist Peter Collier, Division of Chemistry, US Department of Agriculture, began investigating food and drug adulteration.

1883   Harvey Wiley, Chief Chemist of Bureau of Chemistry assigned staff to study food and drug adulteration.

1891   Act passed requiring inspection of animals for diseases before slaughter.

1897   The Tea Importation Act was passed, providing for inspection of all tea entering US ports.

1902   Congress made appropriations to establish pure food standards.

1906   Congress regulated the economic integrity, safety and labelling of food.

1907   Bureau of Chemistry began administration of the Food and Drugs Act.

1913   Gould Amendment required that definite quantity information appear on food packages.

1938   Federal Food Drug and Cosmetic Act passed, forming the basis of modern US food safety regulation.

1958   Delaney Clause 'no food or color additive may be deemed safe that has been found to induce cancer when ingested by man or animals'.

1966   Fair Packaging and Labelling Act required consumer products in interstate commerce to be honestly and informatively labelled; Food and Drugs Administration to enforce provisions that affect foods, drugs, cosmetics and medical devices.

Source: Darby, WJ (1993) *Journal of Nutrition*, vol 123, pp277–78 with additional information from Miller, Salford (1993) *Journal of Nutrition*, vol 123, pp279–84

number of committees, such as the Food Advisory Committee (FAC), the Committee on Toxicity (COT) and the Committee on Medical Aspects of Food Policy (COMA). Committee members are liable to prosecution under the Official Secrets Act if they disclose documents without authority. The Ministry says, 'Members of these committees represent academic, health, analytical and food technology interests... Members of the committees are appointed in a personal capacity as experts in their own fields, and not as representatives of particular interests.' In the late 1980s, the FAC, MAFF's key food policy committee, had eight food industry-related members, – four academic and medical members, two local authority members and one consumer advocate member. By early 1994, the food industry-related group had dropped to seven and consumer numbers had risen to two.

Box 9.5
British Food Laws and Regulations

In the United Kingdom, as elsewhere, legislation controlling food has two principal aims – the protection of the health of the consumer and the prevention of fraud. In addition there is legislation which has an economic motive (tariffs, taxes, quotas, etc) which provide economic benefits to an industry but which do not usually involve the technologist. The two principal aims are achieved in the UK by a combination of primary legislation (the Acts) and more detailed secondary legislative measures (the regulations or orders). The Acts usually contain general prohibitions, which, when interpreted and enforced by the courts, provide general consumer protection. The scientific and technical requirements of food production require more detailed controls. The Acts therefore also contain the authorisation for Ministers, subject to specific parliamentary approval, to issue detailed regulations.

Dukes, D J (1993) *Food Legislation of the UK – A concise guide,* Third Edition, Butterworth-Heinemann, Oxford, p2

The Food and Drugs Act of 1955 consolidated all previous measures. It established the two principles of protection of health and prevention of consumer exploitation and allowed Ministers to formulate a much wider range of subordinate legislation. Under the Act, a series of Regulations was introduced covering sweeteners, antioxidants, emulsifiers and stabilisers, colouring matter, miscellaneous additives, lead, preservatives, chloroform, solvents, and arsenic in food, as well as bread and flour, food hygiene (general), and food labelling. A list of permitted additives was drawn up for each group and the amount allowed in foods set for preservatives, antioxidants and some miscellaneous additives.

The Food Act 1984 was a consolidating Act that combined previous legislation, some dating back to the nineteenth century, without developing anything new. The main objective was to ensure that food is wholesome and fit for human consumption. Standards were laid down in the relevant regulations and kept under review by the Food Advisory Committee.

Other laws relating to food control included the Food and Environment Protection Act 1985 which gave statutory backing to emergency arrangements (previously voluntary) to protect people from food which had been rendered unsuitable by harmful substances: the Control of Pesticides Regulations 1986 controls the use of pesticides; the Pesticide (Maximum Residue Levels in Food) Regulations 1988 specified residue levels in crops, food and feeding stuff; the Health and Safety at Work etc Act 1974 placed general duties on employers to provide for the welfare of their employees and may be regarded as complementary to the specific duties of the Food Act 1984; The Factories Act laid down requirements for employers to provide washing facilities and a place to eat. Other laws, such as those dealing with trade descriptions, also cover food as well as EU generated laws such as that for Materials and Articles in Contact with Food.

The public outcry over a series of food safety scares in Britain in the 1980s (see Annex 3) led to the adoption of a new Food Safety Act in 1990 that replaced the Food Act 1984. It was an enabling Act to plug the gaps in the 1984 law and to provide a framework to cope with the considerable changes in food technology and with food safety and other directives emanating from the European Commission in Brussels. It gave a wide range of powers to the Government for the following:

➤ promulgation of emergency control orders in cases involving unfit food or drink, where there is an imminent risk of injury;

➤ enforcement where enforcement officers can issue improvement or prohibition notices or can seize stocks, not just samples; Trading Standards Officers can enter premises, not just test the product;

➤ compulsory registration of food manufacturers, restaurants, sandwich bars or other food retailers; previously only ice-cream, sausages and pickled food needed a licence;

➤ ministers to make a wide range of regulations on food composition and to fulfil EU directives;

➤ ministers to control novel foods and prescribe special designations for milk;

➤ to bypass the immediate offender to prosecute the real offender, whose defence is to take all reasonable precautions or due diligence;

➤ to enable food authorities to provide training courses for people handling food for sale;

➤ to make it an offence to sell food which is not of the 'nature substance or quality' demanded by the purchaser, or falsely describing, advertising or presenting food.

The bypass provision allows a prosecutor to proceed against any person whose act or default, he believes, led to the commission of the offence rather than, as before, to require the person who sold the food to bring that other person to court.

In 1994, the Government announced deregulation of the food hygiene laws as part of its plans to implement the EC Food Hygiene Directive which it played a significant part in developing. The existing food hygiene regulations were replaced with less prescriptive rules which:

➤ place a general requirement to carry out food preparation in an hygienic way;

➤ put a duty on food businesses to assess and control potential food hazards;

➤ allow for voluntary guides to good hygiene practice to be produced by different industry sectors;

➤ introduce new requirements for food hygiene training.

Sources: Clutterbuck, C (1990) *Trade Unions and Food*, Fact pack 10, unpublished and MAFF

A long-standing complaint from many consumer organizations is that consumer and producer-industry responsibilities should be split, and not contained within the same ministry. There has been considerable change in the UK since the Food Safety Act 1990, although it has not gone as far as critics want. The major development in MAFF was a restructuring to establish separate food safety and agricultural commodities, trade and food production divisions, thus distinguishing more clearly between producer and consumer interests. The Ministry has also introduced consumer panels, which meet four or five times a year, at which consumer representatives can raise issues with the Ministry.

In 1991, Australia created a National Food Authority (NFA) in Canberra to:

facilitate the provision of a safe and wholesome food supply for the community and to promote national and international trade in food. The food standards it maintains and develops are taken up by the states and territories [of Australia] as food laws.

The NFA is the latest phase of a process of centralization and co-ordination in Australia. When the country became a federation known as the Commonwealth of Australia in 1901, this new form of government joined six self-governing colonies together in matters of trade, foreign affairs and defence. Gradually during this century, particularly during the past two decades, the federal government has gained more influence over the lives of Australians. However, food standards and legislation had been largely left to the individual states. This caused great inconvenience as each state enacted its own rules and regulations. For example, kangaroo meat could be sold for human consumption in South Australia but not in Victoria, and low fat sausages were introduced in some states but not in others.

The new NFA is a statutory Commonwealth (ie, Federal) body whose most important role is in public health and safety:

People want to know that: the food they eat is safe; the use of additives is carefully controlled; labels are accurate and informative; and, they are protected by food standards.

Other functions of the Authority include the following:

➤ The NFA conducts a national survey of pesticides and contaminants in food. The 'Market Basket' survey has shown the Australian food supply to be 'exceptionally safe'. The results have been published for consumers in a pamphlet entitled *How Green is my Trolley?*
➤ The NFA co-ordinates the recall of foods which may pose a risk to public health.
➤ It publishes the *Food Composition Tables* which list the nutrients in the Australian food supply. These are useful for dieticians, health workers, teachers and food industry personnel. A consumer version – *Food for Health* – is also available.
➤ The Authority continually updates food standards as new products appear. These are published in *The Australian Food Standards Code*. Much more of the work of the Authority, which consists of 5 commissioners and around 50 support staff, concerns the continuing review of new products and proposals from industry and other groups to vary the food standards. The Authority advertises these proposals in the national press and the results of its deliberations are published in *The Food Standard* which has wide distribution. Occasionally, open hearings are held which canvass views on controversial issues – for example, on a proposal to change vitamin and mineral restoration and fortification regulations, and the pros and cons of functional foods.

How to regulate

Regulatory authorities vary in the ways that they regulate. Some adopt *laissez-faire* or liberal policies, others are far more conservative. This can be seen in the two opposing approaches to managing the risks posed by the addition of chemicals to foodstuffs. In Norway, for example, manufacturers are allowed to add only chemicals chosen from a short list of chemicals which have been used for a long time. They are believed to have no adverse effects on health. All other chemicals, including many colouring agents, are banned. The alternative view allows the use of chemicals unless they are known to have adverse effects on health. In practice, this means that more or less any chemical which is considered to be harmless is allowed until adverse effects show up. Then it is removed.

Most countries' practice lies somewhere between these viewpoints. The USA and Australia, for example, allow the use of a small number of chemicals (in Australia about 300).

In the USA, new chemicals are tested and monitored extensively by the FDA.

The FDA's Delaney Clause forbids the use of any substance in food which has been shown to cause cancer in experimental animals. In these experiments, animals are exposed to doses of chemicals far in excess of the amounts to which humans will ever be exposed through their diets. Doses over 15,000 times the maximum human exposure are common.

Newspaper headlines such as 'Cyclamates cause cancer' and 'Too much caffeine bad for hearts' can create the impression that many foods contain deadly amounts of 'artificial' substances – especially in the many people who never read much beyond the headlines. Yet toxicologists and cancer experts agree that huge doses of many compounds will induce cancer processes. Dose rates are what make most substances healthy or harmful – the higher the dose the more likely the substance is to be harmful. One effect of this well-meaning FDA policy has probably been to cause a greater degree of panic and loss of confidence among the public about the food supply than might be justified.

To be effective, regulators need flexibility to keep up with advances in science and in society. This, however, probably requires a broadening out of those involved in framing regulations and better understanding of the issues by the public. The people sitting on advisory bodies drafting regulations tend to come from the scientific and industrial circles concerned with the particular topics and may lack this broader perspective.

In Britain, when the Food Safety Act 1990 consolidated food safety requirements throughout the food chain, it affected food handling, retail and manufacturing practice and the relationships between the different actors. This was partly due to the introduction of the statutory defence of 'reasonable precaution and due diligence'. This makes each person or body in the food chain supplying products responsible for their safety, but should problems arise and they have taken all reasonable precautions and exercised due diligence, then they have not committed an offence. Precisely what due dili-

gence means will be tested in the courts, but the nature of due diligence is likely to vary, depending on the size of the enterprise. Small shops, for example, might be expected to buy from reputable suppliers and maintain certain levels of hygiene, while large supermarkets may be expected to do much more to ensure the safety of the food they offer for sale. In fact, major retailers have gone right back down the line with their suppliers and original producers to inspect their operations in order to cover themselves. Although businesses that have exercised due diligence and are subsequently prosecuted may have an acceptable defence before the courts, this does not apply to defendants who manufactured or imported the food.

Pressures to change

The different actors in the system put various kinds of pressure on law-makers and regulatory bodies to change the framework in which they operate. Sometimes changes respond to public concerns, such as the 1990 Food Safety Act in the UK, and at other times changes are more a response to the concerns of industry. One example of the latter is compositional standards. Also known as *vertical standards*, they define legally what must be contained in certain products, such as jams, sausages, cheese and chocolate, for them to be called by a certain name, and some relate to the traditional ways of producing such products. The FDA and the EU have yielded to industry pressure over the enforcement of traditional compositional standards which are gradually being abolished. They are being replaced by so-called horizontal standards which are mainly concerned with safety and labelling.

A switch from vertical to horizontal standards allows manufacturers to make new products which mimic traditional products but contain different ingredients. Traditional products no longer need to contain as much of certain ingredients as consumers might expect, as long as the ingredients are declared on the label. In the UK, when the compositional standards for meat products, such as pies and tinned products, were abolished in 1984, the meat content of some of these fell but not the price.

Soon it may be possible to buy products like tomato sauce which look and taste like tomato sauce but which contain no tomatoes. All the ingredients would be judged 'safe', but would the product be what most people expect? They will have to read the label to find out, but the label may not be designed to tell them.

The science of designing labels to ensure that only that which the seller wishes to extol is given, and that to which he/she wishes to give less prominence is hidden away, has led to specialist label compilers, and hence, the use by trading standards officers of psychologists to explain to the courts the significance of label designs, many of which cannot be said to give meaningful information

Roberts, David (1991) 'Compositional Standards and Labelling', *British Food Journal*, vol 93, no 6, p16

The move to horizontal standards shifts the definition of what is a food away from experience and tradition (as enshrined in compositional standards and common-sense understanding of the meaning of words such as meat or ham) to food companies, marketers and sellers (see Box 9.6). Horizontal standards remove the need for manufacturers to explain what they are doing as they market something new or to sell what is really in the product. Few, if any, are likely to label as such products that contain significant amounts of bone, gristle and mechanically recovered meat or show their production processes in their advertisements. Although the EU Food Labelling Directive actually requires a statement of treatment, where its absence might mislead, few comply with it and the Commission directorate responsible has watered down its position on labelling mechanically recovered meat.

The intense pressure on regulatory agencies such as the FDA is highlighted by the saga of the health claims on labels (see also Chapter 7, box 7.3). Prior to the mid-1980s, health claims about the therapeutic benefits of certain food constituents were banned by the FDA. In the mid-1980s, after lobbying from manufacturers and health agencies (such as the National Cancer Institute), these restrictions were relaxed. The result was a flood of dubious claims about the health-giving, disease-preventing effects of particular products (eg, this product contains dietary fibre which has been shown to prevent bowel cancer, therefore buy X). By 1990, almost half of the products sold in the USA were estimated to include a health or nutrition claim. For the public, it adds to the confusion.

The FDA's new regulations for compulsory nutrition information use the same old highly technical and numeric language in the nutrition information panel which does little to deal with public confusion (see Figure 9.1).

Serving size 1/2 cup (114g)			Servings per container 4	
Amount per serving				
Calories 90			*Calories from fat 30*	
			*Per cent daily value**	
Total fat 3g				5%
Saturated fat 0g				0%
Cholesterol 0mg				0%
Sodium 300mg				13%
Total carbohydrate 13g				4%
Dietary fibre 3g				12%
Sugars 3g				
Protein 3g				
Vitamin A	80%	*	Vitamin C	60%
Calcium	4%	*	Iron	4%

* *Per cent daily values are based on a 2000 calorie diet. Your daily values may be higher of lower depending on your calorie needs:*

	Calories	2000	2500
Total fat	Less than	65g	80g
Saturated fat	Less than	20g	25g
Cholesterol	Less than	300mg	300mg
Sodium	Less than	2400mg	2400mg
Total carbohydrate		300g	375g
Fibre		25g	30g

Calories per gram:

*Fat 9   *   Carbohydrate 4   *   Protein 4   **

Figure 9.1 The US Food and Drug Administration's new food label for the 1990s. Some labels may list more nutrients than this sample

Box 9.6
What's in a Word – Food or Fraud?

We like to think that we know what we mean when we use words. Unfortunately, this is probably not true when it comes to food. There are two problems. One is related to adulteration – we ask for something and get a product that is rather different from what we expect. In Britain, for example, when Liverpool trading standards officers sampled mince (ground beef) at 25 shops in the city, they found that 68 per cent of the samples were adulterated. One sample of 'minced Scotch beef' actually contained 70 per cent beef and 30 per cent lamb and pork. The worst mince contained 43 per cent fat, 20 per cent lamb and 20 per cent pork!

Selling foods described in such a misleading way is fraud. The second problem is not so clearcut but can be equally misleading. What you think you are buying – based on normal, everyday experience – may be rather different from what you are actually getting, since the law may allow a name for a substance that common sense would not accept. Take meat as an example again.

In the EU, there is no real definition of meat for compositional purposes, but each member state has its own definition. In the UK, for example, there is a legal definition of what is meat, which itself is pretty far removed from what most people would expect to buy if they went to a butcher's shop. Legally, meat in the UK can include the flesh, including fat, and the skin, rind, gristle and sinew in amounts that are naturally associated with the flesh. Meat when used to describe the contents of a processed product, however, can also include head meat (muscle meat and associated fatty tissue only), heart, kidney, liver, pancreas, tail meat, thymus and tongue. So any product listing meat as an ingredient could be made from any of these bits.

More confusingly still, sausages called pork have to be 65 per cent meat, only half of which need be lean, the rest could be fat. Beef sausages need contain only 50 per cent 'meat'. Moreover, a pork sausage can contain up to 20 per cent of meat other than pork; and a beef sausage can contain up to 50 per cent of other meats. It is not surprising, then, that countries where sausages must be 100 per cent meat are not very keen on allowing British sausages to be called by the same name. Yet under mutual recognition, foods sold in one EU country must be accepted in another.

Apart from words used to describe a product's ingredients being misleading to customers, different definitions can also affect industry. For example, much of the meat canning industry in the UK has moved to other EU member states where the definition of 'meat' is still broader and where blood plasma, bone protein and masses of connective tissue can all be used to make tinned meat products. This puts UK manufacturers at a disadvantage as they cannot use these cheaper ingredients.

Such examples are but the tip of a giant iceberg when it comes to understanding just what it is that someone is selling. When compositional standards are removed, so that anything can be called jam – for example, as long as the fruit content is put on the label, there is an especial need for clear, accurate labels. There is also a need for clear definitions of words that consumers have commonsense definitions of, like meat and ham, for use in food products. For trading standards officers like Dave Roberts from Shropshire in England, concerned about truthful, accurate descriptions and fairness for customers and industry alike, such confusion over labelling facilitates the perpetration of fraud. And in English law, at least, it is not just what is said, but also what might be omitted from a label that can make it false.

Sources: Dukes, quoted in Box 9.5 and Roberts , David (1989) 'Lean times for meats' *Municipal Journal*, 6 October

Scant attention has been given to the kinds of nutrition information consumers want from food labels, such as clearer details of ingredients, additives and overall nutritional status – for example, how much of the energy in the food comes from saturated and unsaturated fat or sugars. And, as usual, nutrition dominates the label to the exclusion of other legitimate consumer interests, such as the ecological status of the food product or methods by which it was produced. One improvement, however, is a dietary value column which tries to explain how much of an individual's daily requirement a serving of the food provides.

In Japan, the situation is more chaotic as manufacturers have been allowed to make wild claims about the supposed effects of their products (eg, 'X calms you', 'Y has aphrodisiac properties'). Most of these claims appear to be based on loose interpretations of traditional herbal remedies. Perhaps this is why Japan leads in the development of so-called 'functional' foods which are supposed to contain ingredients which have specific effects on particular body systems (see Chapter 4, Box 4.10). For example, it is claimed that some diets are

anti-hypertensive and, when combined with 'mild' anti-hypertensive drugs, that they can exert powerful health-restoring effects.

The American policy on health and nutrition claims was reversed in 1992 when the FDA regulated for only a small number of precise nutrition claims. Terms like 'low fat' and 'lean' were defined to give the public reliable, consistent information. They also defined what a food must be if it made certain health claims (see Box 9.7).

It seems that the tension between demand for simple, traditionally formulated products free of advertising hype and demand for 'functional foods' is likely to keep regulators very busy. Both may be reasonable paths, but are regulatory authorities able to use appropriate participatory decision-making skills and techniques to allow reasonable consensus between industry, scientists and consumers?

Box 9.7
FDA Defines Key Words

The FDA in the USA has defined a range of key words used on food labels that 'claim' something for the product about its healthiness or certain other properties. Some examples include the following:

Key words	*What they mean*
Fat free	Less than 0.5 gram fat per serving
Low fat	3 gram fat (or less) per serving
Lean	Less than 10 gram fat, 4 gram saturated fat and 95 milligram cholesterol per serving
Light (lite)	One third less calories or no more than half the fat of the higher calorie, higher fat version; or no more than half the sodium of the higher sodium version
Cholesterol free	Less than 2 milligram cholesterol and 2 gram (or less) saturated fat per serving

To make health claims about...	*The food must be*
Heart disease and fats	Low in fat, saturated fat and cholesterol
Blood pressure and sodium	Low in sodium
Heart disease and fruits, vegetables and grain products	A fruit, vegetable or grain product low in fat, saturated fat and cholesterol, that contains at least 0.6 gram soluble fibre, without fortification, per serving

Source: FDA & American Heart Association, (1993) *How to read the new food label*, AHA, Dallas, TX

Regulating the regulators

Much depends on who makes the rules and regulations in the food system. They are normally agreed on the advice of specialist advisory bodies to governments, such as the FAC in the UK, or international organizations such as Codex. These bodies, as we have seen, usually draw on specialists in a particular area, such as labelling, novel foods, safety, and so on, who work in academia and industry. Relatively few, if any, represent consumers or enforcement officials.

Broader factors than simply the composition of the committees may also be at play, however, as in the case of the FDA's new labelling regulations which seem to ignore the findings of important research into the information processing needs of consumers. Certainly, the power and influence of different actors plays a role. Different groups of actors and individuals often perceive any new proposal as a threat to their well-being and control. Related factors include the experience and expertise of organizations in dealing with information. Public servants are often recruited to food and nutrition authorities because of their background in food science and technology, nutrition, microbiology or law. They tend to foster the habits and viewpoints of any one of these sciences – rarely do they have backgrounds in more than one discipline and more rarely do they span disparate areas such as nutrition science and the law.

Such specialized academic backgrounds make it difficult for public servants to appreciate the wide, diffuse agendas of ordinary shoppers and food consumers. Consumers' views can usually be dismissed as based on some form of ignorance. Moreover, it is unusual for any group of consumers to maintain prolonged influence over any government department.

A related factor concerns the ways that we make the rules and laws which govern food system practices, at least in the Anglo-American world. Debate about issues tends to be on an adversarial basis. Skirmishes between different points of view are organized in the form of media releases and rebuttals, court cases, and scientific and industrial meetings. Although discussion of issues such as food labelling may occur in depth for a long time within such groups, often preaching to the converted, there is rarely any prolonged debate between groups which hold opposing views.

In general, Westerners have lost those social structures which facilitate consensus building and conflict resolution. In contrast, many traditional societies still have such structures and processes. In Maori and Aboriginal tribes, for example, long discussions take place over several days (or longer), which bring conflicting parties together under conditions which allow them to negotiate their differing expectations and views. These 'meetings' end only when consensus is reached, and the different parties (and their social networks) are bound by the agreements that have been reached. Perhaps food regulation authorities need to experiment with better ways of reaching a consensus, using the range of social marketing skills used by industry, which pays market researchers to aim their products in ways which will link into consumers' concerns. There is a need to find ways to allow fairly equal negotiation between expertly defined 'needs' and viewpoints and those felt by shoppers and other food consumers which, as discussed earlier, span non-scientific values such as equity, ethics, ecology and well-being.

Enforcement

The complexity of laws and regulations can make life difficult for those whose job it is to apply them locally – the enforcement officials. They need clearly framed, explicit and precise guidelines to be able to do their job effectively for, whatever the rules and regulations say, and with whatever good intent, without enforcement mechanisms people can ignore them with impunity. And even with enforcement, some are tempted to stretch them as far as they can.

Responsibility for legally enforcing 'fair trading' practices – for example, policing weights and measures regulations, and environmental health requirements such as hygiene, normally fall upon various authorities within states. The staff and departments involved vary widely between countries. In the UK, for example, these are the responsibility of local authority Trading Standards Departments (usually concerned with composition and labelling) and Environmental Health Departments (health and hygiene). Elsewhere, food hygiene is often local-authority based, but outside the UK it is rare to find weights and measures at anything other than the central or regional level. Enforcement agencies check what the different actors are doing to ensure that they comply with the regulations and also they investigate complaints from the public. Precisely how they operate varies even within the UK, depending on whether they are in England and Wales, Scotland or Northern Ireland. In England and Wales, the operation varies between metropolitan areas and non-metropolitan areas. Public analysts, usually employed by county councils, analyse samples taken by the enforcement officers.

For people marketing food, the label and the way the food looks are a part of selling it. Consumers and legislators, however, tend to see the label as there to inform them about the food. For those whose job it is to enforce the law and to ensure that the public get a fair deal, the move from compositional standards to 'informative labelling' just changes the nature of the job, it does not make it easier. Unfortunately, they are rarely sufficiently involved in shaping the rules and regulations, and are not asked for guidance on what can, and cannot, be enforced.

Overall, food laws, rules and regulations are complex and ever-changing to meet changing technical, social and economic demands. Clarifying and making explicit the issues involved and resolving conflicts is the job of those involved in food policy, which we discuss briefly next.

Policies for food

Food policy encompasses the collective efforts of governments to influence the decisionmaking environment of food producers, food consumers, and food marketing agents in order to further social objectives.

Timmer, C Peter, Falcon, Walter P and Pearson, Scott R (1983) *Food Policy Analysis* World Bank, published by the Johns Hopkins University Press, Baltimore MD, p9

Laws change. They are made by politicians in power. Politics is about power, benefits and ultimately implementing policies to secure particular interests. Throughout the food system, as in all other areas, there are actual or potential conflicts of interest. Our governments are responsible for developing policies that reconcile the conflicts of interest in the best interests of the community – local, national and global. Debate and conflict about what is desirable and undesirable, about which directions are good and bad, and whose interests should be protected, are the meat and drink of political life.

Policies influence events. Policy making is at the heart of government. It is what interest groups seek to influence. Often, policies relate to immediate problems with little thought for the overall aims and goals of society. Politicians and administrators may keep going as best they can to defend the interests of those with power and control. But such people do not form an homogeneous group. They act in various sub-systems that make up the food system. Today, as we have seen in the various chapters of this book, policies in the food area must deal with a wide range of issues, for example:

➤ overproduction in the developed countries and underproduction in many developing countries;

- a strategy for a healthy diet;
- the ecological sustainability of agriculture;
- concentration of power in the food industry and retailing;
- reduced local access to food as small retailers disappear;
- the terms of trade between commodity producers and industrial producers;
- the direction and aims of research and development.

Policy concerns

Food is an exceptionally political commodity...the idea of there being a COLLECTIVE wellbeing in addition to an individual one will lead to an array of policy actions designed to protect or foster the interests of the weak and vulnerable, referee the interactions both within and between groups all along the food chain, insulate one group from the actions of others, and redistribute the benefits to achieve a more equitable pattern than the market system can manage on its own.

McInerney, John (1983) 'A Synoptic View of Policy-making for the Food Sector', in *The Food Industry: Economics and Policies*, Burns, Jim McInerney, John and Swinbank, Alan eds, Heinemann, London, pp163 and 172

Food policy, of course, is an aspect of socio-economic policy. Dealing with some issues would require policy changes beyond the food system. However, effective policy making requires an understanding of how the various pieces involved in the jigsaw puzzle that makes up the food system operate. John McInerney, in *The Food Industry*, points to five areas with which the conventional economic approach to policy concerns itself:

1. efficiency – or avoiding waste of resources; inefficiency leads to a loss to the community with benefits of consumption being less than they might have been;
2. growth – in which society gains progressively more benefits from its resource use;
3. stability – limiting fluctuations in supply, prices or consumption;

4. sustainability – the ability to maintain consumption levels into the indefinite future;
5. equity – how fair the economic workings of society are to all concerned.

In market economies, the main pattern of distribution depends very much on who owns what and how the different markets work.

Policy implementation

Policy changes generally are designed to affect one or more of these economic criteria. To turn policies into actions that bring about change, however, requires more than a policy statement. Three factors are crucial for implementing policies:

1. Institutions able to enact the policies. These may be national ministries, international trade bodies, and a range of sectional bodies like trade and consumer associations, or trade unions.
2. Policy instruments which influence the system. These can be direct, such as price-fixing mechanisms and redistributive taxation, or indirect, such as voluntary operating codes, which use pressure and persuasion to change practices. Both types tend to do one or more of three things:
   - initiate or encourage things to happen which otherwise would not;
   - inhibit or prevent things from happening which otherwise would;
   - redistribute goods, services or power in the system.
3. Information – without this we cannot know how to attempt to act or what are the effects of the action.

Food policy

Food policy concerns far more than agriculture, and there is more to agriculture than producing food. Today, agriculture accounts for less than half of the average cost of final foodstuffs. In the UK, for example, agriculture accounted for a little over 15 per cent of food costs in the late 1980s. In 1991 in the USA, only 22c in every dollar spent on food went to the farm, down from 32c in 1971 (see Figure 9.2). Nestlé, a major multinational food processor, spent only 28 per cent of its turnover on raw materials in 1991.

Production in any sector is geared to producing commodities, be they raw materials, such as wheat or cotton, or finished products, such as bread and textiles, for exchange in a market system for money. Processors, manufacturers and distributors must produce for a market at a profit and more and more of the final cost of a food goes to them.

For policy makers, a central question is just who gets what benefits from the system. Power, influence, and control of resources largely determine who gets what benefits from food. One example in recent years, as we saw in Chapter 6, is the shift in relative power between food manufacturers and food retailers in the UK, which has resulted in a larger share of the profit pie from food going to the retailers.

It was not really until the 1980s that food policy was taken as a serious issue of concern to the rich countries. Mostly, it had been seen as something relevant only for poor developing countries with difficulty feeding their populations, which is the focus of *Food Policy Analysis*, quoted at the start of this section. In 1981, however, the OECD in Paris published a short book called *Food Policy*. It saw food policy as describing:

> *a balanced government strategy regarding the food economy, which takes account of the interrelationships within the food sector and between it and the rest of the national and international economy.*

Despite its age and bureaucratic language, it is still a useful document that has not been followed up sufficiently in subsequent OECD work. A modified version of their definition of

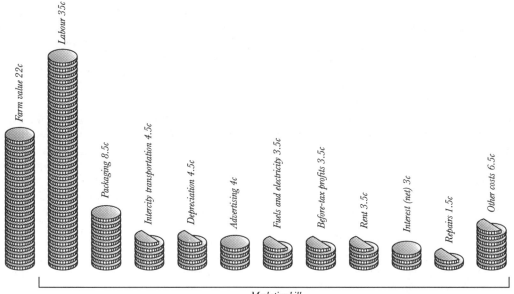

Source: Elitzak, Howard (1992) 'Marketing Bill Is Largest Chunk of Food Expenditures' *Food Review*, vol 12 no 2 July–September, p13

Figure 9.2 Breakdown of US food costs per dollar spent in 1991 (includes food eaten at home and away from home). 'Other costs' includes property taxes and insurance, accounting and professional services, promotion, bad debts, and miscellaneous items

the food economy also provides a good definition of the food system as:

> *the set of activities and relationships that
> interact to determine what, how much, by
> what method and for whom food is produced
> (and distributed). The food economy is a
> sub-set of the overall national and interna-
> tional economy.*

By food is meant 'the end products that consumers buy to eat or drink', yet the
relationships that affect what happens when
the food is taken home and transformed into
meals are also important – who selects and
prepares, and how their choice is shaped, as
we discussed in Chapter 7.

The changing playing field

Agriculture has traditionally been the only part
of the food system consciously considered in
policy-making in the industrialized countries.
It could benefit from a broader approach to
food policy, without which it will be increasingly squeezed. This is because, as we saw in
Chapter 5, farmers tend to buy inputs from,
and sell their products to, very much larger
businesses operating in markets with a few
major players. This means that agriculture is
vulnerable to shocks and pressures from these
businesses as well as from the inherent instability of agricultural production itself. A broader
approach to food policy could make the relative position of agriculture better appreciated.
Indeed, current technological developments
could transform the whole food system:

> *The changing technical base of food process-
> ing is creating new entry points for chemical
> and pharmaceutical companies, which will
> lead to the significant restructuring of the
> industry. The ultimate prize is domination
> and proprietary ownership of the scientific
> knowledge and process engineering technolo-
> gy required to control the complex biological
> reactions and microbial activities involved in
> food manufacture...*
> *The cumulative effect of this tendency is*

> *to obscure the specificity or 'identity' of rural
> commodities, reinforcing the long-run move-
> ment of substitutionism to reduce the share of
> agriculture and land in value added generat-
> ed by the food system....*
> *The ability to regulate and directly trans-
> form biological processes will give a major
> stimulus to the computerized monitoring and
> robotization of agricultural activities. The
> application of modern biotechnologies marks
> the decisive break with traditional farming
> knowledge. This already had been eroded by
> the advance of mechanical and chemical inputs.
> Now, however, genetically engineered plants
> and animals and controlled environment pro-
> duction demand a radically new information
> base. The farmer will give way to the 'bio-
> manager' and observation will be replaced by
> 'software'. Biotechnology and information
> technologies therefore go hand-in-hand to create
> the new production process in agriculture.*

Goodman, David Sorj, Bernardo and Wilkinson,
John (1987) *From Farming to Biotechnology – A
Theory of Agro-Industrial Development*, Basil
Blackwell, Oxford, pp138–39 and 184

Over the past decade or so in the food system
developed by the industrialized countries, we
have seen a continuation of the trends identified in the OECD report:

➤ food becoming a smaller part of the overall economy, with agriculture accounting
for an increasingly smaller share of
employment and value added within the
food economy;
➤ non-food components of economic activity increasing relative to food components,
with basic nutrition per capita intakes,
calories per head, increasing only slowly;
➤ a greater variety of foods available and
some internationalization of consumption
patterns;
➤ differing bargaining positions within food
economies, from competitive to monopo-
listic with a general trend to fewer and
larger companies, with a few often
accounting for a large share of output in
some OECD countries;

➤ the 'distance' (space, time and psychological) between producer and consumer lengthening with specialization;

➤ food being less a group of commodities consumed for nutritional value and more as something eaten for its novelty, status and entertainment value, as well as convenience, reliability and safety.

Limits to eating

Within the OECD countries, industrialization of the food system has continued, while production methods and consumption patterns have become more international. But in these rich markets where basic food needs have been met, the crucial problem for the food industry is the limited quantity of food that people need. There is a limit to what people can sensibly eat. This is a central issue for consumer-based market economies and underlies discussions about sustainability and green and other issues today. For while you can increase the number of clothes, shoes, TVs or other electric gadgets in a household, it is hard to increase the quantity of food consumed, especially if it is consumed as basic food commodities.

The limits to what we can eat put pressures on those involved in food businesses, as we saw in Part 2, to find ways to increase the return that they get from these basic commodities – to increase 'added value', in the industries' euphemistic term, for increasing the profitability of what they sell. That means turning 5lb of grain, say, into 1lb of meat, which costs more to buy than the basic grain. Today, beef and chicken are value-added forms of grain. Next, by processing the meat, it can be sold for even more. Another approach to increasing returns has been to mix cheap ingredients, like fat, sugar and starch, into a wide range of products using food technology.

Such pressures have helped to bring about the development of food patterns and products which, when consumed in such quantities and with the lifestyles of those in the industrialized, urbanized world, lead to the very problems which have prompted concern about food, health and sustainability. This is particularly the case with a system which in the rich countries suffers from chronic overproduction.

Many factors then – including the changing size, age structure, mobility, location and lifestyles of the population – affect consumption trends for different commodities. They also affect the ways in which these commodities are eaten. The growth of real disposable income affects the composition of foods in people's diets, but also leads to a saturated market in which the quantity consumed hardly increases at all and the percentage of family budgets spent on food declines (see Box 9.8).

As real incomes have increased in the industrialized countries, so too has the range of goods and services available. People also have more options on how to spend their time. Since food is consumed frequently, the consumption of convenience food and the acquisition of household technology have together greatly reduced the time spent on household food preparation which has been shifted into factories. Shopping habits have also changed. Fewer people buy provisions daily, and more do bulk shopping in supermarkets once a week or less.

The market squeeze increases the demand for technological change to enable firms to get a competitive edge, which in turn generates further competitive pressures throughout the market. This, coupled with agricultural policies, has led to greatly increased productive capacity with no real market, as in the dairy sector, and is a major reason for overproduction in the food system of the richer countries.

The range and volume of information available to producers and consumers has also increased, helping the internationalization of production and consumption – for example, in soft drinks, fast foods and some convenience foods. Computerization allows improved control throughout the food system with important potential implications for reducing storage, wastage, improving delivery scheduling, and therefore for resource use and costs.

The interlocking and changing nature of the food system means that policies directed at one part of it have increasingly important effects elsewhere. They can conflict with objectives pursued in other parts of the food system or in the wider economic system.

Obviously, many policies affect food today, such as

➤ agricultural policy, both national and international;
➤ consumer and social policy in cost, safety, consumer information and nutrition;
➤ economic policy in management and industry;
➤ energy policy;
➤ international relations and trade policy;
➤ employment, transport, education and environmental policies.

Changes in some of the latter, for example, are likely to increase the immediate costs of food production but enhance its sustainability.

This is why policy formulation is complicated and difficult, and makes it unlikely, perhaps, that a completely coherent policy covering all aspects of food will emerge. A far greater coherence between different areas affecting the food system is possible, however, avoiding the situation where policies in different areas are in complete opposition to each other. This was the case, for example, with nutritional recommendation to cut fat intakes in the 1980s in the UK, which was in opposition to the farming policies which encouraged fat production.

Policies for change

Few industrialized countries have tried to develop broadly-based food policies until recently. Interest grew in such food policies in the 1970s and 1980s, following public concern about the links between diet and disease, appalling famines, food mountains and food safety.

Norway has made perhaps the most wide-

Box 9.8
Competitive Pressures

The combination of slow growth in per capita demand, coupled with a relatively slow growth in population, results in increased pressure on the capital and other resources employed in the food economy, increasing the degree of competition.

To survive in such a market, businesses from farm to mouth try to:

➤ improve efficiency (and lower unit costs);
➤ diversify product ranges and markets;
➤ remain in the food economy and accept lower returns on resources than elsewhere or transfer resources to activities providing higher returns outside food, or persuade governments to subsidize the various parts of the food economy.

The results of these pressures, as identified by OECD in 1981 remain true.

In virtually all sectors of the food economy these demand pressures have resulted in rationalisation, concentration, oligopolisation and a reduction in the overall level of resource use and the substitution of capital for labour and land....there has been a diversification of resource use – in agriculture through part time farming, including, for example, the use of land for leisure purposes;... in food processing and manufacturing through the creation and advertising of new products, and in expansion into non-food areas; in food retailing in terms of spatial competition and, again, selling non-food products; in the post farm sectors generally, in some export of capital and development of multinational enterprises.

OECD (1980) *Food Policy* OECD, Paris, p40

ranging attempt, following acceptance by the Government of a report in 1975 which provided a blueprint for the development of a food and nutrition policy. The main objectives were to ensure adequate food supplies, the use and preservation of the natural resource base, regional development, equality of agricultural incomes with the average incomes of industrial workers, reasonable consumer food prices and adequate nutrition, and to take account of just international relations (see Box 9.9).

In the USA, Britain and most other industrialized countries, concern focused more on links between diet and the pattern of disease and premature deaths. Various reports have listed evidence of links with heart disease, bowel disease, and so on. In the USA, major public health initiatives were mounted to reverse some consumption trends, in line with dietary recommendations; these seem to have been quite effective in leading to a drop in deaths from coronary heart disease.

In Britain, a major Government-sponsored study by the National Advisory Committee on Nutrition Education (NACNE) in 1983 suggested ways to influence food production and consumption which would bring overall health benefits through greater life expectancy and reduced risk of disease. In an action which is typical of the British approach to openness (or 'transparency'), it was initially suppressed by the government department responsible, although the leaked contents and finally the published report came as no sur-

Box 9.9
Norway's Food Policy

Norway set four main goals for its comprehensive food policy adopted in the mid-1970s:

1. encourage a health-promoting diet, reducing overall consumption of fats, especially saturated fats, and replacing them with polyunsaturated fats, whole grains and vegetables;
2. promote domestic food production and reduce food imports, increasing national self-sufficiency from 39 per cent of total calories to 52 per cent by 1990;
3. promote agricultural development in the country's less advantaged, outlying regions with due regard for preserving the environmental resource base;
4. contribute to world food security, promoting food production and consumption in poor countries.

By the early 1990s, much progress had been made, especially on food self-sufficiency, regional development and the contribution to world food security aspects. Dietary changes were most evident in fat consumption, with average total fat consumption dropping from over 40 per cent to under 34 per cent of total calories; deaths due to heart disease had fallen correspondingly.

Perhaps more interesting from a food policy development perspective was the experience it has produced in the difficulties, strategies and policy instruments needed to implement a broad-ranging food policy. The Interministerial Council which was set up to co-ordinate policy implementation and which had the power to do so was relatively passive. The older established advisory body, the National Nutrition Council, became the *de facto* force pushing the policy along, using information as a key weapon.

While much has changed in Norway, the complex interactions between the different actors continue. Both official and consumer groups are continuing their efforts to meet all the goals. The whole process, however, has led to former opponents of the policy, such as the food and agriculture industries, becoming much more willing to listen to arguments from nutritionists and health authorities.

Sources: Milo, Nancy (1990) *Nutrition Policy for Food-Rich Countries, A Strategic Analysis,* Johns Hopkins University Press, Baltimore MD; Cannon, Geoffrey (1992) *Food and Health: The Experts Agree* Consumers Association, London

prise. Although the Government never officially accepted the report, it led to great debate and continuing public interest in diet and health. The COMA report of 1984 on diet and cardiovascular disease, like the NACNE report, also introduced target numbers for change and was accepted by the Government. Improving the nutritional status of the population to reduce the damaging effect of diets rich in fats, sugars and highly refined foods has received most attention, with further targets set in 1990 as part of the Health of the Nation initiative.

Nutrition, however, is only one consideration in the development of a food policy. Both national and international equity issues, the health of other people involved in our food system, various agricultural practices, operations by transnational corporations, pricing systems, the structure of technical and food aid, and so on, are relevant. All need open discussion and debate.

Without informed public debate, policies remain implicit and are dictated behind the scenes by the outcome of pressure from the various actors who are trying to control the food system and capture the benefits arising from it. That control, however, cannot be seen in a simple national framework, since food is an international business, as we have seen throughout this book.

An open information system can also enable politicians and policy makers, perhaps only after public pressure, to correct unexpected side-effects that frequently arise from policies. Some unexpected results are caused by actions outside the national economic system and others by national policies which have an impact elsewhere than intended. Openness facilitates feedback which allows modifications to be made.

Information flows and gatekeepers

There are various barriers to openness and public debate, so necessary for effective policy-making and implementation. One is the development of bureaucracies by all institutions and organizations. In both public and private sectors, the bureaucracies have their own agenda in continuing and can either inhibit or promote change. Much information is generated by and within these structures and should flow through and out from them

Susan George, in a long critique of the World Bank's policies, notes that among its employees is a large group of ecologists, many of whom are extremely dedicated. Yet the Bank has pursued policies which are often detrimental to the environment. Part of the reason for this lies in the structure of the organization itself, as in the case of the Fordist factory. Decision-making is hived off to the very top of the organization. Information has to travel up through a series of barriers (eg, departmental heads) before decisions are made by those at the top who have the power but not the responsibility to act on the information. These are authoritarian institutions.

Another intangible influence is the ideology of the organization and its key personnel. In this sense, ideology refers to basic assumptions about the world – for example, that it is a just world, that the market is the best arbiter of policy, or that there is no such thing as society, only individuals whose needs are paramount. Information which conflicts with these basic world views is often ignored, watered down, or denigrated. If the structure of the organization is based on the supposed wisdom of those at the top, views differing from those held by the people in power are unlikely to influence it.

Finally, the personality of key staff is all important. Norman Dixon in a study of military leadership has shown that in hierarchical organizations, authoritarian personalities tend to rise to the top where they exert power on behalf of themselves and only rarely in the interests of the organization's aims. Thus, organizations which have fossilized are often led by people who are unable to take decisions and who tend to deny or repress information which diverges from their views.

Feminists would point to one factor which unites practically all powerful organizations. They are almost all run by men. This means that male ideologies about competition and success, and ritualistic behaviours, as Margaret Mead's *Male and Female* (Victor

Gollancz, 1949) showed, pervades them in contrast to more feminine views of harmony and co-operation. Given this male preponderance, organizations would need a great deal of input from women's viewpoints, which most do not seem to have.

Most government and international organizations are dominated by male-centred ideologies and élites and some are antithetical to the purposes of liberal government. Michael Pusey's study of senior officials in the Australian public service, for example, presented the absurd but true finding that government service was dominated by people whose personal ideologies had little room for government intervention. Pseudo-religious belief in the market tends to substitute for a lack of decision-making ability, which can only be expected of those whose actions are removed from the reality of their effects. During the past 20 years, we have been plagued not so much by big government as by bad or non-existent government.

Such organizational maladies can be found in all sectors, including food. Sometimes organizations do make major shifts if their survival is threatened. In a sense, these can be seen in some of the large firms discussed in Part 2 of this book. For government organizations, this is less the case, but these are affected by the changing ideologies of those in power. We should not expect all individual employees to act heroically if they discover adverse effects of their company's, organization's or government's policies and actions on public well-being. In fact, they are unlikely to do so because they act according to the mind-set engendered by their working environments. To food technologists, public concern about 'additives' is largely

unfounded as they will have found that most additives have little or no effects on most consumers. Consumers, on the other hand are more concerned about what the additives *might* do to them, to their children and the environment, and may view risk differently.

People's thoughts and behaviours are influenced by the groups to which they belong. Retail and trade associations are mainly concerned with the welfare of their members – retailers will have quite different opinions about importing cheap processed foods than, say, farmers. Each group has its own ways of working, its own views of what is correct conduct, and its own interests, and each group is largely isolated from close contact with other groups. Some are secretive, from the in-flight catering association to the grain trade association, as we know from trying to get information from them in writing this book.

Throughout the food system, the degree of openness to outside influences varies enormously. Each group and the key people within the groups act as information gatekeepers or brokers, facilitating, filtering or blocking information flow. To an extent, all information brokers selectively filter the information they process according to the interests they represent. Most technical specialities require very specialized forms of information – for example, the information from bar codes. Irrelevant information about other aspects of food would interfere. However, some groups such as advertisers, trade associations and lobbyists selectively filter or control information so as to advance (ostensibly) the interests of their members or clients. There appear to be few societal or moral constraints on these forms of information filtering.

A Food Policy Fit for the Future

Despite these barriers, the opportunities for disseminating information about all aspects of food are greater than ever before. New technologies, greater scientific understanding, the particular experience of different groups of people can all provide feedback on what is happening to our food and help us to define the food policies that improve it. As we said in

the introduction, we see the challenge of food policy as being to produce a safe, secure, sufficient, sustainable and nutritious diet for all, equitably.

In the final chapter, we discuss some general approaches that may help to achieve policies which could help us to reach that goal as we move into a new millennium.

10

Conclusion – Food Policies for a New Millenium

Study the rich and powerful, not the poor and powerless...not nearly enough work is being done on those who hold the power and pull the strings.

George, Susan (1976) *How the other half dies – the real reasons for world hunger* Penguin, (revised edition 1977) p289, copyright © Susan George, 1976, 1977. Reproduced by permission of Penguin Books Ltd

There is no definite food future, only a range of possible futures. How we deal with current issues will create the future, determine how well the food system functions and how far it meets everyone's food needs. Adopting as the basic aims of food policy the provision of a safe, secure, sustainable, sufficient, nutritious diet for all, equitably and relating the various issues to these aims, provides a guide with which to evaluate current and proposed changes. The question is then whether the institutions, instruments and information needed to meet the challenges ahead are adequate and, if not, how they should change. We discuss this briefly in the last part of this chapter.

Today's food system, as we have seen, has been developed by and for the people of the rich industrialized world. We put the world in our mouths. Our foods are drawn from all round the globe, produced using plants, animals and technologies developed in one country and transferred to another, and brought to the table by companies that operate world-wide. Consumers are offered an ever growing abundance of foods to buy, more conveniently presented but ever less under their own control. It is a system now spreading throughout the world. But the present system does not work for everyone. Hunger and malnutrition persist in many poor countries and even in some rich countries. Indeed, the largest food aid programme in the world, costing almost US$29 billion in 1991, is within the USA and directed at its own poor citizens. The poor lack choice and are nutritionally disadvantaged.

As we have seen in this book, food producers and consumers everywhere are connected through an increasingly global food system. This links some of the poorest people on earth with the richest via corporations whose turnovers dwarf the GNP of many countries (see Figure 10.1). In 1991, Unilever's turnover of over US$40 billion or Nestlé's US$35 billion was more than the GNP of over 110 of the 161 countries for which figures were given in the 1992 World Bank Atlas. The economic activities, policies and bargaining power of these companies cannot be ignored as today's board-room decisions may affect people all over the planet. These companies want an international trading system that suits their needs.

The food system that links the farm, factory, shop and consumer has been constructed by the political and economic institutions of the rich of the world in their interests. Now it is being restructured economically and transformed technically to serve the needs of rich markets world-wide. Within it, different actors are trying to gain the upper hand, to shape and gain the most benefits from change. This produces great dynamism. Changes are coming about in Europe through the development of a single European market and the changing nature of Eastern Europe, in trade as the rules are renegotiated in GATT, and in the relations between the rich industrialized countries and the poor developing countries.

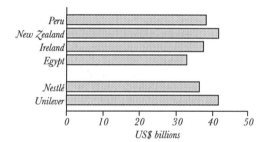

Source: 1992 Company Annual Reports, and World Bank (1992) *The World Bank Atlas* World Bank, Washington DC

Figure 10.1 Turnover and GNP figures for major food firms and selected countries in 1991 in billions of US$

Food Policy Goals

We believe that changes are good if they help to meet the goals we laid out for food policy and that they should be measured against these goals. Trade-offs will be necessary, sometimes sacrificing short-term gains for long-term benefits, sometimes making life more uncomfortable for the successful in the food business and bureaucracies in order to open up the debate about future directions to a wider public. The aim is to create a food system that fulfils the following criteria.

Sustainable

Greater scientific understanding of the effects of human action on this planet have made sustainability a critical question for current and future food practices. It is, to say the least, questionable whether the earth can continue to maintain human populations in their projected numbers – 10–12 billion by 2050 – in the affluent consumer lifestyles of the industrialized countries. Pessimists say that we must reach a solution in the next 40 years, optimists believe that the human race will survive until about 2100. Clearly this is the challenge – to ensure the long-term sustainability of the systems that nourish us, our children and our children's children.

For food, this raises issues about the nature of agricultural production, mentioned briefly in Chapter 5, and the energy and pollution costs of food distribution networks. Only about 10 per cent of the fossil fuel energy used in the world's food system is used in production, according to a 1993 estimate. The other 90 per cent goes on distribution and marketing – for example, in transporting food long distances or in expensive, non-recyclable packaging. The desirability of shifting great quantities of basic and processed foodstuffs around the world compared with more local and regional production needs to be reconsidered.

Other changes are needed in the way that research and development priorities are set and work financed to encourage research into making more efficient use of the inputs used in farming and into organic and other systems that use lower inputs which are sustainable indefinitely. Biotechnology offers a very powerful tool to affect food futures; it could be used to benefit large companies and rich consumers, or small farmers and poor people world-wide. Some research, such as research into substitution and factory production of feedstocks, could lead to the replacement of weather-dependent, farm-level food production, with enormous impact on livelihoods. Other research, such as some genetic engineering of plants, produces materials that are available only from particular firms and thus supports monopolies, while research into low input agriculture, for example, could be of widespread benefit to many poor producers.

Secure

Everyone needs food security – the knowledge that they will have enough food throughout life. Obviously, security has to be built on an ecologically sustainable food production system. Yet it involves so much more, reflecting the socio-economic distribution of wealth and power in societies. Food security is needed at an individual level, by households, by nations and globally. Carry-over grain stocks that are available for trade and as international food aid can ensure global food security. Nations need the capacities to produce or buy food to feed their citizens, plus the relevant transport and distribution infrastructure. Households need the means of producing their own food or income from employment, social security or other support systems to supply themselves with food. Individuals need the skills, household structures and relationships that are necessary to ensure that they receive sufficient food. Poverty is the root cause of food insecurity. However, food insecurity can be exacerbated by ecological, political and economic changes and can be caused by wars and civil conflicts.

Safe

Food should be safe to eat and not contaminated with bugs or pollutants that harm the eater or producer, either in the short or long term. As the food system becomes more complex, the scale larger, and the food chains longer, and as more people live in urban areas, the ability of unsafe food to affect many people increases.

Ensuring food safety is a broadly-based but controversial responsibility. Is the consumer responsible for adequate cooking to kill off the effects of poor hygiene practices during production, or is the processor or retailer responsible for providing uncontaminated food? Obviously, all three bear responsibility for their actions, but where does the balance lie?

Different actors view the risks involved differently. Essentially, the debate is about who should bear the risks of innovation, the producer or the consumer. Extensive testing of additives, for example, places more costs on the manufacturer, while little testing risks the health of (some) consumers. We believe that procedures which estimate such risks adequately and place the risks on those most able to bear them, are most important. Open, democratic forms of decision-making which involve consumers' representatives and enforcement agencies as well as food producers and manufacturers, are also necessary.

Sufficient and nutritious

Not only must the food supply be sustainable, secure and safe but it must also provide sufficient food, in quantity and variety, for a nutritious diet. Whether individuals eat sufficient nutritious food – which is also a long-term safety issue – depends on whether the foods that would make up such a diet are available, accessible and culturally acceptable to them. It also depends on having the knowledge and skills to choose foods that provide a sufficient nutritious diet.

Ironically, perhaps, the greater the number of items to choose from the harder it becomes to choose such a diet. Nutritional sufficiency, with its long term health benefits, may be hard to achieve with cultural and social pressure promoting particular foodstuffs. People need the knowledge required to make sound choices, the opportunity, and the economic possibilities to do so.

Equitable

The rich do not starve, the poor do. The rich, however, may be malnourished through over-consumption. Hunger and malnutrition are unacceptable in a civilized world. They are signs of inequality in the world which are reflected in the food system and exacerbated by its failures. To develop an equitable system, action is needed at global, regional, national and community levels to end hunger and malnutrition, whether of affluence or poverty. At each level policies have to be judged on how far they enable people to exercise their entitlement to adequate food supplies. Changes are required that will empower people who are hungry and malnourished to feed themselves well.

Equity in the food system is about more than simply meeting people's entitlements to food. It is about the livelihood of billions of people world-wide. This function of the food system contributes directly to the economic and social well-being (or otherwise) of many communities. If people are drawn into a market system but have little money, they are unable to gain access to the necessities of life. They lose hope in the future and may become mentally or physically sick as a result. Tragic examples are the state of many indigenous peoples, especially those living in urban areas, and the chronic unemployed.

We need a sense of perspective when we consider individual physical health and community health concerns. Which is more important, a small increase in an individual's life expectancy due to lowered fat intakes, or the maintenance of a stable employed rural population producing food? In many develop-

ing countries and in commodity exporting countries such as Australia and New Zealand, much of the national wealth (and the income of families) is generated by food exports, and much of the populations' health is dependent on them. Many developing countries need income from food exports to buy the necessities of life.

This argument for the food system as a basis of community health depends on the fairness of income distribution within and between countries. The trend in industrialized countries, which is likely to be replicated world-wide, has been to squeeze out the many small food businesses in favour of fewer, larger companies. Is this situation fair or desirable if it causes massive poverty and social disruption? It is a major role for governments to ensure that wealth is shared so that the extremes of poverty are not a by-product of the activities of a few major actors in the system. But for that to happen, governments must be concerned to promote the public good of community welfare, and the welfare of the planet as a whole, and not simply the short-term desires of individuals and companies.

Achieving the Goals

In 1992, in the World Declaration and Plan of Action for Nutrition (WDPAN) following the International Conference on Nutrition, governments expressed many desirable sentiments and goals (see Box 10.1). Unfortunately, these goals have no legal force. Technically, they can almost certainly be achieved. But ensuring a well-nourished world is primarily a political and economic problem concerning power and control over a range of resources and the distribution of benefits arising from their use. The plan assumes a commonality of interest and intent to improve nutritional well-being in the global food system that is almost certainly mistaken. The food system, in reality, as we have seen in this book, is not driven by nutritional but by market and competitive needs.

Clearly, the food system is a sub-set of the wider social and economic system which is in a great state of flux following the end of the Cold War. A global economic restructuring is underway at the same time as the development of new biotechnologies which could transform food production. Most key actors in the food system are businesses which happen to deal in the commodities that people eat instead of drive or wash with. They employ all the tools that are necessary to move their products and increasingly have sought market segmentation, and new and more profitable products. The challenge that governments face is to ensure that the actors in this market meet people's nutritional needs. Nutritional well-being was acknowledged in WDPAN as a crucial measure of 'the development of soci-

Box 10.1
International Conference on Nutrition
World Declaration and Plan of Action on Nutrition

When 159 countries declared their 'determination to eliminate hunger and to reduce all forms of malnutrition' at the UN's first ever International Conference on Nutrition in December 1992, they also agreed a plan of action. The plan has four overall objectives:

➤ ensuring continued access by all people to sufficient supplies of safe foods for a nutritionally adequate diet;
➤ achieving and maintaining the health and nutritional well-being of all people;
➤ achieving environmentally sound and socially responsible development to contribute to improved nutrition and health;
➤ eliminating famines and famine deaths.

To achieve these objectives, it included a number of major policy guidelines for countries:

➤ commitment to promoting nutritional well-being;
➤ strengthening agricultural policies;
➤ environmentally sound and sustainable development;
➤ growth with equity – the need for both economic growth and equitable sharing of benefits by all segments of the population;
➤ priority given to the most nutritionally vulnerable groups;
➤ focus on Africa;
➤ participation of the people;
➤ focus on women and gender equality;
➤ development of human resources;
➤ population policies;
➤ health policies;
➤ promoting nutritional well-being through strengthened economic and technical co-operation among countries;
➤ allocating adequate resources.

It recognized that improved nutrition cuts across many sectors of government, and proposed nine strategies and actions:

1. incorporating nutritional objectives, considerations and components into development policies and programmes;
2. improving household food security;
3. protecting consumers through improved food quality and safety;
4. preventing and managing infectious diseases;
5. promoting breast-feeding;
6. caring for the socio-economically deprived and nutritionally vulnerable;
7. preventing and controlling specific micronutrient deficiencies;
8. promoting appropriate diets and healthy lifestyles;
9. assessing, analysing and monitoring nutrition situations.

The conference called on governments to produce follow-up plans of action by the end of 1994 involving the private sector and non-governmental organizations, and for international agencies to develop means to support carrying out the plan.

The central test of the WDPAN's success will come in the follow-up to the conference in specific national plans. If they contain clear policy goals and targets, identify institutions and policy instruments for reaching them and information systems that permit public monitoring and evaluation of the results and ways of turning this feedback into revised policies and practices, then we have a chance of moving more rapidly towards a well-fed world.

Source: International Conference on Nutrition (1992) *World Declaration and Plan of Action for Nutrition* FAO/WHO, Rome and Geneva

eties' and 'of progress in human development'. Using nutritional well-being as a measure of policy effectiveness can help to determine the types of institutions, policy instruments and information systems that are needed to balance the competing interests in favour of nutritional well-being.

As we discussed in the last chapter, these three elements – institutions, policy instruments and information – are essential to achieve policy goals. It is here that the greatest changes must come to meet the goals set.

Institutions

Both government and private institutions affect the food system, although generally we think of government institutions in terms of policy. This is too narrow a view. Since the early 1980s, there has been a shift away from government regulation of activities to allowing the market to determine the direction of change and distribution of benefits. Just how the balance should be struck is a highly political issue and should feature clearly in political debate. Whichever party prevails, we believe that the measure of success of whatever approach is taken to food is how well the food system meets the goals set out above. If it fails to do so, then the institutions involved need to change, and to employ appropriate policy instruments and to generate the information needed to reach the goals.

A major flaw in much of the market based discussion has been its neglect of the real world and its harking back to an idealized, economic model of free markets. That model, as its early

proponents recognized, required several major conditions for it to function well. These included the immobility of national capital – that is, money that stayed in its country of origin – a market-place consisting of many small players involved in the buying and selling of goods and services, and, finally, perfect information so that all buyers and sellers had the information they needed to make good decisions.

This is a long way from the real world today, with just a few major players dominating activities in most food sectors, a completely international capital market operating 24 hours a day, seeking returns on funds invested from anywhere in the world, and a far from fully informed set of people in the market-place.

In the post Cold War world, with the collapse of state directed economies and spread of the market economics, the institutions involved need a complete rethink. Given the system has global, regional, national and local community dimensions, appropriate institutions are needed at each of these levels to interact effectively. They need a set of enforceable laws, rules and regulations within which to operate, a primary basis for which should be a guarantee of the right to food.

Clearly, such a system is some way off, but the basis for developing it does exist, from the international UN and other agencies, which would need to be much more democratic and transparent in their operation, to national ministries and local government departments that affect food. The new WTO could become a body promoting sustainable trade, taking far greater notice of environmental considerations in regulating trade relationships than seems likely at present. National institutions, as the

Norwegian and other experience shows, can be mobilized behind clear policies to affect how the food system operates.

Securing a right to food is not simply a matter for governments, however, but concerns the new international actors on the scene, the non-governmental private institutions that now set their own policies world-wide and operate to fulfil them – the transnational corporations. They consist of mainly publicly, but also privately, owned companies that operate in different parts of the food system. States, however, have given companies certain legal status, rights and immunities. These need to be reassessed and the institutions perhaps put on a new legal footing. Private companies, for example, do not have to disclose very much information about what they do, while the limited liability status of companies means, for example, that they do not have to take any responsibility for the social, environmental or economic impact of their actions. They can treat these costs as external to their operations. They may range from the ecological consequences of production processes or packaging methods that leave society to foot the bill for their treatment or amelioration, to the social disruption caused by plant relocation in pursuit of a greater return on investment from elsewhere.

New rules and regulations could require companies to be open about their activities and to bear the risk for change, rather than society, and could shift their activities in directions that are more likely to produce a sustainable food system to feed everyone. Private agents that act at a global level need policing, just as private citizens who are acting in a national context need to be checked. But this requires international agencies and enforcement mechanisms.

Policy instruments

The range and scope of the policy instruments used also need reviewing. From patents and plant protection laws to fiscal measures, such as export subsidies, all should be mea-sured against the objectives set out above. If the effects of existing measures do not enhance these objectives, other measures should be adopted which do. In farming, for example, the use of a single policy instrument – pricing – has failed to protect the small farmers it was aimed at in the EU. Indeed, much of the money spent never reaches them and has had the effect of increasing production and leading to surpluses. World-wide, farmers have shown repeatedly that they can and will increase production of different crops if the pricing structure and availability of other inputs are such that it is attractive for them to do so.

Information

The level of education and information available to people throughout the food system has a great effect on its functioning. The aim throughout should be to communicate clearly and fully about what is happening so that people at all levels can make good decisions. This means running institutions with a far greater transparency and openness than tends to be the case now. It also means ensuring that people have sufficient education and information to be able to secure a good diet for themselves, and a sustainable food system for everyone. Schools, the media, advertising, labelling and all the other means through which information is transmitted have to be used. However, the information should be balanced, honest and clear. This is often not the case, with advertising heavily weighted towards the promotion of a narrow range of biased messages that are aimed to increase sales, not improve diets, with education lacking the resources to teach, and marketing skills only available to the larger actors in the system.

Getting the mix right

A mix of institutions, instruments and information is needed to implement particular policies, to obtain the feedback that is required to monitor their effectiveness and to make any necessary alterations. Policies that might bring nutritional benefits to those most in need may be opposed by others in the system who feel that their interests are threatened. Already, for example, some lobby groups in the industrialized countries have opposed funding for agricultural assistance to developing countries whose farmers were perceived as competitors. Legal controls on marketing of breast milk substitutes might benefit mothers and children, but might affect the producers of the substitutes, as would curbs or levies on food advertising, especially that aimed at children, to pay for education to match its influence.

In the new food system, those actors between the farmer and consumer are increasingly powerful, well-organized and connected, and have access to a wide range of information. Farmers and especially consumers act more as isolated individuals. But it is when consumers act as citizens, by voting for particular policies and by being organized through a wide variety of public interest non-governmental organizations (NGOs), that they have most effect. Then consumers can spread information and concern about food issues, exert political pressure, keep food high on the national agenda, help to mobilize the political will that is needed to reach, for example, WDPAN's goals, and to shape the framework in which the major actors operate. It is an increasingly important form of public participation that provides a degree of countervailing power to the major vested interests in the food system. Indeed, as the Chairman, Dr Ramalingaswami, of the International Conference on Nutrition was quoted as saying, 'When we all go home, it will be the NGOs that will continue the pressure on governments'. There is, however, a need to distinguish between the interests of the different actors in the system, the interest groups that lobby for them and the citizen-based NGOS (see Box 10.2).

For WDPAN to deliver its goals on nutrition, and for food to be given its proper place in the changing world, requires a vigorous national follow-up across a wide range of government sectors, involving a broad range of private and academic interests as well as public, and loud and continuing pressure from 'public interest NGOs to keep the fine words and aspirations well up on the public agenda'. It also means finding ways to enable citizen-based NGOs to develop international networks to match the national networks, to ensure that the various issues in the food system are openly and publicly debated.

Box 10.2
PINGOs and BINGOs

Although the term non-governmental organisation (NGO) is used rather broadly, it stems from those voluntary, public groups formed by ordinary citizens that concern themselves with hunger, malnutrition, the environment, consumer rights and the like. At the International Conference on Nutrition where NGOs had a very welcome involvement in both the preparation and the conference itself, these groups called themselves 'PINGOs' – public interest NGOs. This was to distinguish themselves from other non-governmental organizations which represent powerful business and industrial actors in the food system. These are operated and funded by increasingly large business enterprises and were labelled business and industry ngos or 'BINGOs'.

Changing Perspectives

Societies are changing. The market is penetrating every aspect of human life with a great fragmentation occurring in consumerist society. Television, computers, and satellites mean that inside knowledge soon becomes public – so consumers have a very good idea of the wide range of problems facing people and the planet (from famines and deforestation, to salmonella and listeria food-poisoning from the chill cabinets of the supermarkets). Yet in the USA, for example, it is increasingly difficult to distinguish between advertisements and news on the radio, since everything is said with a similar degree of seriousness. Such an approach degrades humanity, trivializes the important, and elevates the unimportant, such as what brand of this or that you use.

Greater knowledge about the world and food issues can lead to a feeling of powerlessness if there are no effective ways in which people can respond to these concerns. They need to be able to act. As institutions become more powerful, global and remote, and are able to use increasingly sophisticated tools to understand the concerns of individuals and market their products as if they would meet these concerns, individuals and small groups can feel powerless. People need to be able to act locally, in their own circumstances, through NGOs as well as through the political processes that affect change.

New alliances are possible between different actors in the food system which could push food policy goals forward. Consumer groups, for example, may be able to put pressure on farming methods through alliances with retailers, or, as in Australia with Landcare, farmers and environmentalists may also make common cause to safeguard the environment.

Power may be compared, politely, to a dung heap. In a heap dung stinks, does no good and may pollute. But spread out, scattered over the fields, it fertilizes and life can blossom. Power is like dung. The skills, knowledge and technology that we need to have power over our food must be spread for an effective food system. The key challenge is to transform the institutions and design the tools so that the power is shared and spread.

A radical reassessment of needs

The measure of success is how well *all* people are able to feed themselves, not just a few or particular groups. We saw in Chapter 4 that food plays many roles in human life and culture, not simply as fuel for the body. We also saw in that chapter and in Chapter 7, how most consumer behaviour models are based on Maslow's hierarchy of needs. We need a radical rethink about the nature of human needs and how they may be met to create a food system that truly meets the needs of all.

Food, shelter, clothing and suchlike are *not* basic needs, argues Manfred Max-Neef, director of the Development Alternatives Centre in Santiago, Chile, but *satisfiers* of more fundamental human needs.

> *Human needs must be understood as a system; that is, all human needs are interrelated and interactive. With the sole exception of the need of subsistence, that is, to remain alive, no hierarchies exist within the system. On the contrary, simultaneities, complementarities and trade-offs are characteristics of the process of needs satisfaction....*
>
> *We have organised human needs into two categories...on the one hand, the needs of Being, Having, Doing and Interacting; and, on the other hand, the needs of Subsistence, Protection, Affection, Understanding, Participation, Creation, Leisure, Identity and Freedom.*
>
> *...food and shelter must not be seen as needs, but as satisfiers of the fundamental need for Subsistence. In much the same way, education (either formal or informal), study, investigation, early stimulation and meditation are satisfiers of the need for Understanding...*
>
> *A satisfier may contribute simultaneously to the satisfaction of different needs, or*

conversely, a need may require various satisfiers in order to be met. Not even these relations are fixed. They may vary according to time, place and circumstance. For example, a mother breast-feeding her baby is simultaneously satisfying the infant's needs for Subsistence, Protection, Affection and Identity....

...fundamental human needs are finite, few and classifiable; and...are the same in all cultures and in all historical periods. What changes, both over time and through cultures, is the way or means by which the needs are satisfied...one of the aspects that define a culture is its choice of satisfiers. Whether a person belongs to a consumerist or to an ascetic society, his/her fundamental needs are the same... Furthermore, needs are satisfied within three contexts: (1) with regard to oneself...(2) with regard to the social group...and (3) with regard to the environment.

Max-Neef, Manfred (1992) 'Development and human needs' in Ekins, Paul and Max-Neef, Manfred, eds, *Real-Life Economics – Understanding Wealth Creation*, Routledge, London, pp199–200

Although the nine needs – subsistence, protection, affection, understanding, participation, creation, recreation, identity and freedom – are universal, how they are satisfied varies. Satisfying them, Max-Neef suggests, involves four other core needs – being, having, doing and interacting. Subsistence, for example, is satisfied by *being* healthy, adaptable; *having* food, shelter and work; *doing* things such as feeding, procreating and resting, and *interacting* with the living environment and social setting.

These fundamental needs form a dynamic system in which no one need is more important than another or necessarily has to be met before another. They are not a hierarchy. They may be met simultaneously, complement each other, and be traded off against each other. However, if the minimal need for subsistence is not met, the other needs may be blocked and a single intense drive to fulfil it is likely. This can be true of other needs, too – for example, a complete

lack of affection or loss of identity can lead people to extremes of self destruction.

How and with what we seek to fulfil these needs matter. For food, this means that the social, economic and cultural aspects of food, as well as its content (ie, its nutritional value), affect whether it fulfils our needs. These fundamental needs can be met in ways that violate or destroy others. The need for protection, for example, might be met by an arms race, a national security doctrine, or authoritarianism, which can impair the satisfaction of other needs, such as subsistence, affection, freedom, participation and identity. Pseudo-satisfiers only *appear* to satisfy, for example, representative democracy which may only appear to satisfy the need for participation. Some apparent satisfiers of one need inhibit others. Paternalism, for example, may meet the need for protection but inhibits satisfaction of the needs for understanding, participation, freedom and identity. Some satisfiers only satisfy one need – for example, food programmes which meet the need for subsistence, while others have synergistic effects, satisfying one need and stimulating satisfaction of others at the same time. Self-managed production, for example, may satisfy the need for subsistence and stimulate satisfaction of needs for understanding, participation, creation, identity and freedom.

A successful food system

In a successful food system, food is a satisfier not simply of the need for subsistence, but also of various other needs. It has synergistic effects on other needs rather than inhibiting or violating them. This approach recognizes that human beings are multi-faceted creatures that cannot be reduced to a single variable, such as maximizing economic well-being. This is revealed when people talk about what food means to them as individuals and households, and as citizens of a local community, national entity, regional grouping, and one world. They draw in concerns about livelihood, health, freedom from environmental destruction, enjoyment, and the like, and about participating in decisions and actions that shape their lives. In large measure, it is the antithesis of the consumer culture, which defines people by what they can buy and assumes that all needs are fulfilled through what people consume.

Public concerns about food will not go away. They are central to the way that society develops – the way it satisfies fundamental human needs – since the kind of food system that society has plays a key role. With the world's population expected to double by 2050, rapid economic restructuring underway, and major technological advances waiting in the wings, tomorrow's food system will be a dynamic, changing place to be. The choices made now will help to determine whether it will satisfy our needs more abundantly or more narrowly. We hope in this brief guide to the food system to have provided a basis for understanding the processes that underpin the system and to encourage a concern across the food system to ensure a well-fed future for all.

Annexes

Annex 1

The International Framework: A Very Brief Guide

The World Bank and International Development Association, Washington DC Headquarters

The World Bank, or the International Bank for Reconstruction and Development, to give it its official name, was created in 1944 as part of the Bretton Woods Agreement to govern the post-war international finance system. It began operation in 1946 and was geared to help finance the reconstruction in Europe and Japan after the war. It raised capital on the markets and offered long-term loans to credit-worthy countries for productive, revenue producing projects at more or less commercial rates of interest.

By the late 1950s, a growing number of newly independent countries also needed finance for development, not just for the import of capital goods but also for building the infrastructure, such as roads, ports and water supplies, and financing the development of human skills and well-being through education and healthcare. This led to the creation of an affiliate to the Bank, the International Development Association (IDA).

The World Bank became the lender on more commercial terms to developing countries and the IDA makes highly concessional

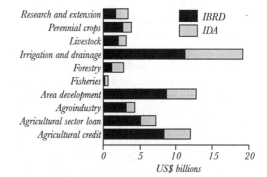

Source: World Bank (1993) *The World Bank Annual Report 1993* World Bank, Washington DC

Figure A1 IBRD and IDA cumulative lending operations by purpose to 30 June 1993

credits, repayable over 50 years with no interest charges but a service fee of 0.75 per cent and a ten-year grace period before repayment begins. By mid-1993, the Bank and the IDA had lent a total of US$313 billion – US$77.8 billion from the IDA, the rest from the World Bank. Some US$67.7 billion, or 22.6 per cent, were lent for agriculture and rural development, a further US$4.8 billion for fertilizer and other chemical industry and US$7.7 billion for population, health and nutrition. About 37 per cent of the lending for agriculture was from the IDA (see the chart for the breakdown by source and category).

The Bank is governed by its member governments. Voting powers depend upon the number of shares held. The majority are held by the major industrial powers, with the USA the largest shareholder at nearly 18 per cent in 1991, which gives them the major say in the Bank's operation. By mid-1993, there were 176 member governments in the Bank and 152 in the IDA.

The Food and Agriculture Organization of the United Nations (FAO), Rome headquarters

The FAO was the first of the UN specialized agencies set up after the Second World War. Its basic aim was 'to raise levels of production and distribution of food and agricultural products'. The thinking behind the FAO goes back to the League of Nations. In 1943, President Roosevelt called a war-time conference of the allied nations on food and agriculture which set up a commission to plan a permanent organization for food and agriculture. The FAO was the result. Set up by 34 nations on 16 October 1945 – now celebrated as World Food Day – it had grown to include 169 member governments by the end of 1993 with a biennial budget for its regular programmes of US$673 million in 1994–95. The regular budget finances administrative costs, services to member governments in agriculture and rural development; it also helps to combat hunger and provides technical assistance. The member governments form the FAO's governing council which meets every two years to approve its budget and biennial work programme. The council also elects a director general for a six-year term.

About half the 6500 staff work in Rome, with the rest in country and regional offices and in field projects. About three-quarters of the field programmes are geared to agriculture, with the focus on increasing crop production. Funds for these activities come from donor countries and the United Nations

Development Programme. It covers agriculture, fisheries and forestry, and runs a global early warning programme for famine.

The World Food Programme (WFP), Rome Headquarters

The WFP was set up in 1963 as the food aid organization of the UN. It provides food aid to support economic and social development projects and meets emergency needs. Most food aid is bilateral – an arrangement between two governments – but the WFP is a multilateral organization that handles around 20–25 per cent of food aid. Its budget has grown from an initial US$87.3 million from 33 countries to pledges and contributions of US$1749 million from over 80 countries (83 + UN agencies) for 1989–90.

A 30-member Committee on Food Aid Policies and Programmes, which usually meets twice a year, governs the WFP. The members are elected for a three-year term, half by the Economic and Social Council (ECOSOC) of the UN and half by the FAO Council.

The World Food Council (WFC), Rome Headquarters

The World Food Council is the highest level political body in the UN dealing with food matters. It was set up in 1974 by the General Assembly as its food policy arm and reports to the Assembly via ECOSOC. It aims to raise the political profile and priority of global food issues through specific proposals in order to increase food production in countries where this is most needed, to reduce poverty, to improve access to food for the poor, and to promote international support for the efforts of developing countries and so produce a better system of world food security. In 1992, there were 36 members, who met at ministerial level annually. The Council's Secretariat in Rome was wound up in December 1993 and proposals made to abolish the Council.

The International Fund For Agricultural Development (IFAD), Rome Headquarters

The idea for IFAD was proposed at the UN World Food Conference in Rome in 1974. IFAD began operations in 1977 with a fund of about US$1000 million. Its lending aims to benefit the poorest people in the poorest countries by supporting projects that increase food production, reducing rural poverty by increasing incomes and producing jobs, and improving the level of nutrition.

Its unusual governing structure divides voting power equally between three groups of countries – the rich industrialized countries, the oil-rich developing countries and the non-oil-exporting developing countries. Financing, however, is split differently with the first group providing US$567 million, the second group US$435 million and the third group US$19 million at the start. Subsequent replenishments of the fund have maintained a similar balance, although there are always arguments, particularly between the industrialized and oil-exporting countries, over the balance.

By the end of 1990, IFAD had lent over US$3200 million to 292 projects in 93 developing countries. About two-thirds of these loans are on highly concessional terms to very low income countries (less than US$300 per capita in 1976). The total involved in these projects was some US$11,800 million, owing to co-financing of some projects by other agencies and national contributions to the projects.

The World Health Organization (WHO), Geneva Headquarters

The WHO is another specialized agency of the UN, which was set up in the late 1940s to provide support to member governments in health activities. In recent years, the WHO has taken a growing interest in dietary patterns and health problems and produced advice on dietary goals. It also runs, jointly with the FAO, the Codex Alimentarius, which works to establish international quality and safety standards and fair trading practices in food (see Chapter 9, Box 9.2).

The WHO is governed by its member governments – 187 at the end of 1993 – meeting at the annual World Health Assembly. Its regular budget for 1994–95 was US$822 million, of which US$11 million was for nutrition and US$4 million for food safety. It also has special project budgets for work on, for example, Aids and tropical diseases, and had a total staff world-wide of about 6300 in 1994.

The World Trade Organization (WTO), Geneva Headquarters

As a result of the Uruguay Round agreement in the General Agreement on Tariffs and Trade (GATT), a World Trade Organization was established which was expected to begin operation in 1995. It will encompass the GATT and all agreements and arrangements made under its patronage. The WTO will be run in a similar manner to the UN specialized agencies, with a Ministerial Conference meeting at least once every two years and a General Council to oversee its operations and to act as a Dispute Settlement Body and Trade Policy Review Mechanism. The WTO has been established more than 45 years after a similar body, an International Trade Organization, failed to become established as part of the UN system to complement the IMF and World Bank, owing to opposition by the US Congress to the transfer of sovereignty on trade matters to such an organization.

United Nations Administrative Committee on Co-ordination, Subcommittee on Nutrition, (ACC.SCN), Secretariat in the WHO, Geneva

This little-known subcommittee is the focal point for harmonizing the policies and activities concerned with nutrition in the UN system. The main ACC consists of the heads of UN agencies, and recommended that the SCN be set up in 1977. The SCN is a co-ordinating mechanism for the exchange of information and technical guidance, and acts to help the UN respond to nutritional problems. The UN members are the FAO, International Atomic Energy Authority (IAEA), IFAD, International Labour Organization (ILO), UN, UN Development Programme (UNDP), UN Environment Programme (UNEP), UN Educational, Scientific and Cultural Organization (UNESCO), UN Fund for Population Activities (UNFPA), UN High Commissioner for Refugees (UNHCR), UN Children's Fund (UNICEF), UN Research Institute for Social Development (UNRISD), UN University (UNU), WFC, WFP, WHO and the World Bank. Bilateral aid donors also take part in meetings. The ACC/SCN produces regular reports and other materials on world nutrition.

Consultative Group on International Agricultural Research (CGIAR), Washington DC Headquarters

The CGIAR is an *ad hoc* grouping of some 40 public and private sector donors that supports a network of international agricultural research centres. The best known are the International Rice Research Institute (IRRI) in the Philippines, and the Wheat and Maize Research Institute, (CIMMYT) in Mexico. These produced the green revolution high-yielding varieties of wheat and rice.

The group was set up in 1971 to support four existing centres – IRRI, CIMMYT, the International Institute for Tropical Agriculture (IITA) in Nigeria and the International Centre for Tropical Agriculture in Colombia – which had been established by the Ford and Rockefeller Foundations. In 1993 there were 18 centres supported by the group but, owing to amalgamations, this number fell to 16 in 1995 (see Table A1).

The CGIAR centres maintain about one-third of the world's plant genetic material, have trained about 45,000 scientists in developing countries and spent about US$235 million in 1991 on the existing centres. That is about a tenth of spending on agricultural research in developing countries, which itself is about a quarter of the world total of around US$10 billion.

Table A1 CGIAR centres in 1993

Centre, location, date founded	Full title (in English)	Main areas of work
CIAT, Colombia, 1967	International Centre for Tropical Agriculture	Improvement of crops and agriculture in lowland tropics of Latin America – rice, beans, cassava, forages and pastures
CIFOR, Indonesia, 1992	Centre for International Forestry Research	Forest conservation and sustainable development
CIMMYT, Mexico, 1966	International Research Centre for Wheat and Maize	Improvement of maize, wheat, barley and triticale
CIP, Peru, 1971	International Potato Centre	Focus on improvement of potatoes and sweet potatoes
IBPGR[a] Italy, 1974	International Board for Plant Genetic Resources	Conservation of gene pools of current and potential crops and forages; research on plant genetic resources
ICARDA, Syria, 1977	International Centre for Agricultural Research in the Dry Areas	Improvement of farming systems for North and West Africa; research on wheat, barley, chickpeas, lentils, pasture legumes, and small ruminants
ICLARM, Philippines, 1977	International Centre for Living Aquatic Resources Management	Improvement in the efficiency and productivity of culture and capture fisheries
ICRAF, Kenya, 1977	International Council for Research on Agroforestry	Initiation and support of research on integrating trees in land-use systems in developing countries
ICRISAT, India, 1972	International Crops Research Institute for the Semi-Arid Tropics	Crop improvement and cropping systems; sorghum, millet, chickpeas, pigeon peas and groundnuts
IFPRI, USA, 1975	International Food Policy Research Institute	Strategies and plans to meet food needs; all aspects of policy analysis
IIMI, Sri Lanka, 1984	International Irrigation Management Institute	Improvement and sustainability of the performance of irrigation systems through better management
IITA, Nigeria, 1967	International Institute of Tropical Agriculture	Crop improvement and land management in humid and semi-humid tropics; farming systems; maize, cassava, cowpeas, plantain, soybeans, rice and yams
ILCA,[b] Ethiopia, 1974	International Livestock Centre for Africa	Farming systems to identify livestock production and marketing constraints in Sub-Saharan Africa; research covers ruminants, livestock and forages
ILRAD,[b] Kenya, 1973	International Laboratory for Research on Animal Diseases	Control of major livestock diseases in Sub-Saharan Africa; theileriosis (East Coast fever) and trypanosomiasis (sleeping sickness)
INIBAP,[c] France, 1984	International Network for the Improvement of Banana and Plantain	Bananas and plantains
IRRI, Philippines, 1960	International Rice Research Institute	Global rice improvement
ISNAR, The Netherlands, 1979	International Service for National Agricultural Research	Strengthening and development of national agricultural research systems
WARDA, Cote d'Ivoire, 1970	West Africa Rice Development Association	Rice improvement in West Africa; rice in mangrove and inland swamps, upland and irrigated areas.

Notes:

a This was renamed the International Plant Genetic Resources Institute (IPGRI) in 1994

b The ILCA and the ILRAD were amalgamated into a new global institute – the International Livestock Research Institute (ILRI) – in 1995

c This is no longer a separate CGIAR centre and its programmes were integrated into those of IPGRI in 1995.

Source: CGIAR, Washington, DC

Annex 2

Basic Functions of Nutrients

Three classes of nutrients supply the body with energy:

➤ Carbohydrates supply glucose (a sugar), the main energy for the brain and the nervous system;
➤ Fats (lipids) supply fatty acids (mostly as triglycerides); these are the main energy source for muscles, including the heart muscle;
➤ Proteins (made of amino acids) are the main structural material of cells, but they may be broken down for energy if the body has run out of glucose, fatty acids or carbohydrate; they can be broken down to supply the brain with glucose.

Generally, energy which is not used in body metabolism or activity is stored in the body as layers of fat. Extreme amounts of stored fat lead to obesity. Fat stored around the abdomen increases the risk of disease such as heart disease. This is particularly true of men. Pre-menopausal women tend to store fat on the thighs and buttocks and do not appear to have as high a risk of cardiovascular disease as men.

Carbohydrates

There are various forms of carbohydrate. Complex carbohydrates are long chains of glucose units. One form of complex carbohydrate, starch, is found in many grains and vegetables. Simple carbohydrates are sugars (eg, glucose, fructose, lactose, dextrose, sucrose). There are several forms, but they are chemically similar to glucose and can be converted to glucose in the body. They are found mostly in fruits and milk.

Dietary fibre includes many different chemical compounds. It was once thought to be chemically inert and to have only mechanical properties, stimulating bowel function and preventing constipation and diseases of the intestinal tract such as diverticulosis. During the 1980s and 1990s, various forms of fibre have been shown to be chemically active in the gut, fermenting to produce compounds such as volatile fatty acids. Among other effects, these may help to lower blood cholesterol levels. Most Westerners do not consume enough dietary fibre (it is found mainly in vegetables and cereal grains). It is believed to be one factor in the prevention of bowel cancer and other bowel diseases.

Fats

There are many types of fat. They provide heat and shock insulation for the body when laid down in layers under the skin. They also provide energy and perform other functions in the cells. Polyunsaturated fats include essential fatty acids which are involved in the composition and functioning of cell membranes. Low intakes of such fats are believed to bring about cardiac arrhythmias (leading sometimes to sudden death). Another form of fat, saturated fat, is believed to raise blood cholesterol levels (especially low density lipoproteins) which are associated with an increased risk of heart disease and other diseases.

Protein

Proteins are made of amino acids. Although protein is known as a body-building nutrient – for example, muscles are made of it – there are thousands of types of protein and many work in the cells in various ways. Enzymes, for example, are proteins which assist a specific chemical reaction in the cell, either synthesizing or breaking down chemical materials. The instructions for making proteins are found in the cell nucleus in the genes.

There are 20 amino acids which form proteins. Our bodies can make eleven of them, but we need to obtain the other nine from our diet. These are found mainly in meats, fish and dairy products, but also in legumes (eg, baked beans and chickpeas), cereals and vegetables.

Vitamins and minerals

Vitamins and minerals are required in much smaller amounts than carbohydrates, fats or protein, but they play essential roles. They are often called micronutrients and they facilitate many processes throughout the body.

Vitamins were originally divided into fat-soluble and water-soluble vitamins. Fat-soluble vitamins tend to be stored in the body and can be ingested in large quantities at intervals. Water-soluble vitamins tend to be excreted every day in urine and so have to be consumed more often; however, water-soluble vitamins can be stored in the body, especially in the liver.

Fat-soluble vitamins

Vitamin A is a pigment of the eye important in vision, especially night vision. It participates in the modelling of bones during growth and in the mending of breaks, and it also helps to maintain the body's many surfaces (skins and linings of lungs, digestive tract, urinary tract, vagina and eyelids). Along with vitamins C and E and other factors, vitamin A helps to repair and prevent genetic damage which could lead to cancer. It is a factor in the production of

sperm and in the maintenance of pregnancy.

Vitamin D is a regulator of the calcium concentration in the blood and a factor in bone formation and repair.

Vitamin E protects compounds which are susceptible to destruction by oxidation. It protects red blood cells from bursting as they pass through the lungs and increasingly appears to be important in preventing heart disease and genetic damage from free radicals such as ionic oxygen.

Vitamin K is necessary for blood clotting and is implicated in bone metabolism and in the prevention of osteoporosis.

Water-soluble vitamins

Thiamin, Riboflavin and Niacin help enzymes to facilitate the release of energy from nutrients in every cell of the body.

Vitamin B6 is necessary for the metabolism of protein.

Vitamin B12 helps cells to divide, especially blood cells and cells of the intestinal lining. Vitamin B12 may be deficient in the diets of non-meat eaters. Folate deficiency is common in many countries and is associated with neural tube defects if deficiency occurs during the early stages of pregnancy

Biotin and pantothenic acid are also B vitamins. They are rarely deficient in humans but are required in the diets of laboratory animals.

Vitamin C maintains the body's connective tissue. It is important for wound healing and is part of the 'glue' that holds cells together.

Major minerals

Calcium is important for bones and teeth. It plays roles in muscle contraction, nerve transmission, immune function and blood clotting.

Phosphorus is important for bones and teeth, in genetic processes and in energy transfer.

Magnesium is involved in bone mineralization, protein synthesis, muscular contraction and transmission of nerve impulses.

Sodium, chlorine and potassium are electrolytes that maintain fluid balance and the balance of acids and bases inside and outside cells. People in rich countries ingest 10–15 times more sodium than they require, in the form of common salt. Excess sodium intake has been linked to elevated blood pressure, a risk factor for cardiovascular disease in a substantial proportion of the population. Chlorine is also included in the hydrochloric acid in the stomach which is necessary for digestion. Potassium facilitates protein synthesis and the maintenance of nerves and muscles.

Sulphur is a component of certain amino acids and part of the vitamins biotin and thiamin and the hormone insulin.

Trace elements

Iodine is a component of the thyroid hormone which helps to regulate growth and metabolic rate in the body. It is necessary for proper growth of the early foetus.

Fluorine is involved in the formation of bones and teeth; it helps to make them resistant to mineral loss.

Selenium, together with vitamin E, protects body compounds from oxidation and so may prevent genetic damage.

Iron is part of the red blood cell protein haemoglobin which carries oxygen around the body, and of the muscle protein myoglobin which makes oxygen available for muscle work. It is necessary for the use of energy in every cell. Iron deficiency is fairly common in women of reproductive years owing to menstrual blood losses and/or poor intakes. It can be a cause of fatigue, poor concentration and anaemia. Severe iron deficiency three to six months after birth is believed to cause irreversible cognitive impairment.

Zinc is a working part of many enzymes (biological catalysts) and of the hormone insulin. It is involved in the making of genetic material and proteins, immune reactions, the transport of vitamin A, taste perception, wound healing, the making of sperm, and the development of the foetus.

Copper is necessary for the absorption and use of iron in the formation of haemoglobin. It is a part of several enzymes and helps to form the protective covering of nerves.

Cobalt is part of vitamin B12 which is involved in cell division.

Chromium is associated with insulin and required for the use of glucose.

Molybdenum is important in enzymes, the facilitators of many cell processes.

Manganese, vanadium, tin, nickel, silicon and others are necessary for many biological functions in animals, but their roles in humans are unclear.

Sources: Truswell, A S (1986) *ABC of Nutrition* BMA, London; Whitney, E N and Sizer, FS (1989) *Essential Life Choices: Health Concepts and Strategies* West Publishing, New York, pp66-85

Alcohol

Alcohol is not a nutrient but it is a common dietary constituent in wines, beers and spirits. Moderate use is not harmful, but alcohol abuse has disastrous effects on health, and on family and social life. Alcoholism is probably the world's most widespread drug problem.

Prohibition has been used for centuries in Islamic countries, and in the USA during the earlier part of this century, as a way of controlling its adverse effects. The health benefits of prohibition are controversial. There is some evidence that in the USA and during the German occupation of France during the Second World War, many lives were saved because of the reduction in the availability of alcohol. However, prohibition can lead to the criminalization of the sources of supply and distribution.

In recent years there have been greater attempts to raise the public's awareness of the harmful effects of alcohol abuse as well as measures to ban the promotion of alcohol by the advertising industry. In Western countries, there is a narrow gap between the acceptance of restrictions for the public good and infringement of personal freedoms, which are often seen to be 'puritanical' forms of 'social engineering'.

Diary of Food Safety Scares in the UK, 1988–89

Prelude

May 1988 — In three days 120 peers are poisoned at the House of Lords. Eggs and hygiene were held to be responsible. Not reported in the press.

25 May–10 June — 70 ill owing to salmonella at a Leeds Hospital.

13 June — Department of Health and Social Security (DHSS) officials meet Ministry of Agriculture, Fisheries and Food (MAFF) officials to discuss outbreaks of food poisoning in two hospitals 'caused by eggs'. Not reported in the press.

June–July — Outbreaks at a Manchester Hospital and Cardiff Remand Centre. Further outbreaks and deaths at hospitals in Birmingham.

14 August — Report of Institute of Environmental Health Officers says that the cook-chill problem extends way beyond Birmingham hospitals.

26 August — DHSS issue a draft statement warning against eating uncooked eggs. British Egg Producers are furious because they had not been consulted.

12 October — Business people at two banquets in the City, including the Lady Mayoress and two Old Bailey judges, are struck down with salmonella. They had eaten poisoned cheese and egg canapés, chilled and flash grilled.

1 November — Dr Richard Lacey (Leeds University) announces further evidence of chilled food being poisoned with listeria.

18 November — Department of Health issues advice that it is prudent to avoid eating raw eggs, yet 'the public should be reassured of any risk to them. For the housewife it is clearly ...very small.'

December 1988

1. Plymouth Area Health Authority (AHA) advises banning eggs.
2. The Plymouth AHA decision is reported by Independent Television News (ITN).
3. ITN interview with Edwina Currie, Junior Minister of Health, who says 'Most of the egg production in this country, sadly, is now infected with salmonella'.
4. MAFF officials are extremely angry at Currie's comments on ITN. They say 'that there is no research to support her claim; 30 million eggs are eaten every day, with only 26 outbreaks of salmonella attributable to them'.
5. Dr David Clark, opposition spokesperson on agriculture says that although Edwina Currie is scaremongering, MAFF is 'incredibly complacent and tried to deceive the public'.
6. Kenneth Clarke, Secretary of State for Health, rejects backbench Tory demands

to sack Currie, repeats the advice of the Chief Medical Officer not to eat raw eggs and that the risk to healthy people is small.

Robin Cook, opposition spokesperson on health, says that Clark 'will let her insult pensioners, caricature Northerners, and threaten child benefits, but to stub the toe of the NFU is going too far'.

MAFF propose a seven-point Code of Practice to the poultry industry, including improved hygiene and disease security. A joint Government and egg industry working party reported that it had traced 12 of the 46 cases definitely linked to eggs to specific flocks.

8. Free-range eggs are found to be far more likely to carry salmonella than eggs from battery hens.

10. John MacGregor, the Minister of Agriculture, says that it is not the case that most eggs are infected with salmonella.

12. The National Farmers Union (NFU) demand compensation for egg producers who fear losses. Simon Gourlay, NFU President, accuses the Prime Minister of procrastination and presents the Government with an action plan.

Richard North, ex-environmental health officer, now consultant, showed that all the 'egg-related' cases occurred in the catering context. There was no record of eggs alone being responsible, but they were being used as a scapegoat by caterers to excuse poor hygiene practices.

13. Mrs Thatcher and senior ministers consider a publicity campaign costing £500,000 to allay public anxiety over contamination.

15. Thames Valley Eggs start legal action against Edwina Currie over losses, after her warning about salmonella poisoning.

16. Full page advertisements appear in most papers, 'Eggs – The Facts. Eggs are a valuable and nutritional part of a balanced diet. We eat 30 million a day. The number of reported cases of salmonella poisoning is very small in comparison with the huge number that are consumed. Vulnerable people should cook eggs thoroughly.

Everyone should avoid raw eggs or uncooked foods made from them. Please follow this notice. HM Government.'

Richard Lacey: ' We are facing the largest salmonella epidemic in our history.' There is Government confusion over a compensation scheme.

Edwina Currie resigns, facing 12 legal actions from egg producers.

19. Government Rescue Package sets aside £20 million to compensate egg producers following Currie's remarks.

Richard Lacey, ' Eggs kill one person a week due to salmonella.'

24. John MacGregor, Minister of Agriculture, refuses to name manufacturers who have produced salmonella-infected animal feed.

It is confirmed that a quarter of protein manufacturers inspected during the past year failed, yet none were prosecuted.

27. Supermarkets claim that not one egg out of 100,000 tested was infected. Lacey says that this is not surprising, as only one in 7,000 on average is infected anyway.

January 1989

6. The egg industry is 'likely' to announce an agreement on an egg quality assurance scheme, following two years' discussion.

11. All Party Select Committee on Agriculture ask Edwina Currie to reconsider not giving evidence to them. They receive MAFF Officials.

12. Sir Donald Acheson, Government Chief Medical Officer, gives evidence to the Parliamentary Select Committee, that salmonella in eggs is a serious public health problem, but Edwina Currie was wrong to say that most egg production was infected. Nevertheless, the increase in cases amounts to 'an epidemic'.

In one series of tests, a quarter of precooked cook-chill meats bought from leading supermarkets were found to contain Listeria.

13. MAFF confirms that material containing

human effluent can be fed to animals, according to the Protein Processing Order 1981.

Egg producers are worried about further losses due to the evidence given to the Select Committee.

Egg production at farms known to be infected was allowed to continue, according to the Ministry of Health. MAFF officials did not pass this information on to the Select Committee when they met them two days before.

18. The National Consumer Council and the Institute of Environmental Health Officers call for an end to MAFF's responsibility for food safety.

19. The Government is to take extensive new legal powers to inspect poultry farms and egg-producing plants, and enforce registration in monitoring.

20. John MacGregor confirms the cost of the compensation scheme at only £4 million.

The body of a boy suspected of dying from food poisoning is paraded through his school.

23. MAFF is holding up the action to prevent the sale of infected eggs, despite a request from the DHSS. The use of the existing legislation – Zoonosis Order 1975 – would suffice.

24. John MacGregor collapses in Brussels, found later to be the result of a duodenal ulcer.

The Labour Party stages a debate on health and safety standards on food. At times, only five Labour MPs are present.

25. The Select Committee criticizes MAFF for cutting back spending on research into the causes of food poisoning.

26. Currie admits her gaffe over infected eggs – or does she? She says that she intended to explain that 'a significant number of egg-laying hens in many of the egg-laying flocks in this country were infected with salmonella'.

27. Evidence about the poisoning at the House of Lords becomes public.

28. Egg farmers found to have flocks infected with salmonella are banned from selling

raw eggs, until their farms are given the all-clear by the use of the Zoonosis Order.

February 1989

2. MAFF is found to have known about the salmonella epidemic a year ago, but was unable to persuade egg producers of its seriousness or to warn the public. A Veterinary Committee met in February 1988 to discuss the 100 farms that were infected in 1987.

7. Pressure grows on Currie to give evidence to the Select Committee. Currie writes to its Chair to ask 'to be let to get on with my life'.

8. Convenience foods, including cook-chill foods, are reported to be as convenient to germs, especially listeria, as to people.

Pregnant women are warned not to eat soft cheeses.

9. Currie appears before the Select Committee, but refuses to clear up the central mystery – why it took seven weeks to clarify her controversial statement.

An official report, compiled by MAFF, DHSS and British Egg Industry, estimates that there may be up to two million salmonella infections a year in England and Wales. The estimate is based on the assumption that only 1 per cent of outbreaks is reported.

John MacGregor reports that of 12,000 eggs tested since November 1988, a few were infected with salmonella.

10. Neil Kinnock, Leader of the Opposition, clashes with the Prime Minister, Margaret Thatcher, after disclosing the results of the official report. Thatcher announces that two million leaflets will be placed in supermarkets and surgeries giving tips about avoiding food poisoning from salmonella and listeria.

11/The MAFF announces a ban on the sale
12. of unpasteurized milk following consultation. Cheeses made with unpasteurized milk are to be included.

13. John MacGregor confirms that draft legis-

lation for a Food Bill is well advanced. It will include new offences for unwholesome food, training for handling, and increased government powers to seize.

John MacGregor meets a French counterpart to say that the UK is not planning to ban sales of soft cheeses made from unpasteurized milk or to impose a ban on French soft cheeses. He still warns of the danger of listeria from soft cheeses.

14. France demands official explanations over a possible UK ban on the import of certain cheeses, amid growing confusion over possible restrictions.

Kenneth Clarke does not reduce his intake of unpasteurized goat cheese.

15. The full story emerges that while eggs may have been the vehicle for infection, the scale of the outbreaks is due to poor hygiene and food production methods. The *Independent* carries details of how, rather than improving standards for cook-chill foods which would prove expensive, the DHSS let eggs take the blame. Evidence of outbreaks between May and November 1988 (see Prelude) emerge.

16. Labour proposes new rules for cook-chill foods, which would require chilled foods to be kept below 3 °C.

17. Egg producers call for a judicial inquiry into the way that Government departments handled the food-poisoning problem. They believe that eggs were blamed for outbreaks that would not have occurred if there were stricter standards for food preparation in the catering industry.

22 Major House of Commons debate. Labour accuse the Conservative Government of backing production and shareholders to the detriment of consumer interests, and accuse Mrs Thatcher of acting like a 'hyperactive fairy queen'.

Kenneth Clarke announces the formation of a new Committee, on Microbiological Food Safety, to try to establish whether there is a link between the increased rate of food-borne illness, changes in agriculture and food production, food technology, distribution, retailing, catering and food handling at home. He also announces new laws for setting maximum temperatures for manufacture and storage of pre-cooked supermarket food.

1 March 1989

The report of the All Party Agriculture Select Committee is published. The main conclusions of the report are as follows:

1. Former Junior Health Minister Edwina Currie is denounced in strong terms for her 'most egg production' comment that was never withdrawn, although later it was qualified and judged to be wrong and alarming.

2. Kenneth Clarke is accused of naivety for relying on the opinion of the Chief Medical Officer and not correcting Edwina Currie's mistake sooner. Money for food and hygiene education should be established.

3. MAFF is criticized for 'tardiness over several years' in meeting the threat of salmonella.

4. The industry is criticized for 'a head-in-the-sand attitude' and 'dragging their heels' in seeking to safeguard consumers.

5. Eggs from farms that were responsible for salmonella outbreaks should not have been allowed to enter the food chain.

6. They concluded 'it was a failure of government, and not just a single Minister, not to make the proper correction sooner'.

Dr Tim Lang, Director of the London Food Commission in welcoming the report, said on BBC News, 'You'd be wrong to think of MAFF as a cosy club for farmers. It is a cosy club for farmers, the agrochemical and food industry'.

Source: Clutterbuck, C (1990) *Trade Unions and Food*, Fact pack 9, unpublished

Annex 4

The UK's Leading Food Brands in 1991

Table A4.1 The top 50 leading food brands by sales in 1991 in the UK (£million)

Coca-Cola	203.0	Mars bar	55.3
Nescafé	188.5	Clover spread	51.9
Whiskas	178.1	Kellogg's Frosties	50.5
Silver Spoon sugar	144.5	Del Monte fruit juices and drinks	49.6
PG Tips tea	140.7	Katto-meat canned cat food	49.1
Flora margarine	125.3	Müller	49.1
Tetley tea bags	116.6	Gold Blend spread	49.0
Heinz baked beans	111.4	Birds Eye peas	48.6
Robinson's squash	106.4	Ty-phoo tea	48.2
Walker's crisps	105.6	Birds Eye fish fingers	46.4
Pedigree Chum	105.3	Golden Wonder pots	46.3
Heinz soup	95.9	Maxwell House coffee	43.3
Mr Kipling cakes	92.6	Hula Hoops snacks	42.4
Kellogg's Corn Flakes	85.2	St Ivel Shape yoghurt	41.7
Kit Kat	84.9	Co-op 99 tea	41.0
Tate & Lyle granulated sugar	84.9	Golden Wonder crisps	40.8
Anchor butter	80.2	Heinz tomato ketchup	39.7
Ribena	72.5	John West canned salmon	39.3
Bird's Eye Menu Master	72.3	Felix cat food	39.0
Bird's Eye Steakhouse Burgers	71.6	Schweppes mixer	38.8
Pepsi-Cola	62.8	Twix	38.1
Ski yoghurt	60.7	Bisto gravy maker	37.9
St Ivel Gold spread	60.5	Lucozade	37.2
Weetabix	58.6	Kellogg's Crunchie Nut	36.8
Lurpak butter	55.9	Kellogg's Rice Crispies	35.5

Source: Seymour Cook (1992) The European Food Report 1992 Food Research International, London, p89

Annex 5

Consumer Charters

Various bodies have produced consumer charters. The International Organization of Consumer Unions (IOCU, now called Consumers International) has produced a list of consumer rights and responsibilities of consumers. These are as follows:

➤ *The right to basic needs* means the right to basic goods and services which guarantee survival. It includes adequate food, clothing, shelter, health care, education and sanitation.
➤ *The right to safety* means the right to be protected against products, production processes and services which are hazardous to health or life. It includes concern for consumers' long-term interests as well as their immediate needs.
➤ *The right to be informed* means the right to be given the facts needed to make an informed choice or decision. Consumers must be provided with adequate information enabling them to act wisely and responsibly. They must also be protected from misleading or inaccurate publicity material, whether it is included in advertising, labelling, packaging, or other means.
➤ *The right to choose* means the right to have access to a variety of products and services at competitive prices and, in the case of monopolies, to have an assurance of satisfactory quality and service at a fair price.
➤ *The right to be heard* means the right to advocate consumers' interests with a view to their receiving full and sympathetic consideration in the formulation and execution of economic and other policies.

It includes the right of representation in governmental and other policy-making bodies as well as in the development of products and services before they are produced or set up.

➤ *The right to redress* means the right to a fair settlement of just claims. It includes the right to receive compensation for misrepresentation of shoddy goods or unsatisfactory services and the availability of acceptable forms of legal aid or redress for small claims where necessary.
➤ *The right to consumer education* means the right to acquire the knowledge and skills to be an informed consumer throughout life. The right to consumer education incorporates the right to the knowledge and skills needed for taking action to influence factors which affect consumer decisions.
➤ *The right to a healthy environment* means the right to a physical environment that will enhance the quality of life. It includes protection against environmental dangers over which the individual has no control. It acknowledges the need to protect and improve the environment for present and future generations.

The IOCU also balances this by encouraging citizens:

➤ to be critically aware in the market;
➤ to be alert and ready to question the price and quality of goods and services;
➤ to be assertive and act to ensure that fairness prevails;

➤ to be aware of the impact of their consumption on other citizens as well as the environment, and to act accordingly;

➤ to organize together as consumers to develop the strength and influence to promote and protect their interests.

Source: IOCU, London, 1994

In other cases, national organizations have produced charters. The following charter, aimed at a UK audience, was produced by the London Food Commission in 1988.

A Food Quality Charter

Britain needs a new National Food Policy. All people have the right to good quality, enjoyable food which everyone can afford. Food should be produced in a manner which consumers and workers can trust and be proud of. It is high time there was once more a Ministry of Food.

1. *Public health protection.* The quality of food and food laws and regulations should be set to the highest standard, with priority being given to protecting the health of consumers and food workers.
2. *Access to information.* Government and industry should provide full and open access to all the relevant information about food, especially safety and technical data. Food should bear full ingredient and nutrition labelling.
3. *Democratic decision-making.* Decision-making on food matters should involve widespread public consultation. Evidence presented to government committees about food quality matters should be given in public, not behind closed doors. The

Ministry of Agriculture should be turned into a Ministry of Food.

4. *Enforcement.* Food laws should be strengthened and rigorously enforced. More resources and powers should be given to food law enforcement officers.
5. *Resources.* An independent fund for research into and monitoring of food safety and quality should be set up with funds provided by a levy on the food and chemical industries.
6. *Need.* The public has a right to good quality and convenient food. The addition of unnecessary substances to food and the use of unnecessary processes should be minimized.
7. *Hygiene.* Food should be produced and treated in the most hygienic manner. Legislation should be strengthened and strictly enforced. There should be mandatory hygiene training for all food handlers. Additives and processing techniques should not be used to conceal inadequate hygiene.
8. *Environment.* Food policies should conserve natural resources and protect the environment for future generations.
9. *EEC.* The EEC should harmonize its food laws according to the highest standards, not to the lowest common denominator.
10. *Responsibility to the people of the world.* No food or food-processing technology which is banned in the UK should be exported to other countries. The UK should take a lead in setting high food standards. Surplus or substandard food should not be dumped on world markets.

Source: The London Food Commission (1988) *Food Adulteration and How to Beat It*, The London Food Commission, Unwin Hyman Ltd, London

Annex 6

Further Reading

There are many academic journals in all the fields covered by this book which give up-to-date work going on in their area, as well as many textbooks on ecology, agriculture, food science, and so on, which give much more technical detail about specific areas. We would direct you to the sources quoted throughout the text as a first step. They will take you further into particular subject areas and into some of the debates raging in the food system.

A broad, readable guide to the history of food, to which we are greatly indebted, is *Food in History* by Reay Tannahil, Penguin, London, 1988. More detailed studies exist for specific areas and periods – for example, *The Englishman's Food* by J. C. Drummond and Anne Wilbraham, Jonathan Cape, London, 1969.

Urban-rural ecological links are discussed in 'Ecological footprints and appropriated carrying capacy: what urban economics leaves out' by William E Rees in *Environment and Urbanization*, vol 4 no 2, October 1992, pp121–30. A useful overview of social and cultural aspects of food is *Food and Nutrition: Customs and Culture* by P Fieldhouse, Croom Helm, London, 1986. This is now a little dated but illustrates the important significance of food in social and cultural life. A really engrossing read about things culinary is by M Harris, *Good to eat: riddles of food and culture*, Allen & Unwin, London, 1986 – but it is not for weak stomachs! Margaret Visser's works are a readable mix of history, anthropology and sociology of food as their titles suggest: *Much Depends on Dinner – The Extraordinary History and Mythology, Allure and Obsessions, Perils and Taboos*

of an Ordinary Meal, Penguin, London, 1989, and *The Rituals of Dinner – The Origins, Evolution, Eccentricities and Meaning of Table Manners*, Viking, London, 1992.

In addition to the nutrition texts referred to in Annex 2, see also the major WHO report *Diet, nutrition, and the prevention of chronic diseases*, WHO Technical Series Report 797, Geneva, 1990. For an insight into what dietary change in one area means, see S J Fallows and J V Wheelock 'Means to dietary change: the example of fat' *Journal of the Royal Society of Health*, vol 103, no 5, 1983, pp186–91. A short but eminently readable book on the often obtuse area of sensorimetrics is *The Bliss Point*, by R L McBride, Sun Books, Melbourne, 1990.

The specialist trade press serving the different groups of actors, such as *Farmers Weekly* and *The Grocer* in the UK or *The Packer* and *Supermarket News* in the USA, gives a detailed picture of day-to-day developments in their areas and even the advertisements are worth reading, for they show many of the trends. Also not to be missed are company annual reports and the financial pages and newspapers, such as the *Financial Times* for an idea of what companies are about. For a guide to the problems of Australian farmers see *The Foodmakers* by S Sargent, Penguin, Ringwood, Victoria, 1985.

Many good sources of information are not easily available, for they are costly, low circulation reports from specialist companies or documents from conferences where the various actors in the system talk among themselves. These are often worth seeking out if you can.

The FAO produces an annual general publication *The State of Food and Agriculture* as well as many more specialized ones. The quarterly *Australian Commodities* produced by ABARE in Canberra provides forecasts and reviews of commodity issues. For some areas, such as traders, information is less easy to get and Dan Morgan's *Merchants of Grain*, Viking, New York, 1979 quoted in the text, still provides the best background to the grain business, while Brewster Kneen's work *Trading Up: How Cargill, the World's Largest Grain Trading Company, is Changing Canadian Agriculture*, NC Press, Toronto, 1990, also quoted in the text, is a critical look at one major company's influence and the issues behind it.

Trade Union Quality in the Factory of the Future by Paul Elshof and Ed Wetzel, SOMO, Amsterdam, 1992, reviews industry trends from the workers' point of view. A useful introduction to all aspects of consumer behaviour as seen by marketers is *Consumer Behaviour: Implications for Marketing Strategy* by D I Hawkins, R J Best and K A Coney, Homewoord, Boston, Irwin, 1992, 5th edn. This book deals with issues, such as the influence of values, market segmentation, buying behaviours and much more. John Sutton's *Sunk Costs and Market Structure: Price Competition, Advertising, and the Evolution of Concentration* MIT, Cambridge MA, 1991, is heavy on the econometrics but contains reviews of over a dozen food sectors including sugar, flour, bread, soft drinks, coffee and breakfast cereals.

Despite its age, *The Food Industry – Economics and Policies*, edited by Jim Burns, John McInerney and Alan Swinbank, Heinemann, London, 1983, is a good overview. Food law in Britain is also covered in *Food Legislative System of the UK* by S J Fallows, Butterworths, London 1988. The OECD produces many useful reports, including the annual monitoring and outlook report on *Agricultural Policies, Markets and Trade*. Leslie Sklair's *Sociology of the Global System*, Harvester Wheatsheaf, Hemel Hempstead, 1991 gives a fuller account of trends in globalization discussed in the paper from which we quoted in Chapter 8.

A useful description of nutritional problems in affluent societies and their implications for nutrition policy is given by W P T James, A Ferro Luzzi, B Isaksson and W B Szostak, in *Healthy Nutrition*, WHO Regional Publications No 24, Copenhagen, 1988. For a quick guide to the policy proposals and guidelines linking food and health policy in most industrialized countries between 1961 and 1991, see Geoffrey Cannon, *Food and Health: The Experts Agree*, Consumers Association, London, 1992. The journal *Food Policy* is a forum for discussion about food policy issues. Finally, do observe what is happening in the pubs, cafés, restaurants, shops, supermarkets, food labels, newspapers, TV advertisements and among your family and friends as they eat. This too will tell you much about changes in the food system.

Index

Other titles available from Earthscan

Regenerating Agriculture – Policies and Practice for Sustainability and Self-Reliance

Jules Pretty

For the past half-century, the aim of agriculture has been solely to increase food production, relying on high inputs of fertilizers and pesticides and intensive industrial methods. Along the way it has damaged environmental, social and economic systems. Now, finding an alternative approach is one of the greatest challenges that it faces. *Regenerating Agriculture* shows that there are viable alternatives. Agriculture can conserve resources without loss of productivity, while maintaining rural employment and minimizing external costs. Using case studies from both industrial and developing countries, it shows that environmentally, economically and socially beneficial forms of agriculture are available and viable.

'Wonderful, extremely well written. It blends analysis, argument and example very well. This book will challenge you.' *Ed Mayo, The New Economics Foundation*

320pp £12.95 paperback ISBN 1 85383 198 0 £29.95 hardback ISBN 1 85383 227 8

The Earthscan Reader in Sustainable Development

Edited by John Kirkby, Phil O'Keefe and Lloyd Timberlake

Such a huge number of books, journals and papers have been devoted to defining, assessing and implementing 'sustainable' development that students and other readers face information overload. Now, the most authoritative writings have been carefully assessed and collected together in *The Earthscan Reader in Sustainable Development*. The contributions included span five years of the debate, and cover all the principal themes; for ease of use, the essays have been split into key subject areas. These seminal essays will provide readers with a unique overview of the subject, as well as the long-awaited course material for students of environmental studies, economics, geography, politics, planning and the social sciences.

384pp £14.95 paperback ISBN 1 85383 216 2 £35.00 hardback ISBN 1 85383 223 5

Wasted – Counting the Cost of Global Consumption

Michael Redclift

Without the creation of more sustainable livelihoods, sustainable development will remain an elusive goal. Yet given the huge differences in levels of consumption between North and South, how is this to be brought about? Taking the 1992 Rio summit as its point of departure, *Wasted* examines what we need to do now to live within our limits. In the process of consuming, we also create sinks; and these back pockets of the global biogeographical system are no longer empty. The fate of the global environment is intimately connected with our rates of consumption, particularly in the energy-profligate North. In an absorbing analysis, Michael Redclift argues that the way we understand the environment conditions our responses to it, and discusses policies for increased sustainability grounded in recent research and practice.

224pp £12.95 paperback ISBN 1 85383 355 X £32.50 hardback ISBN 1 85383 360 6